Pitman Research Notes in Mathematics Series

Submission of proposals for consideration
Suggestions for publication, in the form of outlines and representative samples, are invited by the Editorial Board for assessment. Intending authors should approach one of the main editors or another member of the Editorial Board, citing the relevant AMS subject classifications. Alternatively, outlines may be sent directly to the publisher's offices. Refereeing is by members of the board and other mathematical authorities in the topic concerned, throughout the world.

Preparation of accepted manuscripts
On acceptance of a proposal, the publisher will supply full instructions for the preparation of manuscripts in a form suitable for direct photo-lithographic reproduction. Specially printed grid sheets can be provided and a contribution is offered by the publisher towards the cost of typing. Word processor output, subject to the publisher's approval, is also acceptable.

Illustrations should be prepared by the authors, ready for direct reproduction without further improvement. The use of hand-drawn symbols should be avoided wherever possible, in order to maintain maximum clarity of the text.

The publisher will be pleased to give any guidance necessary during the preparation of a typescript, and will be happy to answer any queries.

Important note
In order to avoid later retyping, intending authors are strongly urged not to begin final preparation of a typescript before receiving the publisher's guidelines. In this way it is hoped to preserve the uniform appearance of the series.

Addison Wesley Longman Ltd
Edinburgh Gate
Harlow, Essex, CM20 2JE
UK
(Telephone (0) 1279 623623)

Titles in this series. A full list is available from the publisher on request.

151 A stochastic maximum principle for optimal control of diffusions
U G Haussmann
152 Semigroups, theory and applications. Volume II
H Brezis, M G Crandall and F Kappel
153 A general theory of integration in function spaces
P Muldowney
154 Oakland Conference on partial differential equations and applied mathematics
L R Bragg and J W Dettman
155 Contributions to nonlinear partial differential equations. Volume II
J I Díaz and P L Lions
156 Semigroups of linear operators: an introduction
A C McBride
157 Ordinary and partial differential equations
B D Sleeman and R J Jarvis
158 Hyperbolic equations
F Colombini and M K V Murthy
159 Linear topologies on a ring: an overview
J S Golan
160 Dynamical systems and bifurcation theory
M I Camacho, M J Pacifico and F Takens
161 Branched coverings and algebraic functions
M Namba
162 Perturbation bounds for matrix eigenvalues
R Bhatia
163 Defect minimization in operator equations: theory and applications
R Reemtsen
164 Multidimensional Brownian excursions and potential theory
K Burdzy
165 Viscosity solutions and optimal control
R J Elliott
166 Nonlinear partial differential equations and their applications: Collège de France Seminar. Volume VIII
H Brezis and J L Lions
167 Theory and applications of inverse problems
H Haario
168 Energy stability and convection
G P Galdi and B Straughan
169 Additive groups of rings. Volume II
S Feigelstock
170 Numerical analysis 1987
D F Griffiths and G A Watson
171 Surveys of some recent results in operator theory. Volume I
J B Conway and B B Morrel
172 Amenable Banach algebras
J-P Pier
173 Pseudo-orbits of contact forms
A Bahri
174 Poisson algebras and Poisson manifolds
K H Bhaskara and K Viswanath
175 Maximum principles and eigenvalue problems in partial differential equations
P W Schaefer
176 Mathematical analysis of nonlinear, dynamic processes
K U Grusa
177 Cordes' two-parameter spectral representation theory
D F McGhee and R H Picard
178 Equivariant K-theory for proper actions
N C Phillips
179 Elliptic operators, topology and asymptotic methods
J Roe
180 Nonlinear evolution equations
J K Engelbrecht, V E Fridman and E N Pelinovski
181 Nonlinear partial differential equations and their applications: Collège de France Seminar. Volume IX
H Brezis and J L Lions
182 Critical points at infinity in some variational problems
A Bahri
183 Recent developments in hyperbolic equations
L Cattabriga, F Colombini, M K V Murthy and S Spagnolo
184 Optimization and identification of systems governed by evolution equations on Banach space
N U Ahmed
185 Free boundary problems: theory and applications. Volume I
K H Hoffmann and J Sprekels
186 Free boundary problems: theory and applications. Volume II
K H Hoffmann and J Sprekels
187 An introduction to intersection homology theory
F Kirwan
188 Derivatives, nuclei and dimensions on the frame of torsion theories
J S Golan and H Simmons
189 Theory of reproducing kernels and its applications
S Saitoh
190 Volterra integrodifferential equations in Banach spaces and applications
G Da Prato and M Iannelli
191 Nest algebras
K R Davidson
192 Surveys of some recent results in operator theory. Volume II
J B Conway and B B Morrel
193 Nonlinear variational problems. Volume II
A Marino and M K V Murthy
194 Stochastic processes with multidimensional parameter
M E Dozzi
195 Prestressed bodies
D Iesan
196 Hilbert space approach to some classical transforms
R H Picard
197 Stochastic calculus in application
J R Norris
198 Radical theory
B J Gardner
199 The C^*-algebras of a class of solvable Lie groups
X Wang
200 Stochastic analysis, path integration and dynamics
K D Elworthy and J C Zambrini

201 Riemannian geometry and holonomy groups
S Salamon

202 Strong asymptotics for extremal errors and polynomials associated with Erdös type weights
D S Lubinsky

203 Optimal control of diffusion processes
V S Borkar

204 Rings, modules and radicals
B J Gardner

205 Two-parameter eigenvalue problems in ordinary differential equations
M Faierman

206 Distributions and analytic functions
R D Carmichael and D Mitrovic

207 Semicontinuity, relaxation and integral representation in the calculus of variations
G Buttazzo

208 Recent advances in nonlinear elliptic and parabolic problems
P Bénilan, M Chipot, L Evans and M Pierre

209 Model completions, ring representations and the topology of the Pierce sheaf
A Carson

210 Retarded dynamical systems
G Stepan

211 Function spaces, differential operators and nonlinear analysis
L Paivarinta

212 Analytic function theory of one complex variable
C C Yang, Y Komatu and K Niino

213 Elements of stability of visco-elastic fluids
J Dunwoody

214 Jordan decomposition of generalized vector measures
K D Schmidt

215 A mathematical analysis of bending of plates with transverse shear deformation
C Constanda

216 Ordinary and partial differential equations. Volume II
B D Sleeman and R J Jarvis

217 Hilbert modules over function algebras
R G Douglas and V I Paulsen

218 Graph colourings
R Wilson and R Nelson

219 Hardy-type inequalities
A Kufner and B Opic

220 Nonlinear partial differential equations and their applications: Collège de France Seminar. Volume X
H Brezis and J L Lions

221 Workshop on dynamical systems
E Shiels and Z Coelho

222 Geometry and analysis in nonlinear dynamics
H W Broer and F Takens

223 Fluid dynamical aspects of combustion theory
M Onofri and A Tesei

224 Approximation of Hilbert space operators. Volume I. 2nd edition
D Herrero

225 Operator theory: proceedings of the 1988 GPOTS–Wabash conference
J B Conway and B B Morrel

226 Local cohomology and localization
J L Bueso Montero, B Torrecillas Jover and A Verschoren

227 Nonlinear waves and dissipative effects
D Fusco and A Jeffrey

228 Numerical analysis 1989
D F Griffiths and G A Watson

229 Recent developments in structured continua. Volume II
D De Kee and P Kaloni

230 Boolean methods in interpolation and approximation
F J Delvos and W Schempp

231 Further advances in twistor theory. Volume I
L J Mason and L P Hughston

232 Further advances in twistor theory. Volume II
L J Mason, L P Hughston and P Z Kobak

233 Geometry in the neighborhood of invariant manifolds of maps and flows and linearization
U Kirchgraber and K Palmer

234 Quantales and their applications
K I Rosenthal

235 Integral equations and inverse problems
V Petkov and R Lazarov

236 Pseudo-differential operators
S R Simanca

237 A functional analytic approach to statistical experiments
I M Bomze

238 Quantum mechanics, algebras and distributions
D Dubin and M Hennings

239 Hamilton flows and evolution semigroups
J Gzyl

240 Topics in controlled Markov chains
V S Borkar

241 Invariant manifold theory for hydrodynamic transition
S Sritharan

242 Lectures on the spectrum of $L^2(\Gamma\backslash G)$
F L Williams

243 Progress in variational methods in Hamiltonian systems and elliptic equations
M Girardi, M Matzeu and F Pacella

244 Optimization and nonlinear analysis
A Ioffe, M Marcus and S Reich

245 Inverse problems and imaging
G F Roach

246 Semigroup theory with applications to systems and control
N U Ahmed

247 Periodic-parabolic boundary value problems and positivity
P Hess

248 Distributions and pseudo-differential operators
S Zaidman

249 Progress in partial differential equations: the Metz surveys
M Chipot and J Saint Jean Paulin

250 Differential equations and control theory
V Barbu

251 Stability of stochastic differential equations with respect to semimartingales
X Mao
252 Fixed point theory and applications
J Baillon and M Théra
253 Nonlinear hyperbolic equations and field theory
M K V Murthy and S Spagnolo
254 Ordinary and partial differential equations. Volume III
B D Sleeman and R J Jarvis
255 Harmonic maps into homogeneous spaces
M Black
256 Boundary value and initial value problems in complex analysis: studies in complex analysis and its applications to PDEs 1
R Kühnau and W Tutschke
257 Geometric function theory and applications of complex analysis in mechanics: studies in complex analysis and its applications to PDEs 2
R Kühnau and W Tutschke
258 The development of statistics: recent contributions from China
X R Chen, K T Fang and C C Yang
259 Multiplication of distributions and applications to partial differential equations
M Oberguggenberger
260 Numerical analysis 1991
D F Griffiths and G A Watson
261 Schur's algorithm and several applications
M Bakonyi and T Constantinescu
262 Partial differential equations with complex analysis
H Begehr and A Jeffrey
263 Partial differential equations with real analysis
H Begehr and A Jeffrey
264 Solvability and bifurcations of nonlinear equations
P Drábek
265 Orientational averaging in mechanics of solids
A Lagzdins, V Tamuzs, G Teters and A Kregers
266 Progress in partial differential equations: elliptic and parabolic problems
C Bandle, J Bemelmans, M Chipot, M Grüter and J Saint Jean Paulin
267 Progress in partial differential equations: calculus of variations, applications
C Bandle, J Bemelmans, M Chipot, M Grüter and J Saint Jean Paulin
268 Stochastic partial differential equations and applications
G Da Prato and L Tubaro
269 Partial differential equations and related subjects
M Miranda
270 Operator algebras and topology
W B Arveson, A S Mishchenko, M Putinar, M A Rieffel and S Stratila
271 Operator algebras and operator theory
W B Arveson, A S Mishchenko, M Putinar, M A Rieffel and S Stratila
272 Ordinary and delay differential equations
J Wiener and J K Hale
273 Partial differential equations
J Wiener and J K Hale
274 Mathematical topics in fluid mechanics
J F Rodrigues and A Sequeira
275 Green functions for second order parabolic integro-differential problems
M G Garroni and J F Menaldi
276 Riemann waves and their applications
M W Kalinowski
277 Banach C(K)-modules and operators preserving disjointness
Y A Abramovich, E L Arenson and A K Kitover
278 Limit algebras: an introduction to subalgebras of C*-algebras
S C Power
279 Abstract evolution equations, periodic problems and applications
D Daners and P Koch Medina
280 Emerging applications in free boundary problems
J Chadam and H Rasmussen
281 Free boundary problems involving solids
J Chadam and H Rasmussen
282 Free boundary problems in fluid flow with applications
J Chadam and H Rasmussen
283 Asymptotic problems in probability theory: stochastic models and diffusions on fractals
K D Elworthy and N Ikeda
284 Asymptotic problems in probability theory: Wiener functionals and asymptotics
K D Elworthy and N Ikeda
285 Dynamical systems
R Bamon, R Labarca, J Lewowicz and J Palis
286 Models of hysteresis
A Visintin
287 Moments in probability and approximation theory
G A Anastassiou
288 Mathematical aspects of penetrative convection
B Straughan
289 Ordinary and partial differential equations. Volume IV
B D Sleeman and R J Jarvis
290 *K*-theory for real *C**-algebras
H Schröder
291 Recent developments in theoretical fluid mechanics
G P Galdi and J Necas
292 Propagation of a curved shock and nonlinear ray theory
P Prasad
293 Non-classical elastic solids
M Ciarletta and D Ieşan
294 Multigrid methods
J Bramble
295 Entropy and partial differential equations
W A Day
296 Progress in partial differential equations: the Metz surveys 2
M Chipot
297 Nonstandard methods in the calculus of variations
C Tuckey
298 Barrelledness, Baire-like- and (LF)-spaces
M Kunzinger
299 Nonlinear partial differential equations and their applications. Collège de France Seminar. Volume XI
H Brezis and J L Lions

300 Introduction to operator theory
T Yoshino
301 Generalized fractional calculus and applications
V Kiryakova
302 Nonlinear partial differential equations and their applications. Collège de France Seminar Volume XII
H Brezis and J L Lions
303 Numerical analysis 1993
D F Griffiths and G A Watson
304 Topics in abstract differential equations
S Zaidman
305 Complex analysis and its applications
C C Yang, G C Wen, K Y Li and Y M Chiang
306 Computational methods for fluid-structure interaction
J M Crolet and R Ohayon
307 Random geometrically graph directed self-similar multifractals
L Olsen
308 Progress in theoretical and computational fluid mechanics
G P Galdi, J Málek and J Necas
309 Variational methods in Lorentzian geometry
A Masiello
310 Stochastic analysis on infinite dimensional spaces
H Kunita and H-H Kuo
311 Representations of Lie groups and quantum groups
V Baldoni and M Picardello
312 Common zeros of polynomials in several variables and higher dimensional quadrature
Y Xu
313 Extending modules
N V Dung, D van Huynh, P F Smith and R Wisbauer
314 Progress in partial differential equations: the Metz surveys 3
M Chipot, J Saint Jean Paulin and I Shafrir
315 Refined large deviation limit theorems
V Vinogradov
316 Topological vector spaces, algebras and related areas
A Lau and I Tweddle
317 Integral methods in science and engineering
C Constanda
318 A method for computing unsteady flows in porous media
R Raghavan and E Ozkan
319 Asymptotic theories for plates and shells
R P Gilbert and K Hackl
320 Nonlinear variational problems and partial differential equations
A Marino and M K V Murthy
321 Topics in abstract differential equations II
S Zaidman
322 Diffraction by wedges
B Budaev
323 Free boundary problems: theory and applications
J I Diaz, M A Herrero, A Liñan and J L Vazquez
324 Recent developments in evolution equations
A C McBride and G F Roach
325 Elliptic and parabolic problems: Pont-à-Mousson 1994
C Bandle, J Bemelmans, M Chipot, J Saint Jean Paulin and I Shafrir
326 Calculus of variations, applications and computations: Pont-à-Mousson 1994
C Bandle, J Bemelmans, M Chipot, J Saint Jean Paulin and I Shafrir
327 Conjugate gradient type methods for ill-posed problems
M Hanke
328 A survey of preconditioned iterative methods
A M Bruaset
329 A generalized Taylor's formula for functions of several variables and certain of its applications
J-A Riestra
330 Semigroups of operators and spectral theory
S Kantorovitz
331 Boundary-field equation methods for a class of nonlinear problems
G N Gatica and G C Hsiao
332 Metrizable barrelled spaces
J C Ferrando, M López Pellicer and L M Sánchez Ruiz
333 Real and complex singularities
W L Marar
334 Hyperbolic sets, shadowing and persistence for noninvertible mappings in Banach spaces
B Lani-Wayda
335 Nonlinear dynamics and pattern formation in the natural environment
A Doelman and A van Harten
336 Developments in nonstandard mathematics
N J Cutland, V Neves, F Oliveira and J Sousa-Pinto
337 Topological circle planes and topological quadrangles
A E Schroth
338 Graph dynamics
E Prisner
339 Localization and sheaves: a relative point of view
P Jara, A Verschoren and C Vidal
340 Mathematical problems in semiconductor physics
P Marcati, P A Markowich and R Natalini
341 Surveying a dynamical system: a study of the Gray–Scott reaction in a two-phase reactor
K Alhumaizi and R Aris
342 Solution sets of differential equations in abstract spaces
R Dragoni, J W Macki, P Nistri and P Zecca
343 Nonlinear partial differential equations
A Benkirane and J-P Gossez
344 Numerical analysis 1995
D F Griffiths and G A Watson
345 Progress in partial differential equations: the Metz surveys 4
M Chipot and I Shafrir
346 Rings and radicals
B J Gardner, Liu Shaoxue and R Wiegandt
347 Complex analysis, harmonic analysis and applications
R Deville, J Esterle, V Petkov, A Sebbar and A Yger
348 The theory of quantaloids
K I Rosenthal
349 General theory of partial differential equations and microlocal analysis
Qi Min-you and L Rodino

350 Progress in elliptic and parabolic partial differential equations
A Alvino, P Buonocore, V Ferone, E Giarrusso, S Matarasso, R Toscano and G Trombetti
351 Integral representations for spatial models of mathematical physics
V V Kravchenko and M V Shapiro
352 Dynamics of nonlinear waves in dissipative systems: reduction, bifurcation and stability
G Dangelmayr, B Fiedler, K Kirchgässner and A Mielke
353 Singularities of solutions of second order quasilinear equations
L Véron
354 Mathematical theory in fluid mechanics
G P Galdi, J Málek and J Necas
355 Eigenfunction expansions, operator algebras and symmetric spaces
R M Kauffman
356 Lectures on bifurcations, dynamics and symmetry
M Field
357 Noncoercive variational problems and related results
D Goeleven
358 Generalised optimal stopping problems and financial markets
D Wong
359 Topics in pseudo-differential operators
S Zaidman
360 The Dirichlet problem for the Laplacian in bounded and unbounded domains
C G Simader and H Sohr
361 Direct and inverse electromagnetic scattering
A H Serbest and S R Cloude
362 International conference on dynamical systems
F Ledrappier, J Lewowicz and S Newhouse
363 Free boundary problems, theory and applications
M Niezgódka and P Strzelecki
364 Backward stochastic differential equations
N El Karoui and L Mazliak
365 Topological and variational methods for nonlinear boundary value problems
P Drábek
366 Complex analysis and geometry
V Ancona, E Ballico, R M Mirò-Roig and A Silva
367 Integral expansions related to Mehler–Fock type transforms
B N Mandal and N Mandal
368 Elliptic boundary value problems with indefinite weights: variational formulations of the principal eigenvalue and applications
F Belgacem
369 Integral transforms, reproducing kernels and their applications
S Saitoh
370 Ordinary and partial differential equations. Volume V
P D Smith and R J Jarvis
371 Numerical methods in mechanics
C Conca and G N Gatica
372 Generalized manifolds
K-G Schlesinger
373 Independent axioms for Minkowski space-time
J W Schutz
374 Integral methods in science and engineering Volume one: analytic methods
C Constanda, J Saranen and S Seikkala
375 Integral methods in science and engineering Volume two: approximation methods
C Constanda, J Saranen and S Seikkala
376 Inner product spaces and applications
T M Rassias
377 Functional analysis with current applications in science, technology and industry
M Brokate and A H Siddiqi
378 Classical and quantic periodic motions of multiply polarized spin-particles
A Bahri
379 Analysis, numerics and applications of differential and integral equations
M Bach, C Constanda, G C Hsiao, A-M Sändig and P Werner
380 Numerical analysis 1997
D F Griffiths, D J Higham and G A Watson
381 Real analytic and algebraic singularities
T Fukuda, T Fukui, S Izumiya and S Koike
382 Boundary value problems with equivalued surface and resistivity well-logging
T Li, S Zheng, Y Tan and W Shen
383 Progress in partial differential equations Pont-à-Mousson 1997 Volume 1
H Amann, C Bandle, M Chipot, F Conrad and I Shafrir
384 Progress in partial differential equations Pont-à-Mousson 1997 Volume 2
H Amann, C Bandle, M Chipot, F Conrad and I Shafrir
385 The linear theory of Colombeau generalized functions
M Nedeljkov, S Pilipovic and D Scarpalézos
386 Recent advances in differential equations
H-H Dai and P L Sachdev
387 Progress in holomorphic dynamics
H Kriete
388 Navier–Stokes equations: theory and numerical methods
R Salvi
389 Strongly irreducible operators on Hilbert space
C Jiang and Z Wang
390 Mathematical methods in scattering theory and biomedical technology
G Dassios, D I Fotiadis, K Kiriaki and C V Massalas

G Dassios

University of Patras, Greece

D I Fotiadis

University of Ioannina, Greece

K Kiriaki

National Technical University of Athens, Greece

and

C V Massalas

University of Ioannina, Greece

(Editors)

Mathematical methods in scattering theory and biomedical technology

Proceedings of a workshop
dedicated to Professor Gary Roach

LONGMAN

Addison Wesley Longman Limited
Edinburgh Gate, Harlow
Essex CM20 2JE, England
and Associated Companies throughout the world.

*Published in the United States of America
by Addison Wesley Longman Inc.*

First published 1998

AMS Subject Classifications (Main) 35-xx, 92C10, 92C50
(Subsidiary) 78-xx, 92-08

ISSN 0269-3674

ISBN 0 582 36804 9

Visit Addison Wesley Longman on the world wide web at
http://www.awl-he.com

British Library Cataloguing in Publication Data

A catalogue record for this book is
available from the British Library

Printed in Great Britain by Henry Ling Ltd.,
at the Dorset Press, Dorchester, Dorset.

Contents

Preface

Scattering theory deals with the interaction of a spatial discontinuity, prescribed as the scatterer, with a prescribed wave field, known as incident field. In particular, direct scattering theory is concerned with the effects that a given scatterer has on the propagation of an incident field, and inverse scattering is focused on the recovery of physical and geometrical characteristics of the scatterer and provides information about the way that the scatterer disturbed the incident wave. Inverse scattering theory occupies today a leading place in science and modern technology. It extends from non invasive medical diagnostic methods, to non destructive evaluation and testing, remote sensing, radar systems, robot vision and pattern recognition. Actually, inverse scattering theory extends the ability of a human being to hear and see things beyond his physiological limitations. The wide spectrum of activity in scattering theory for acoustic, electromagnetic, elastic, thermoelastic, viscoelastic and other waves justify the increasing demand for effective methods to answer technological and scientific questions as they are described above.

Biomedical technology is a modern subject with ancient roots and covers a very wide territory, i.e. biomaterials, new diagnostic equipment, computer modeling of human organs, software for analysis of medical data, imaging systems, telemetry systems, biomedical sensors, expert systems, new biological products, biomechanics of the human body, etc. The most important contribution of modern biomedical technology lies in its promotion of a better understanding of physiological and pathological mechanisms of human being suffered by harm of organs or/and tissues. The analysis and principles of mechanics, applied mathematics, methods of computer and engineering sciences developed in the age of technological explosion can be adopted to deal with the complex problems of health science and technology.

The papers in this volume address the state-of-the-art and future directions in Applied Mathematics in both Scattering Theory and Biomedical Technology. They are papers contributed by several authors in a workshop organized at the Metsovo Conference Center, Metsovo, Greece from June 30th to July 1st 1997. This workshop brought together some of the world's foremost experts in the field with researchers working in Greece. The two days workshop featured twenty lectures. Sixteen of the contributed papers appear in this volume.

The workshop was dedicated to the retirement of Professor Gary F. Roach of Strathclyde University on the occasion of his retirement. Prof. Roach's monumental contribution, as well as his influence to applied mathematics is well known to the international community and we are pleased to acknowledge it here.

George Dassios

Dimitrios I. Fotiadis

Kiriakie Kiriaki

Christos V. Massalas

Opening Remarks for Workshop on Mathematical Methods in Scattering Theory and Biomedical Technology

Ralph Kleinman

(Prof. R. Kleinman passed away at his home in Delaware on February 19, 1998)

This workshop is dedicated to Gary Roach who has recently been promoted to emeritus status. I have explained to Gary that we would break with tradition at this meeting and rather than have each participant present a paper to which Gary has to listen attentively and ask a penetrating question, this time Gary would present all the talks and we would ask the questions. So if you would all please pass your transparencies to the front I'll give them to Gary so he can get ready for the first talk. No, No, never mind, this way we only have to clear our heads of wine and ouzo when we speak and lapse into a euphoric coma for the rest of the time. But not you Gary. Most of you have probably wondered why Gary retired so prematurely since he is still so young. The thought may even have crossed Isabel's mind. Well, I can reveal that in fact his retirement is not premature at all since he has had some dramatic episodes which have aged him beyond his years. No, I am not referring to that incident when he fell off a mountain and was in a body cast for a year. That after all was only a year. And it didn't even cure him of mountains. No, what really aged him were some of his flying experiences. Not when he was piloting a plane. I mean while he was a passenger as Kouli's husband Theodore was flying through the streets of Athens at warp speed. And what youth Theodore left, I confess, I helped squander. More than 10 years ago Gary and I both attended a Potential Theory Conference in Prague well before the velvet revolution. We arrived in different planes, but I saw him in the baggage claim area although he didn't see me. I walked behind him, tapped him on the shoulder and said in my most menacing voice "May I see your papers, please?" I think Gary aged 5 years in the few seconds it took him to recognize me. So his retirement is not untimely, but perhaps you didn't know that he already has another job. He has found his true calling in making bonnets. In fact he is so good at it that he has been selected as chief bonnet maker. As I understand it, the duties of a bonnet maker are to eat and drink outrageously well and dispense large sums of money to the poor of Glasgow. Our task during this meeting is to convince Gary to expand the brief of these charitable bonnet makers to include indigent applied mathematicians. The way research funding is going we may need to line up with Glasgow's poor.

But despite all of this, you may still ask why Gary retired. A brief look at his career provides part of the answer. Consider the following litany of achievement.

Teaching:	He has taught the full spectrum of analysis courses, guided 20 Ph.D. Students.
Research:	His interests in functional analysis and operator theory as applied to problems in scattering theory, inverse problems, time dependent problems and integral equations has resulted in over a hundred papers in prestigious journals.
Books:	His famous book on Green's functions is a classic and his new book on scattering will become. And he has edited 8 other books on various research topics and translated Jürgens deep book on integral equations.
Administration:	He has served as Department Chair, Dean and has been on every conceivable University Committee as well as a host of external commissions. And he has built the applied analysis group at Strathclyde which has put Strathclyde on the mathematical map.
Editorships:	He has served on a number of editorial boards including managing editor of Mathematical Methods in the Applied Sciences.
Conferences:	He has been plenary speaker at major conferences and has organized the highly successful series of Ross Priory Conferences.
Professional - Honorary Societies:	He is a fellow of 4 Societies including the Royal Society of Edinburgh, has been President of the Edinburgh Math Society.
Honors:	He has received not one but two honorary Doctorats, from Lodz and Manchester. He has visited, lectured, and has collaborators all over the worlds - Canada, Greece, Germany, the U.S., New Zealand, South Africa, Poland, France and the list goes on.

In short why did Gary retire? Because, simply, he has done it all with style and grace and good humor. On a personal note, this year marks the 25th year of our collaboration which I look on as a high point in my professional life and which has grown into warm friendship and more. Gary and Isabel have become family.

I'd like to share with you one incident in that collaboration. It was during one of the Dundee Differential Equation Conferences, as I recall, but I don't remember the year. Gary and I played hokey after the talks one afternoon and were trading ideas in my room when we discovered a way to solve explicitly scattering problems for an arbitrarily shaped scatterer. The result was so astounding that we stopped and literally floated outside to walk a bit and enjoy the excitement, the euphoria, the heady feeling of discovery. Never mind that afterwards we found the flaw in our reasoning. It is for those few blissful moments that we struggle so with problems just a little too difficult for us to solve, and the thought of experiencing that rush again I'm sure will keep Gary up to his ears in research for many years to come.

It is a little early in the day for a toast but Gary, we honor your past accomplishments and we respect your decision to withdraw from the academic rat race. But your credentials as a working research mathematician are still in good order and they won't expire for many years. So on behalf of all the participants I wish to thank Kouli, George and Christos and all those who made it possible to have this opportunity to not only recognize Gary's past achievements, but provide the seeds for future collaboration. So let us begin the workshop and get those seeds planted. Gary - old friend - we salute you and wish you well.

Metsovo, 30 June 1997

SECTION A:
Direct and Inverse Problems in Scattering Theory

DAVID L COLTON

A simple method for solving electromagnetic inverse scattering problems[1]

1 Introduction

It gives me particular pleasure to dedicate this paper to Professor Gary Roach on the occasion of his retirement from the University of Strathclyde. I first met Gary in 1972 when I was a visitor at the University of Glasgow and then as a colleague at Strathclyde from 1975 to 1978. I have always valued our relationship from both a personal and professional point of view and I hope he finds the following of some interest.

In a recent paper [1], Andreas Kirsch and I introduced a new method for solving inverse scattering problems in the resonance region. The attractive features of this method are that no low–frequency or high–frequency approximations are made, its implementation only requires the solution of a linear integral equation, and it is not required to know what boundary conditions are imposed on the scattering object. For both penetrable and impenetrable media the method yields the support of the medium. In many cases in medical imaging and non–destructive testing this information is sufficient and hence our method avoids the need to solve a time consuming nonlinear optimization problem.

The purpose of this paper is to give a tutorial presentation of this "simple method" for solving the inverse scattering problem. For the sake of simplicity I will concentrate my attention on the scattering of electromagnetic waves by an inhomogeneous infinite cylinder for both the TM mode and the TE mode. I will avoid giving any proofs, referring the reader to the existing literature for the proofs of any statements that are made. Generalizations of the simple method to the inverse scattering problem

[1]This research was sponsored by the Air Force Office of Scientific Research.

for perfect conductors, orthotropic media, Maxwell's equations in R^3 and scatterers imbedded in a piecewise constant background media can also be found in the literature [1], [3], [7], [4].

2 The Simple Method for TM Polarized Waves

The scattering of electromagnetic waves polarized perpendicular to the axis of an infinite cylinder where the electric field only has a component in the direction of the axis of the cylinder is called the transverse magnetic or TM mode in scattering theory. If we assume that the cylinder is homogeneous in the axial direction and the incident electric field is a time harmonic plane wave propagating in the direction $(d_1, d_2, 0)$ where $(0, 0, d_3)$ lies along the axis of the cylinder then the electric field u satisfies

$$\Delta_2 u + k^2 n(x) u = 0 \quad \text{in} \quad R^2 \tag{2.1}$$

$$u(x) = e^{ikx\cdot d} + u^s(x) \tag{2.2}$$

$$\lim_{r\to\infty} \sqrt{r} \left(\frac{\partial u^s}{\partial r} - iku^s \right) = 0 \tag{2.3}$$

where $k > 0$ is the wave number, $x \in R^2$, $d = (d_1, d_2)$ is a unit vector and $r = |x|$. We assume that the index of refraction n is a piecewise smooth function where $m := 1 - n$ has compact support $\bar{D}$ with smooth boundary ∂D and the only discontinuities of n are across ∂D. Under these conditions we have that

$$u^s(x) = \frac{e^{ikr}}{\sqrt{r}}\, u_\infty(\hat{x}, d) + O(r^{-3/2}) \tag{2.4}$$

as r tends to infinity where $\hat{x} = x/|x|$ and u_∞ is the *far field pattern* of the scattered wave u^s. The *inverse scattering problem* we are interested in is to determine ∂D from $u_\infty(\hat{x}, d)$ for $\hat{x}, d$ on the unit circle $\Omega := \{x : |x| = 1\}$.

The basic idea of the "simple method" for solving the inverse scattering problem is to try and find a superposition of plane waves such that the scattered field corresponding to this superposition is a constant multiple of a point source $\Phi(\cdot, y)$ located at a point $y \in D$. Since the scattered field is uniquely determined by its far field pattern [2], it suffices to have the far field pattern corresponding to this superposition agree with a constant multiple of the far field pattern of $\Phi(\cdot, y)$. In particular, we can set

$$\Phi(x, y) := \frac{i}{4} H_0^{(1)}(k|x-y|) \tag{2.5}$$

where $H_0^{(1)}$ is a Hankel function of the first kind of order zero. Then, since $\Phi(\cdot, y)$ has a far field pattern Φ_∞ given by

$$\Phi_\infty(\hat{x}, y) = \frac{e^{i\pi/4}}{\sqrt{8\pi k}} e^{-ik\hat{x}\cdot y} \tag{2.6}$$

we want to find a function $g(\cdot, y) \in L^2(\Omega)$ such that

$$\int_\Omega u_\infty(\hat{x}, d)\, g(d)\, ds(d) = \Phi_\infty(\hat{x}, y) \tag{2.7}$$

is satisfied. This basic equation is called the *far field equation.* A short calculation using Rellich's lemma shows that (2.7) is solvable if and only if there exist functions w and v satisfying the *interior transmission problem*

$$\left.\begin{aligned} &\Delta_2 w + k^2 n(x) w = 0 \\ &\Delta_2 v + k^2 v = 0 \end{aligned}\right. \quad \text{in } D \tag{2.8}$$

$$\left.\begin{aligned} &w - v = \Phi(\cdot, y) \\ &\frac{\partial}{\partial \nu}(w - v) = \frac{\partial}{\partial \nu}\Phi(\cdot, y) \end{aligned}\right. \quad \text{on } D \tag{2.9}$$

where ν is the unit outward normal to ∂D and v is a *Herglotz wave function* with kernel g, i.e. v is a solution of the Helmholtz equation $\Delta_2 v + k^2 v = 0$ of the form

$$v(x) = \int_\Omega e^{ikx\cdot d}\, g(d)\, ds(d)\,. \tag{2.10}$$

In order to proceed further, we must show the existence of a unique solution to the interior transmission problem (2.8), (2.9). To this end, we make the additional assumption that there exists a positive constant c such that $\operatorname{Im} n(x) \geq c$ for $x \in D$. It suffices to show the existence of a unique weak solution of (2.8), (2.9) as defined below. To motivate this definition, we note that if a solution $v, w \in C^2(D) \cap C^1(\bar{D})$ to (2.8), (2.9) exists, then from Green's formula and (2.8) we have that

$$w(x) + k^2 \int_D \Phi(x,\xi)\, m(\xi)\, w(\xi)\, d\xi = v(x)\,, \quad x \in D \tag{2.11}$$

for $m := 1 - n$ and (2.9) will be satisfied provided

$$-k^2 \int_D \Phi(x,\xi)\, m(\xi)\, w(\xi)\, d\xi = \Phi(x,y)\,, \quad x \in \partial B \tag{2.12}$$

where B is a ball centered at the origin with $\bar{D} \subset B$. This last statement follows from Rellich's lemma and the unique continuation principle.

Definition: *Let H be the linear space of Herglotz wave functions and $\bar{H}$ the closure of H in $L^2(D)$. Then a pair w, v with $w \in L^2(D)$ and $v \in \bar{H}$ is said to be a weak solution of the interior transmission problem if w and v satisfy (2.11) and (2.12).*

With this definition it is now possible to establish the following basic result [6]:

Theorem: *For every point $y \in D$ there exists a unique weak solution to the interior transmission problem.*

We can now turn our attention to a description of the "simple method" for determining the support $\bar{D}$ from a knowledge of the far field pattern. From our assumptions, it suffices to determine ∂D. The basic idea of the inversion scheme is to determine g such that the far field equation (2.7) is satisfied. If this can be done, the Herglotz wave function v with kernel g satisfies the interior transmission problem (2.8), (2.9). We would then like to conclude that, on ∂D, v becomes unbounded as y tends to ∂D and hence so does $\|g\|_{L^2(\Omega)}$. If this is true then ∂D is determined by those points y such that the L^2–norm of $g(\cdot, y)$ becomes unbounded. In practice, we can do this by solving the far field equation for y on a rectangular grid known a priori to contain D and then look for those values of y where $\|g\|_{L^2(\Omega)}$ is large.

Unfortunately, the far field equation (2.7) is a (improperly posed) linear integral equation of the first kind and in general we cannot conclude that a solution $g = g(\cdot, y)$ exists. Even if a solution does exist it is not clear from (2.8), (2.9) that v (and hence

$\|g\|_{L^2(\Omega)})$ becomes unbounded as y tends to ∂D. However, using the above theorem on the existence and uniqueness of a weak solution to the interior transmission problem, we can prove the following result [6]:

Theorem: *For every $\epsilon > 0$ and $y \in D$ there is a solution $g = g(\cdot, y) \in L^2(\Omega)$ of the inequality*

$$\left\| \int_\Omega u_\infty(\cdot, d)\, g(d)\, ds(d) - \Phi_\infty(\cdot, y) \right\|_{L^2(\Omega)} \leq \epsilon$$

such that

$$\lim_{y \to \partial D} \|g\|_{L^2(\Omega)} = \infty \quad \text{and} \quad \lim_{y \to \partial D} \|v_g\|_{L^2(D)} = \infty$$

where v_g is the Herglotz wave function with kernel g.

To determine the boundary ∂D of the support of $m = 1 - n$ we now need to find a regularized solution of the far field equation which approximates the solution g of the above theorem. Methods which have been shown to be numerically successful for doing this will be discussed in the final section of this paper.

3 The Simple Method for TE Polarized Waves

The scattering of electromagnetic waves polarized parallel to the axis of an infinite cylinder where the magnetic field only has a component in the direction of the axis of the cylinder is called the transverse electric or TE mode of scattering theory. If we again assume that the cylinder is homogeneous in the axial direction and the incident electric field is a time harmonic plane wave propagating in the direction $(d_1, d_2, 0)$ where $(0, 0, d_3)$ lies along the axis of the cylinder then the magnetic field satisfies

$$\Delta_2\, u_0 + k^2\, u_0 = 0 \quad \text{in} \quad R^2 \backslash \bar{D} \tag{3.1}$$

$$\nabla \cdot \left(\frac{1}{n} \nabla u \right) + k^2 u = 0 \quad \text{in} \quad D \tag{3.2}$$

$$u_0 = u\,, \quad \frac{\partial u_0}{\partial \nu} = \frac{1}{n_0} \frac{\partial u}{\partial \nu} \quad \text{on} \quad \partial D \tag{3.3}$$

$$u_0(x) = e^{ikx \cdot d} + u^s(x) \tag{3.4}$$

$$\lim_{r\to\infty} \sqrt{r}\left(\frac{\partial u^s}{\partial r} - iku^s\right) = 0 \tag{3.5}$$

where $x \in R^2$, k and d are as previously defined and the following conditions on n are assumed:

(i) $n \in C^2(R^2)$ and $n(x) = n_0$ for $x \in R^2 \backslash D$ where $\operatorname{Im} n_0 = 0$ and $n_0 \neq 1$.

(ii) $\operatorname{Im} n(x) \geq 0$ and the set $D_0 = \{x \in D : \operatorname{Im} n(x) > 0\}$ is nonempty.

It is again assumed that the support $\bar{D}$ is bounded with a connected compliment and that the boundary ∂D is smooth with unit outward normal ν. The asymptotic expansion (2.4) remains valid and we are again interested in the inverse problem of determining ∂D from a knowledge of the far field pattern $u_\infty(\hat{x}, d)$ for $\hat{x}, d \in \Omega := \{x : |x| = 1\}$.

The "simple method" for determining ∂D now proceeds as in the case of TM polarized waves, i.e. we want to "solve" the far field equation (2.7) for y on a rectangular grid known a priori to contain D and then look for those values of y where $\|g\|_{L^2(\Omega)}$ is large. The difference is that we now must investigate a different interior transmission problem than (2.8), (2.9) in order to provide a mathematical justification for doing this. In particular, the interior transmission problem we must now examine is to find functions w and v satisfying

$$\left.\begin{array}{l} \nabla \cdot \left(\frac{1}{n}\nabla w\right) + k^2 w = 0 \\ \Delta_2 v + k^2 v = 0 \end{array}\right. \quad \text{in } D \tag{3.6}$$

$$\left.\begin{array}{l} w - v = \Phi(\cdot, y) \\ \frac{1}{n_0}\frac{\partial w}{\partial \nu} - \frac{\partial v}{\partial \nu} = \frac{\partial}{\partial \nu}\Phi(\cdot, y) \end{array}\right. \quad \text{on } \partial D\,. \tag{3.7}$$

To this end we have the following theorem [5]:

Theorem: *Under the above assumptions on n there exists a unique solution $w \in C^2(D) \cap C^1(\bar{D})$, $v \in C^2(D) \cap C^1(\bar{D})$ of the interior transmission problem (3.6), (3.7) and v can be approximated in $C^1(\bar{D})$ by a Herglotz wave function.*

Using this theorem, one can now establish the analogue of the theorem stated at the end of the last section of this paper [5]. Hence we are again faced with the problem of finding a regularized solution of the far field equation (2.7) and looking for those points y where this regularized solution becomes large. A procedure for doing this will be briefly described in the next section of this paper.

4 The Numerical solution of the Far Field Equation

The above discussion has shown that in order to determine the support of an inhomogeneous anomaly from measured far field data in the case of either TM or TE polarized electromagnetic waves one must solve the far field equation (2.7) and look for those points y where $g = g(\cdot, y)$ becomes large. However, from a numerical point of view, this is problematical since the integral equation (2.7) is improperly posed and furthermore if u_∞ is noisy and g is large the left hand side of (2.7) is very noisy indeed! In particular we must take into consideration the following facts:

(i) Whatever regularization method we use must allow the regularization parameter α to depend on y. In particular, if Tikhonov's method is adopted, the regularized solution is determined from the minimization problem

$$\left\| \int_\Omega u_\infty(\cdot, d)\, g(d, y)\, ds(d) - \Phi_\infty(\cdot, y) \right\|^2_{L^2(\Omega)} + \alpha\, \|g(\cdot, y)\|^2_{L^2(\Omega)} = \text{minimum} \quad (4.1)$$

where $\|g(\cdot, y)\|^2_{L^2(\Omega)}$ becomes unbounded as y approaches ∂D and hence the choice of α depends on y.

(ii) The far field equation is different than the usual linear inverse problem $Az = f$ where the data f on the right hand side is noisy and the operator A is known exactly. On the contrary, in our case the kernel of the operator A is noisy and the right hand side f is known exactly.

In [5] and [6] these problems were dealt with by regularizing the far field equation using Tikhonov's method where the regularization parameter is allowed to depend on y. In particular, for each y, (4.1) is solved where α is chosen by using the generalized discrepancy principle given in [9]. This criterion represents a generalization of Morozov's discrepancy principle (c.f. [2], [8]) to linear operator equations when the operator is not known exactly. As is seen in [5] and [6], a rather surprising result of this approach is that the level curves of the regularization parameter $\alpha = \alpha(y)$ are often a better indication of the location of the support $\bar{D}$ than the level curves of the L^2–norm of the regularized solution. Hence we have the unusual fact that in these cases the regularization parameter is the main object of interest rather than the regularized solution of the integral equation itself!

References

1. D. Colton and A. Kirsch, A simple method for solving inverse scattering problems in the resonance region, *Inverse Problems*, **13** (1996), 383–393.

2. D. Colton and R. Kress, *Inverse Acoustic and Electromagnetic Scattering Theory*, Second Edition, Springer–Verlag, Berlin, 1998.

3. D. Colton, R. Kress and P. Monk, Inverse scattering from an orthotropic medium, *Jour. Comp. Appl. Math.* **81** (1997), 269–298.

4. D. Colton and P. Monk, A linear sampling method for the detection of leukemia using microwaves, *SIAM Jour. Appl. Math.*, to appear.

5. D. Colton and M. Piana, The simple method for solving the electromagnetic inverse scattering problem: the case of TE polarized waves, submitted for publication.

6. D. Colton, M. Piana and R. Potthast, A simple method using Morozov's discrepancy principle for solving inverse scattering problems, *Inverse Problems* **13** (1997), 1477–1493.

7. D. Colton and R. Potthast, The inverse electromagnetic scattering problem for an anisotropic medium, submitted for publication.

8. A. Kirsch, *An Introduction to the Mathematical Theory of Inverse Problems*, Springer–Verlag, Berlin, 1996.

9. A. N. Tiknonov, A. V. Goncharsky, V. V. Stepanov and A. G. Yagola, *Numerical Methods for the Solution of Ill–Posed Problems*, Kluwer, Dordrecht, 1995.

Department of Mathematical Sciences
University of Delaware
Newark, Delaware 19716
U.S.A.

E–mail: colton@math.udel.edu

A CHARALAMBOPOULOS, G DASSIOS AND G KAMVYSSAS

Reciprocity theorems for point source scalar scattering

1. Introduction

In this work the basic reciprocity theorems for acoustic wave fields, in the framework of point source wave scattering, are presented. A reciprocity theorem interrelates, in a specific manner, the field or wave quantities which characterize two admissible states that could occur in one and the same domain in space. Each one of the two states can be associated with its own set of medium-parameters but these parameters are independent of the values of the fields or wave quantities. In other words the case of linear media in their acoustic behaviour is considered.

In this work we consider the acoustic scattering problem in the exterior space of a soft or hard scatterer, wherein there exists one point source generating the incident field. The two states that are connected through reciprocity theorems are characterized by two different specific locations of the point source. The problem is handled in the frequency domain, using the integral equation approach. We also examine the case when the one point source is replaced by a plane wave excitation and finally the classical reciprocity theorems for plane wave incidence are recovered in the limit as the modified point sources recede to infinity.

2. Formulation of the problem

Let S be a smooth, closed and bounded surface constituting a discontinuity surface in an acoustic wave propagation medium of density p. This surface defines a bounded, smooth and closed subset V^c of R^3, which constitutes the scatterer and is responsible for the disturbance of the wave propagation in the acoustic medium. The complement V of the scatterer forms the medium of propagation, within which a point-source field u^i, emanating from $\mathbf{r}_0$ is propagating. We work in the frequency domain, assuming that the time dependence of the quantities characterizing the problem is harmonic given by the term exp{-

iωt}, where ω denotes the angular frequency. The source point $\mathbf{r}_0$ of the incidence field is located outside the smallest sphere that circumscribes the scatterer V^c. Furthermore, we consider the case of a modified point-source field which conforms with the demands according to which we recover the plane wave incidence in the direction $-\hat{\mathbf{r}}_0$ and preserve the normal acoustic intensity at the origin, as the source point $\mathbf{r}_0$ recedes radially to infinity. According to the framework of [1], these demands determine the amplitude and phase of the incident field, which obtains the following form:

$$u^i(\mathbf{r}) = \frac{r_0}{|\mathbf{r}-\mathbf{r}_0|} e^{ik(|\mathbf{r}-\mathbf{r}_0|-r_0)} \tag{1}$$

where *k* stands for the wavenumber.

The incident as well as the scattered field have their sources within a bounded region and consequently the total field

$$u(\mathbf{r}) = u^i(\mathbf{r}) + u^s(\mathbf{r}), \qquad \mathbf{r} \in V - \{\mathbf{r}_0\} \tag{2}$$

satisfies the Sommerfeld radiation condition

$$r\left(\frac{\partial u(\mathbf{r})}{\partial r} - iku(\mathbf{r})\right) \xrightarrow{r\to\infty} 0 \tag{3}$$

uniformly over directions.

The total field *u* solves the Helmholtz equation

$$(\Delta + k^2)u(\mathbf{r}) = 0, \qquad \mathbf{r} \in V - \{\mathbf{r}_0\} \tag{4}$$

and satisfies a specific boundary condition on the boundary S of the scatterer. We consider here two alternatives.

The case of the soft scatterer, where

$$u(\mathbf{r}) = 0, \qquad \mathbf{r} \in S \tag{5}$$

and the case of the hard one, in which

$$\frac{\partial u(\mathbf{r})}{\partial n} = 0, \qquad \mathbf{r} \in S \tag{6}$$

where $\frac{\partial}{\partial n}$ indicates exterior normal differentiation on S.

Following integral equation approach, we obtain the following integral representation for the total field [1]

$$u(\mathbf{r}) = u^i(\mathbf{r}) + \frac{ik}{4\pi}\int_S \left[u(\mathbf{r}')\frac{\partial}{\partial n'}h\left(k|\mathbf{r}-\mathbf{r}'|\right) - h\left(k|\mathbf{r}-\mathbf{r}'|\right)\frac{\partial}{\partial n'}u(\mathbf{r}')\right] ds(\mathbf{r}'), \mathbf{r} \in V - \{\mathbf{r}_0\} \tag{7}$$

where

$$h(x) = \frac{e^{ix}}{ix} \tag{8}$$

is the zeroth-order spherical Hankel function of the first kind.

Standard asymptotic analysis furnishes the form

$$u^s(\mathbf{r}) = g(\hat{\mathbf{r}})\, h(kr) + 0\left(1/r^2\right), \qquad r \to \infty \tag{9}$$

where the scattering amplitude is given by the surface integral

$$g(\hat{\mathbf{r}}) = -\frac{ik}{4\pi}\int_S \left[\frac{\partial}{\partial n'}u(\mathbf{r}') + ik\hat{\mathbf{r}}\cdot\hat{\mathbf{n}}' u(\mathbf{r}')\right] e^{-ik\hat{\mathbf{r}}\cdot\mathbf{r}'} ds(\mathbf{r}') \qquad \hat{\mathbf{r}} \in S^2 \tag{10}$$

where S^2 stands for the unit sphere.

The scattering amplitude plays a very important role in scattering theory as it incorporates all the necessary information concerning the scatterer and its properties (geometrical and physical) that can be exploited from the measurement of the far field. There are many methods to extract the necessary information from the knowledge of the scattering amplitude and to obtain useful results about the parameters of the scattering procedure. One of the most interesting approaches revealing a lot of qualitative and quantitative properties of the scattering procedure is the determination of some reciprocity relations governing the wave propagation under the circumstances of every specific problem. The reciprocity theorems interrelate two admissible states occurring in the same propagation medium, and are useful as they reveal the degenerate states of the problem, the hidden symmetries and provide us with a priori information applied to the specific scattering procedure.

For the plane wave excitation scattering problem, the most useful reciprocity relation is this interrelating the scattering amplitudes of two different scattering problems taking place in the same acoustic wave propagation medium.

More precisely, denoting the dependence of the scattering amplitude on the incident plane wave direction by a second argument we have the well-known formula

$$g(\hat{\mathbf{r}};\hat{\mathbf{k}}) = g(-\hat{\mathbf{k}};-\hat{\mathbf{r}}) \tag{11}$$

indicating that interchanging incidence with observation direction does not alter the measured scattering amplitude. Our goal is to find similar relations for the case of point source excitation acoustic scattering problem connecting now states where the presence of the wave source in the finite space plays a crucial role.

3. Reciprocity in point source excitation acoustic scattering.

The interrelation between different states is going to be established through suitable use of the integral representations of the fields entering the scattering process. In all these treatments a crucial quantity of "inner product-type" makes its appearance. More precisely we define

$$\{\alpha,\beta\}_C = \int_C \left[\alpha(\mathbf{r}';\mathbf{r})\frac{\partial}{\partial n'}\beta(\mathbf{r}';\mathbf{r}) - \frac{\partial}{\partial n'}\alpha(\mathbf{r}';\mathbf{r})\,\beta(\mathbf{r}';\mathbf{r}) \right] ds(\mathbf{r}') \tag{12}$$

for every pair of scalar sufficiently smooth functions α, β and every closed smooth surface C.

In this terminology, the integral representations introduced in previous section can be presented in a simpler way. Indeed Eq. (7) takes the form

$$u(\mathbf{r}) = u^i(\mathbf{r}) + \frac{ik}{4\pi}\{u(\mathbf{r}'), h(k\,|\,\mathbf{r}-\mathbf{r}'|)\}_S, \qquad \mathbf{r} \in V - \{\mathbf{r}_0\} \tag{13}$$

In order to establish reciprocity results, let us consider two scattering processes taking place in the same propagation medium we have already introduced but let us suppose that the two point sources of the two problems are situated at points $\mathbf{r}_0$ and $\mathbf{r}_1$, respectively. In order to distinguish the two problems we add a subscript in all the quantities of the problem indicating the point-source position.

Consequently, we obtain

$$u_{\mathbf{r}_0}(\mathbf{r}) = u^i_{\mathbf{r}_0}(\mathbf{r}) + \frac{ik}{4\pi}\left\{u_{\mathbf{r}_0}(\mathbf{r}'), h\left(k\,|\,\mathbf{r}-\mathbf{r}'|\right)\right\}_S \tag{14}$$

$$\left(u^s_{\mathbf{r}_0}(\mathbf{r}) = \frac{ik}{4\pi}\left\{u_{\mathbf{r}_0}(\mathbf{r}'), h\left(k\,|\,\mathbf{r}-\mathbf{r}'\,|\right)\right\}_S\right) \tag{15}$$

for the first scattering process and

$$u_{\mathbf{r}_1}(\mathbf{r}) = u^i_{\mathbf{r}_1}(\mathbf{r}) + \frac{ik}{4\pi}\left\{u_{\mathbf{r}_1}(\mathbf{r}'), h\left(k\,|\,\mathbf{r}-\mathbf{r}'\,|\right)\right\}_S \tag{16}$$

$$\left(u^s_{\mathbf{r}_1}(\mathbf{r}) = \frac{ik}{4\pi}\left\{u_{\mathbf{r}_1}(\mathbf{r}'), h\left(k\,|\,\mathbf{r}-\mathbf{r}'\,|\right)\right\}_S\right) \tag{17}$$

for the second one.

In order to take advantage of the integral representation properties let us consider the geometrical feature consisting in the scatterer surface S, the two positions $\mathbf{r}_0$, $\mathbf{r}_1$ (of the point sources), too small spheres S_0 and S_1 surrounding the points $\mathbf{r}_0$ and $\mathbf{r}_1$ as well as a large smooth surface J surrounding the whole system of the scatterer and the two spheres.

Let us now determine the "inner product" $\left\{u_{\mathbf{r}_0}, u_{\mathbf{r}_1}\right\}_J$. By exploiting Green's theorem, we can transfer this integral to the interior surfaces obtaining

$$\left\{u_{\mathbf{r}_0}, u_{\mathbf{r}_1}\right\}_J = \left\{u_{\mathbf{r}_0}, u_{\mathbf{r}_1}\right\}_{S_0} + \left\{u_{\mathbf{r}_0}, u_{\mathbf{r}_1}\right\}_{S_1} + \left\{u_{\mathbf{r}_0}, u_{\mathbf{r}_1}\right\}_S \tag{18}$$

Either set of boundary conditions forces $\left\{u_{\mathbf{r}_0}, u_{\mathbf{r}_1}\right\}_S$ to vanish. Indeed

$$\left\{u_{\mathbf{r}_0}, u_{\mathbf{r}_1}\right\}_S = \int_S \left[u_{\mathbf{r}_0}(\mathbf{r}')\frac{\partial}{\partial n'}u_{\mathbf{r}_1}(\mathbf{r}') - \frac{\partial}{\partial n'}u_{\mathbf{r}_0}(\mathbf{r}')u_{\mathbf{r}_1}(\mathbf{r}') \right] ds(\mathbf{r}') = 0 \tag{19}$$

independently of what boundary condition (5) or (6) is satisfied by the total fields $u_{\mathbf{r}_0}$ and $u_{\mathbf{r}_1}$.

As far as the integral on S_0 is considered, we remark that

$$\left\{u_{\mathbf{r}_0}, u_{\mathbf{r}_1}\right\}_{S_0} = \int_{S_0} \left[u^i_{\mathbf{r}_0}\frac{\partial}{\partial n'}u_{\mathbf{r}_1} - \frac{\partial}{\partial n'}u_{\mathbf{r}_0}u_{\mathbf{r}_1} \right] ds' + \int_{S_0} \left[u^s_{\mathbf{r}_0}\frac{\partial}{\partial n'}u_{\mathbf{r}_1} - \frac{\partial}{\partial n'}u^s_{\mathbf{r}_0}u_{\mathbf{r}_1} \right] ds'$$

Due to the regularity of $u_{\mathbf{r}_1}$ and $u^s_{\mathbf{r}_0}$ inside S_0 the second integral of the last term of the above equation vanishes and so

$$\left\{u_{r_0}, u_{r_1}\right\}_{S_0} = \int_{S_0} \left[u^i_{\mathbf{r}_0}\frac{\partial}{\partial n'}u_{\mathbf{r}_1} - \frac{\partial}{\partial n'}u_{\mathbf{r}_0}u_{\mathbf{r}_1} \right] ds' =$$

$$= \lim_{\varepsilon\to 0} \int_{\sigma_\varepsilon} \frac{r_0}{\varepsilon} e^{ik(\varepsilon - r_0)} \frac{\partial}{\partial r'}u_{\mathbf{r}_1} ds' +$$

$$+ \lim_{\varepsilon\to 0} \int_{\sigma_\varepsilon} \frac{r_0}{\varepsilon} e^{ik(\varepsilon - r_0)} u_{\mathbf{r}_1} ds' =$$

$$= 4\pi r_0 e^{-ikr_0} u_{\mathbf{r}_1}(\mathbf{r}_0) \tag{20}$$

where we transferred the integral to a sphere σ_ε of radius ε shrinking to $\mathbf{r}_0$ ($\varepsilon\to 0$).

Similarly, we obtain

$$\left\{u_{\mathbf{r}_0}, u_{\mathbf{r}_1}\right\}_{S_1} = -\left\{u_{\mathbf{r}_1}, u_{\mathbf{r}_0}\right\}_{S_1} = -4\pi r_1 e^{-ikr_1} u_{\mathbf{r}_0}(\mathbf{r}_1) \tag{21}$$

Combining (18)-(21), we obtain

$$\left\{u_{\mathbf{r}_0}, u_{\mathbf{r}_1}\right\}_J = 4\pi r_0 e^{-ikr_0} u_{\mathbf{r}_1}(\mathbf{r}_0) - 4\pi r_1 e^{-ikr_1} u_{\mathbf{r}_0}(\mathbf{r}_1) \tag{22}$$

The "inner product" $\left\{u_{\mathbf{r}_0}, u_{\mathbf{r}_1}\right\}_J$ can be determined alternatively exploiting the asymptotic behaviour as $r\to +\infty$ of the involved fields.

Remark first that

$$\left\{u_{\mathbf{r}_0}, u_{\mathbf{r}_1}\right\}_J = \left\{u^i_{\mathbf{r}_0}, u^i_{\mathbf{r}_1}\right\}_J + \left\{u^i_{\mathbf{r}_0}, u^s_{\mathbf{r}_1}\right\}_J + \left\{u^s_{\mathbf{r}_0}, u^i_{\mathbf{r}_1}\right\}_J + \left\{u^s_{\mathbf{r}_0}, u^s_{\mathbf{r}_1}\right\}_J \tag{23}$$

Using Green's theorem all the previous "inner products" can be transferred to integrals over a sphere S_R of radius R tending to infinity. Let us consider for example $\left\{u^i_{\mathbf{r}_0}, u^i_{\mathbf{r}_1}\right\}_J$

$$\left\{u^i_{\mathbf{r}_0}, u^i_{\mathbf{r}_1}\right\}_J = \int_J \left[u^i_{\mathbf{r}_0} \frac{\partial}{\partial n'} u^i_{\mathbf{r}_1} - \frac{\partial}{\partial n'} u^i_{\mathbf{r}_0} u^i_{\mathbf{r}_1}\right] ds' =$$

$$= \lim_{R\to+\infty} \int_{S_R} \left[u^i_{\mathbf{r}_0} \frac{\partial}{\partial R} u^i_{\mathbf{r}_1} - \frac{\partial}{\partial R} u^i_{\mathbf{r}_0} u^i_{\mathbf{r}_1}\right] ds =$$

$$= \lim_{R\to+\infty} \int_{S_R} \left[ik \frac{e^{ik(R-r_0)}}{R} e^{-ik\hat{\mathbf{R}}\cdot\mathbf{r}_0} \frac{e^{ik(R-r_1)}}{R} e^{-ik\hat{\mathbf{R}}\cdot\mathbf{r}_1} \right.$$

$$\left. - ik \frac{e^{ik(R-r_0)}}{R} e^{-ik\hat{\mathbf{R}}\cdot\mathbf{r}_0} \frac{e^{ik(R-r_1)}}{R} e^{-ik\hat{\mathbf{R}}\cdot\mathbf{r}_1} \right] ds = 0 \tag{24}$$

It can be shown similarly that all terms of the last term of (23) vanish. Consequently,

$$\left\{u_{\mathbf{r}_0}, u_{\mathbf{r}_1}\right\}_J = 0 \tag{25}$$

Combining (22) and (25) we obtain

$$r_0 e^{-ikr_0} u_{\mathbf{r}_1}(\mathbf{r}_0) = r_1 e^{-ikr_1} u_{\mathbf{r}_0}(\mathbf{r}_1) \tag{26}$$

As an immediate result, we infer that the incident fields satisfy the same relation:

$$r_0 e^{-ikr_0} u^i_{\mathbf{r}_1}(\mathbf{r}_0) = r_1 e^{-ikr_1} u^i_{\mathbf{r}_0}(\mathbf{r}_1) \tag{27}$$

Subtracting (27) from (26) we conclude that

$$\frac{e^{ikr_0}}{r_0} u^s_{\mathbf{r}_0}(\mathbf{r}_1) = \frac{e^{ikr_1}}{r_1} u^s_{\mathbf{r}_1}(\mathbf{r}_0) \tag{28}$$

The last equation is the result of our analysis and constitutes the basic reciprocity theorem for point source excitation scattering process. It expresses the fact that interchanging the excitation with the observation point does not alter the scattered field except a multiplicative constant whose presence is due to the modified point source (1) we have selected in order to render point source and plane wave excitation problems compatible.

The previous reciprocity law can provide some other interesting reciprocity results as limiting cases. Indeed, let us modify the one of the two admissible states presented previously. More precisely, let us move the point source situated at $\mathbf{r}_1$ to infinity following the $\hat{\mathbf{r}}_1$ direction. As it is already mentioned the modified form of the incident field is selected appropriately in order for the point source excitation to give place to the plane-wave one after letting point source go to infinity. Consequently, eq. (28) gives

$$\frac{e^{ikr_0}}{r_0}\frac{e^{ikr_1}}{ikr_1}g\left(\hat{\mathbf{r}}_1;\mathbf{r}_0\right)=\frac{e^{ikr_1}}{r_1}u^s_{plane}\left(\mathbf{r}_0;-\hat{\mathbf{r}}_1\right)\Rightarrow u^s_{plane}\left(\mathbf{r}_0;-\hat{\mathbf{r}}_1\right)=h\left(kr_0\right)g\left(\hat{\mathbf{r}}_1;\mathbf{r}_0\right) \qquad (29)$$

where we used asymptotic analysis for $r_1 \to +\infty$.

Expression (29) indicates that the scattering amplitude at the direction $\hat{\mathbf{r}}_1$ of the point source excitation scattering problem with the source situated at $\mathbf{r}_0$ is -modulo a multiplicative constant - equal to the actual value of the scattered field, at the point $\mathbf{r}_0$ of the plane wave excitation problem in which the direction of the incident field is $-\hat{\mathbf{r}}_1$.

Finally, in case that the first admissible state becomes as well a plane wave excitation scattering problem - when point sources recedes to infinity at the $\hat{\mathbf{r}}_0$ direction - then (29) becomes

$$h\left(kr_0\right)g\left(\hat{\mathbf{r}}_0;-\hat{\mathbf{r}}_1\right)=h\left(kr_0\right)g\left(\hat{\mathbf{r}}_1;-\hat{\mathbf{r}}_0\right)\Rightarrow g\left(\hat{\mathbf{r}}_0;-\hat{\mathbf{r}}_1\right)=g\left(\hat{\mathbf{r}}_1;-\hat{\mathbf{r}}_0\right) \qquad (30)$$

4. References

[1] Dassios, G. and Kamvyssas, G. (1995), "Point Source Excitation in Direct and Inverse Scattering: The Soft and the Hard Sphere", *IMA J. Appl. Math.*, **55**, pp. 67-84.

[2] Jones, D.S. (1986), *Acoustic and Electromagnetic Waves*, Oxford University Press, Clarendon.

[3] Colton, D. and Kress, R. (1992), *Inverse Acoustic and Electromagnetic Scattering Theory*, Springer-Verlag, New York.

A A LACEY, D E TZANETIS AND P M VLAMOS

Global existence and finite-time blow-up of a non-local hyperbolic problem modelling Ohmic heating of foods

1 Introduction and Model

One method for sterilizing food is to heat it rapidly electrically. The food is passed through a conduit, part of which lies between two electrodes. A high electric current flowing between the elecrodes results in Ohmic heating of the food which quickly gets hot.

The problem was considered by Please *et al.* (1994) who looked at the stability of models allowing for different types of flow. Both homogeneous and inhomogeneous cases were discussed. More background on this type of process can be found in Biss *et al.* (1989), De Alwiss & Fryer (1992), Fryer *et al.* (1993), Skudder & Biss (1987), Stirling (1987), and Zhang & Fryer (1993).

Heat conduction, both in the direction of flow and normal to it, can be neglected, the potential and temperature vary only with z and time t and satisfy the system

$$\frac{\partial}{\partial z}\left(\sigma\frac{\partial\varphi}{\partial z}\right)=0,\ \rho c\left(\frac{\partial T}{\partial t}+v\frac{\partial T}{\partial z}\right)=\sigma\left(\frac{\partial\varphi}{\partial z}\right)^2,\ 0<z<L;$$

$$\varphi=0,\ T=T_0,\ z=0;\ \varphi=V,\ z=L,$$

where z is the distance along the channel (*i.e.* in the direction of $\mathbf{v}$). (The channel has parallel sides.)

Here V is known if the potential difference across the device is specified (but has to be determined if the process is controlled in some other way). The first equation is integrated,

$$\sigma\frac{\partial\varphi}{\partial z}=J(t),\ \text{so}\ \ V=J\int_0^L\frac{dz}{\sigma}$$

where J is the electric current density (along the channel) and

$$\frac{\partial T}{\partial t}+v\frac{\partial T}{\partial z}=\frac{J^2}{\rho c\sigma}=\left(\frac{V^2}{\rho c}\right)\left(\frac{1}{\sigma}\right)\left(\int_0^L\frac{1}{\sigma}\,dz\right)^{-2}.$$

This is a hyperbolic variant of the non-local parabolic problem considered by Lacey (1995a,b); heat transport is now by convection instead of conduction.

It is convenient to scale distance with the length of the channel, time to make the convective velocity 1, and change the temperature variable so that it becomes 0 at the inlet. The problem can then be written as

$$u_t + u_x = \frac{\lambda f(u)}{\left(\int_0^1 f(u)dx\right)^2},\ 0 < x < 1,\ t > 0; \tag{1}$$

$$u(0,t) = 0,\ t > 0; \tag{2}$$

$$u(x,0) = u_0(x),\ 0 < x < 1; \tag{3}$$

with $f > 0$ being a dimensionless electrical resistivity $(^{\cdot}\ \frac{1}{\sigma})$.

The parameter $\lambda > 0$ can be identified with (square of) applied potential difference; on occasions we shall absorb λ into f.

If the heater forms part of a circuit, so that it is connected in series with a constant resistance, and a fixed EMF is applied across the two, the scaled non-local equation is replaced by

$$u_t + u_x = \frac{f(u)}{\left[a + b\int_0^1 f(u)dx\right]^2},\ 0 < x < 1,\ t > 0 \tag{4}$$

(see Lacey, 1995a). (Results on related, higher-dimensional, non-local parabolic problems can be found in Bebernes & Lacey, Carrillo.)

In the present paper we study the problem (1) - (3) with f taking some particular forms to see what sorts of behaviour might be possible.

For the case of decreasing f (Section 4) much useful information can be gleaned from the corresponding steady-state problem (Section 2)

$$u_x = \frac{\lambda f(u)}{\left(\int_0^1 f(u)dx\right)^2},\ 0 < x < 1; \tag{5}$$

with the boundary condition (2). Such cases are amenable to treatment because a comparison principle holds (Section 3). It is seen that the solution blows up if either of λ or u_0 is large but not if λ and u_0 are both small.

2 Steady States

The stationary problem has solutions $u = w(x)$ with

$$w' = \frac{\lambda f(w)}{\left[\int_0^1 f(w)dx\right]^2}, \quad 0 < x < 1; \quad w(0) = 0.$$

This can be rewritten as

$$w' = \mu f(w), \; w(0) = 0, \quad \text{with } \lambda = \mu \left[\int_0^1 f(w)dx\right]^2.$$

On integrating the ordinary differential equation we see that

$$\lambda = \frac{M^2}{\mu},$$

where $M(\mu) = \max\limits_{[0,1]} w = w(1;\mu)$ and $\mu = \int_0^M dw/f(w)$.

It is clear that $M(\mu) \to 0+$ as $\mu \to 0+$ and, with $f(0) > 0$, $\lambda \to 0+$ as $\mu \to 0+$. Now increasing μ we might distinguish the cases

$$\int_0^\infty \frac{ds}{f(s)} = S < \infty \text{ and } \int_0^\infty \frac{ds}{f(s)} = \infty.$$

For the former $M \to \infty$ and $\lambda \to \infty$ as $\mu \to S-$; there is at least one solution for all $\lambda > 0$. In the latter situation $M \to \infty$ as $\mu \to \infty$. (We shall assume that $f(s) > 0$ for $s \geq 0$.) The behaviour of λ as μ increases depends upon the growth rate of M. From $\lim\limits_{\mu\to\infty}\{M^2/\mu\} = 2 \lim\limits_{M\to\infty} Mf(M)$, $\lambda \to \infty$ as M and $\mu \to \infty$ if $f(M) \gg 1/M$ for $M \to \infty$ but λ is bounded if $f(M) = O(1/M)$ for $M \to \infty$: there is a steady state for all $\lambda > 0$ if $Mf(M) \to \infty$ as $M \to \infty$ but there is some critical λ^* such that there is at least one steady solution for $0 < \lambda < \lambda^*$ but none for $\lambda > \lambda^*$ if $Mf(M)$ is bounded; see Fig. 1.

Freitas (1994) considered the local stability of steady states for parabolic problems.

3 Local Existence and a Comparison Principle.

Using the variable $y = x - t$ in place of x the equation can be written as

$$u_t = g(t)f(u)$$

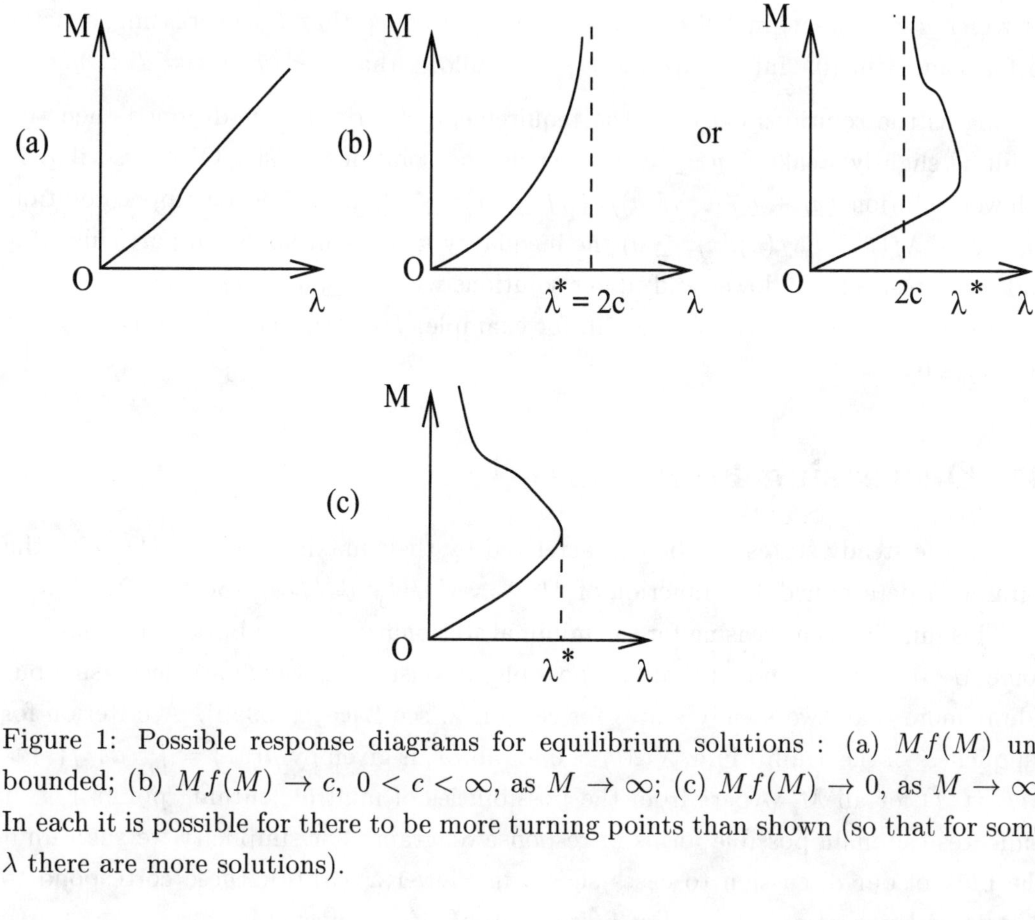

Figure 1: Possible response diagrams for equilibrium solutions : (a) $Mf(M)$ unbounded; (b) $Mf(M) \to c$, $0 < c < \infty$, as $M \to \infty$; (c) $Mf(M) \to 0$, as $M \to \infty$. In each it is possible for there to be more turning points than shown (so that for some λ there are more solutions).

where $g(t) = \lambda/[\int_{-t}^{1-t} f(u)dy]^2$, with $u = 0$ for $y = -t$.

Local existence of a solution (for $f > 0$) and its uniqueness (with f Lipschitz) then follow on employing standard methods, such as Picard iteration. Non-existence can only come about through blow-up with u becoming infinite after some finite time t^*.

More information can be gained from a comparison result which holds when f is decreasing.

Taking two different solutions $\underline{u}$ and $\bar{u}$ with $\underline{u}_0 < \bar{u}_0$ we see that there is some time T for which $\underline{u} < \bar{u}$, $\underline{g} < \bar{g}$ in $0 \le t < T$, $0 < x \le 1$, provided that f is decreasing, strictly so for some s in $(0, \inf_{t\in(0,T)} \{ \sup_{x\in(0,1)} \underline{u}(x,t)\})$. It follows that $\underline{u} < \bar{u}$ at $t = T$: $\underline{u} < \bar{u}$ as long as the solutions exist. If the requirement of strictness is dropped then the result is slightly weaker: $\underline{u} \le \bar{u}$ as long as the solutions exist. Of course if $\underline{u}$ is a lower solution $(\underline{u}_t + \underline{u}_x - \lambda f(\underline{u})/(\int_0^1 f(\underline{u})dx)^2 \le 0)$ and $\bar{u}$ is an upper solution $(\bar{u}_t + \bar{u}_x - \lambda f(\bar{u})/(\int_0^1 f(\bar{u})dx)^2 \ge 0)$ the inequality $\underline{u} \le \bar{u}$ still holds. In particular if $\underline{u}$ and $\bar{u}$ are respectively lower and upper solutions with $\underline{u}_0 \le u_0 \le \bar{u}_0$ then $\underline{u} \le u \le \bar{u}$. (Extra care would need to be taken if, for example, $f(s)$ were not Lipschitz at $s = 0$ and $u_0 \equiv 0$).

4 Decreasing Functions

Firstly, the steady states can be characterized by their maxima, $M = w(1)$, with the parameter determined as a function of M, $\lambda = M^2/[\int_0^M ds/f(s)]$ (Section 2).

This function is increasing for the minimal stationary solution but can decrease for some M. It should be noted that it is possible to construct f which are decreasing but admit more than two steady states for certain λ, see Lacey (1995b). A criterion for uniqueness of the equilibrium, $\lambda'(M) > 0$ for all M, is given by $\mu(M) = \int_0^M ds/f(s) > M/2f(M)$ for all M. Apart from the possibilities of multiple turning points Fig. 1 indicates the main possible forms of response diagram. For simplicity we shall limit the bulk of our discussion to cases shown in Fig. 1(a),(b),(c); these correspond to $\lim_{s\to\infty} sf(s)$ being zero, positive but finite, and infinite, respectively.

We take $v(x,t) \equiv w(x;\mu(t))$ as a prospective comparison function.

¿From $w' = \mu f(w)$ we see immediately that $v_x = \mu f(v)$ and, because $\mu x = \int_0^w ds/f(s)$, $v_t = \dot{\mu} x f(w)$. Then

$$v_t + v_x - \frac{\lambda f(v)}{[\int_0^1 f(v)dx]^2} = f(w)(\dot{\mu}x + \mu - \frac{\lambda\mu^2}{M^2}).$$

Taking

$$\dot{\mu} = \frac{\lambda\mu^2}{M^2} - \mu \tag{6}$$

we get a decreasing upper solution if $\lambda < M(\mu)^2/\mu$ and an increasing lower solution if $\lambda > M(\mu)^2/\mu$. Solutions can be sandwiched by two such v's.

The main types of behaviour which are possible for decreasing f are the following. With $sf(s)$ unbounded, there is always a steady state. Still assuming there is at most one turning point in the response diagram (Fig. 1) there must be, in this case, a unique solution for all λ (Fig.1 (a)). This solution is globally asymptotically stable. With $sf(s) \to c$ as $s \to \infty$, $0 < c < \infty$, (Fig. 1(b)), the unique steady state for sufficiently small λ is globally asymptotically stable ($\lambda \leq c$ for Fig. 1(b) (ii), $\lambda < c = \lambda^*$ for Fig. 1(b) (i)). With $c < \lambda < \lambda^*$ the minimal steady state is stable, the other is unstable; larger u_0 lead to unbounded solutions. For $\lambda = \lambda^*$ (Fig. 1(b) (ii), $c < \lambda^*$) the critical steady state is only stable from below, larger initial conditions result in unbounded u. With $\lambda > \lambda^*$ all solutions are unbounded. Since $\int^\infty f(s)ds = \infty$ the unbounded solutions actually exist for all time. If $sf(s) \to 0$ as $s \to \infty$ (Fig. 1(c)) the minimal solution is stable and the other unstable ($u_0 < w_2$ gives $u \to w_1$, $u_0 > w_2$ gives unbounded u). For $\lambda > \lambda^*$ (or taking u_0 to be large), u is unbounded. If $\int^\infty f(s)ds = \infty$ these unbounded solutions exist for all time (Fig. 2) otherwise they blow up.

Of course, if there are multiple turning points similar results apply. For $sf(s)$ bounded there is a critical λ^* so for $\lambda > \lambda^*$ (or large u_0) u is unbounded. For $sf(s)$ bounded away from zero when s is large there is some λ_* ($0 < \lambda_* \leq \lambda^*$) such that there is a unique steady state for $0 < \lambda < \lambda_*$. This is globally asymptotically stable. If $\int^\infty f(s)ds < \infty$ unbounded solutions blow up, otherwise they exist for all time. Steady states with $\frac{d\lambda}{dM} > 0$ (such as for minimal equilibria) are stable otherwise they are unstable. Fig. 4 indicates what might happen. (*N.B.* if λ_* exists blow-up cannot occur.)

We finally note in this section that blow-up, when it takes place, is global. Since f is decreasing, so f is bounded, and, on characteristics, $\frac{du}{dt} = g(t)f(u)$ blow-up occurs at $t = t^*$ if and only if $g(t) = \lambda/[\int_0^1 f(u)dx]^2$ satisfies $\int^t g(t)dt \to \infty$ as $t \to t^*$. Indeed, blow-up is then uniform on compact subsets of $(0, 1]$. (In fact, for two points $0 < x_1 < x_2 \leq 1$, $|u(x_1,t) - u(x_2,t)| \to 0$ as $t \to t^*$.) A boundary layer develops near $x = 0$ as $t \to t^*$.

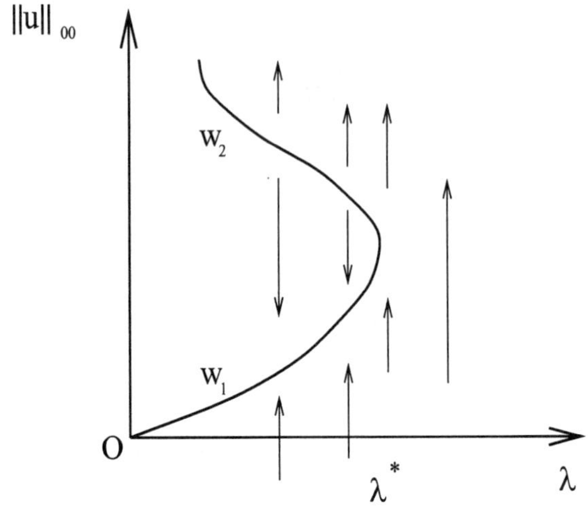

Figure 2: Behaviour of $u_t + u_x = \lambda f(u)/[\int_0^1 f(u)dx]^2$ for $sf(s) \to 0$ as $s \to \infty$ but $\int^\infty f(s)ds = \infty$. ↑ increasing global solution, ↓ decreasing global solution.

A final straightforward, but not Lipschitz, strictly decreasing or positive case uses the Heaviside function so that $f(s) = 1 \;\; s < 1, \;\; f(s) = 0 \;\; s \geq 1$.

¿From the form of f, $\int_0^1 f\,dx = m =$ the measure of the set where $u < 1$. Taking u to be increasing as a function of x (which is guaranteed by increasing $u_0(x)$), as equilibria certainly are, either $u < 1$ for $0 \leq x < 1$, in which case $m = 1$, or there is some $S(t) < 1$ with $u < 1$ for $0 \leq x < S$ and $u \geq 1$ for $x \geq S$, in which case $m = S$. Now $u_t + u_x = \lambda$ or $u_t + u_x = \lambda/S(t)^2$ for these two cases.

There are correspondingly two possible forms of steady state:

$$w < 1 \text{ for } x < 1, \; w' = \lambda \text{ and } w = \lambda x, \; \text{ then } \lambda \leq 1;$$

$$w = \frac{\lambda x}{S^2} \text{ for } x \leq S, \; w = 1 \text{ for } x \geq S, \; \text{ then } S = \lambda \leq 1.$$

We obtain the following results:

if $u_0 < w_2$ then u is a global bounded solution which tends to w_1 as $t \to \infty$;

if $u_0 > w_2$ then u becomes 1 for all $0 < x \leq 1$ after a finite time.

The latter occurrence is a form of blow-up since u_x and u_t become unbounded at this finite time. With $\lambda > 1$ lower solutions can again be used to demonstrate this type of blow-up. The possible behaviour is indicated in Fig. 3.

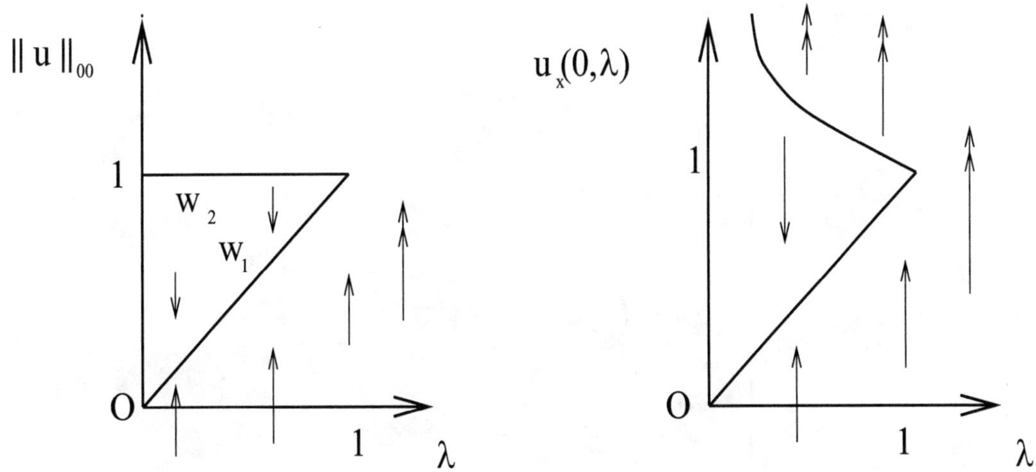

Figure 3: Response diagram for the stationary solution w and behaviour for $u_t + u_x = \lambda H(1-u)/[\int_0^1 H(1-u)\,dx]^2$. $\uparrow$ increasing global solution, $\downarrow$ decreasing global solution, ⇀ increasing solution which blows up through first derivatives becoming unbounded.

5 Increasing Functions

This section differs most significantly from the related one-dimensional parabolic problem, for which it was found that if f is bounded away from zero, the solution u is bounded and global (see Lacey (1995a), also Bebernes & Lacey). In the special case of $f(s) = e^{as}$ it is easily seen that a Lyapunov function exists for the parabolic problem (Freitas, Bebernes & Lacey); this is no longer true for the hyperbolic model.

We can immediately observe that the stationary problem always has a solution (Fig.1(a)). Moreover, because $\mu(M) = \int_0^M ds/f(s) > M/f(M)$ since f is increasing, this steady state is unique.

For convenience we now absorb the positive constant λ into the function f (more correctly we absorb $1/\lambda$ into f) so we study

$$u_t + u_x = \frac{f(u)}{\left[\int_0^1 f(u)dx\right]^2}, \; 0 < x < 1;$$

$$u(0,t) = 0; \;\; u(x,0) = u_0(x), \; 0 < x < 1;$$

where $u_0 \geq 0$ and is continuous and $f(s) > 0, \; f'(s) \geq 0$ for $s \geq 0$.

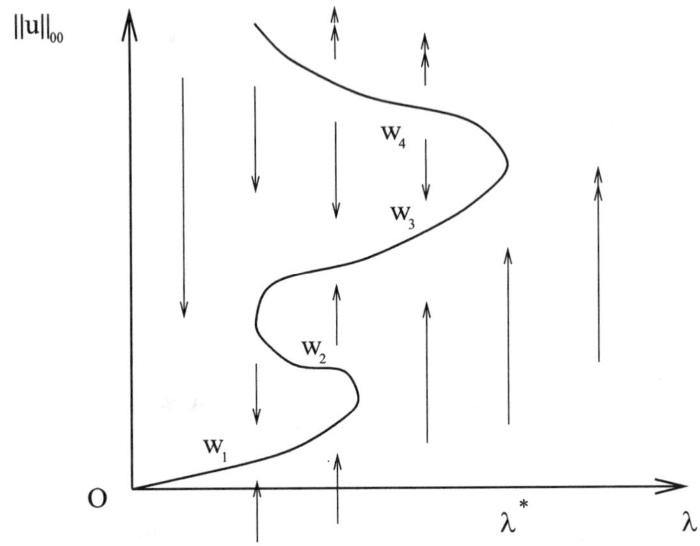

Figure 4: Possible behaviour (if $\int^{\infty} f(s)ds < \infty$) with multiple turning points. ↑ increasing global solution, ↓ decreasing solution, ⇢ solution which blows up.

Writing $M(t) = \max\limits_{[0,1]} u(\cdot, t)$, this maximum satisfies $\frac{dM}{dt} \leq \frac{f(M)}{f(0)^2}$ so blow-up is only possible if $\int^{\infty} ds/f(s) < \infty$. Taking this condition to hold, so blow-up might happen, it is then helpful to make the change of variable

$$v = V(u) = \int_u^{\infty} \frac{ds}{f(s)}.$$

We also assume, without loss of generality, that the problem has been scaled so that $V(0) = v(0,t) = \int_0^{\infty} ds/f(s) = 1$.

Then

$$\left. \begin{array}{r} v_t + v_x + 1/[\int_0^1 F(v)dx]^2 = 0,\ 0 < x \leq 1; \\ \\ v(0,t) = 1;\ \ 0 < v(x,0) = V(u_0(x)) \leq 1,\ 0 < x \leq 1. \end{array} \right\} \tag{7}$$

Here $v(x,0)$ is continuous while $F(v) \equiv f(U(v))$, for $U = V^{-1}$, satisfies $F(v) \geq f(0)$ and is a decreasing function. Blow-up of u corresponds to v becoming zero at some point(s).

It is also helpful to note that the simple form of the equation (7), which can be written as

$$v_t + v_x + g(t) = 0, \text{ or } \frac{dv}{dt} + g(t) = 0 \text{ on } x(t) = \text{const.} + t,$$

$$g(t) = \left[\int_0^1 F(v)dx\right]^{-2},$$

means that

$$v(x,t) = v(x_0, t_0) - \int_{t_0}^{t} g(\tau)d\tau \tag{8}$$

for $x_0 = x - t + t_0$. If, at some time t_0, v is decreasing as a function of x (in particular $v(x, t_0) \leq 1$ for $0 \leq x \leq 1$), then v is a decreasing function of x for all subsequent times. Any later blow-up must then occur at $x = 1$ (perhaps over some interval $[X, 1]$). Of course if u_0 is increasing then $u(x,t)$ is an increasing function of x for all $t \geq 0$ and **any** blow-up must be at $x = 1$.

If f grows "rapidly" (*e.g.* $f(s) = a^{-1}e^{as}$ with $a > 2$ or $f'(s)/f(s) \to \infty$ as $s \to \infty$) then blow-up is possible, even with monotonic initial data. On the other hand, if f grows "slowly" (*e.g.* $f(s) = a^{-1}e^{as}$ with $a < 2$ or $f'(s)/f(s) \to 0$ as $s \to \infty$) then blow-up cannot occur after any time at which u is monotonic in x: blow-up never takes place after $t = 1$; if u_0 is increasing, it does not happen at all. We can obtain solutions which blow up for quite general increasing f.

The form of these solutions means that as long as f grows faster than a power blow-up is possible with Lipschitz data. Blow-up with a smooth initial condition can be achieved provided that $f(u) \sim Cu^p$ with $p > 2$ (or f grows faster than a power).

The estimates can be used to get converses:

If f grows no faster than quadratically then blow-up with a smooth initial condition is impossible. (Blow-up must be at either a point $x^* = a + t^* \leq 1$ corresponding to an initial turning point at $x = s$, contradicted by our present analysis, or $x^* = 1$ with $u'_0(1 - t^*) < 0$, precluded by our earlier results). If $f(u)$ grows no faster than $u^{\frac{3}{2}}$ then blow-up is only possible if u'_0 is unbounded.

The "partial blow-up" obtained here allows the possibility of secondary, or repeated, blow-up for exponentially (or faster) growing f . We could even envisage u blowing up at arbitrarily large times.

To summarize this section, we have seen that the behaviour differs enormously from that of related parabolic problems, unlike cases of decreasing f. Blow-up is now possible for any sufficiently peaked initial data whenever $\int_0^\infty \frac{ds}{f(s)} < \infty$. If, however, f grows "slowly" (*e.g.* as a power) this is the only way in which the solution can cease to exist: if u_0 is smooth or monotonic blow-up does not take place; nor can it happen, whatever u_0 like, after $t = 1$. With rapidly growing f (exponentially or faster) blow-up is possible even for monotonic and smooth initial data. We have not yet ruled out indefinitely repeating blow-up. Even with rapidly growing f if u'_0 is positive then the solution is global. The only cases we saw where blow-up was complete corresponded to a peak in the initial data where it was not Lipschitz; although smooth initial conditions can allow blow-up these solutions could be continued.

We finally note that with these growing f, blow-up can clearly occur for the alternative model $u_t + u_x = f(u)/[a + b\int_0^1 f(u)dx]^2$.

6 Conclusions

We have discussed the non-local, one-dimensional, hyperbolic problem,

$$u_t + u_x = \frac{\lambda f(u)}{\left[\int_0^1 f(u)dx\right]^2}.$$

In Section 4 we saw how blow-up could occur for decreasing functions f and how certain steady states were asymptotically stable. The key tool was a comparison method, obtained in Section 3, employed on trial functions based on steady states from Section 2. It was observed that unique steady states were globally asymptotically stable. The "circuit" model, with denominator $[a + b\int_0^1 f(u)dx]^2$, gives rise to a bounded solution. All these results closely parallel those for the related problem.

Cases of increasing f, Section 5, differ greatly from the parabolic (diffusion-dominated) problem. There blow-up was impossible but here it can happen, although it is precluded for f growing as a power when the problem has a smooth or monotonic initial condition. Should f grow at least exponentially blow-up will be possible even with smooth and increasing initial data. Most cases of blow-up, or thermal runaway, which we demonstrated allowed continuation of a solution; the only one we found for which the blow-up was complete had a quite spiky initial condition.

For $t \geq 1$, u is monotonic in x, which prohibits blow-up for f which grow as a power but might admit it for f growing exponentially. In real situations being modelled, such as sterilization of food, perturbations of solutions can arise from the presence of inhomogenities. Such occurrences, if they are substantial enough, might allow thermal runaway, resulting in local overcooking (or worse) at any time, provided that the resistivity f grows fast enough as a function of temperature for $\int^\infty ds/f(s)$ to be finite.

References

[1] C.H. Biss, S.A. Coombers & P.J. Skudder, 1989, The development and applications of Ohmic heating for continuous processing of particulate foodstuffs, in *Process Engineering in the Food Industry,* R.W. Field & J.A Howell eds, pp 17-27, Elsevier London.

[2] J.W. Bebernes & A.A. Lacey, *preprint*, Global existence and finite-time blow-up for a class of nonlocal parabolic problems.

[3] J.A. Carrillo, preprint, On a non-local elliptic equation with decreasing nonlinearity arising in plasma physics and heat conduction.

[4] A.A.P. De Alwis & P.J. Fryer, 1992, Operability of the ohmic heating process: electrical conductivity effects, *J. Food Eng.*, **15**, 21-48.

[5] P.J. Fryer, A.A.P. De Alwis, E. Koury, A.G.F. Stapley & L. Zhang, 1993, Ohmic processing of solid-liquid mixtures: heat generation and convection effects, *J. Food Eng.*, **18**, 101-125.

[6] P. Freitas, 1994, A nonlocal Sturm-Liouville eigenvalue problem, *Proc. Rog. Soc. Ed,* **124A**, 169-188.

[7] A.A. Lacey, 1995a, Thermal runaway in a non-local problem modelling Ohmic heating. Part I : Model derivation and some special cases, *Euro. Jl. Appl. Maths.*, **6**, 127-144.

[8] A.A. Lacey, 1995b, Thermal runaway in a non-local problem modelling Ohmic heating. Part II : General proof of blow-up and asymptotics of runaway, *Euro. Jl. Appl. Maths.*, **6**, 201-224.

[9] C.P. Please , D.W. Schwendeman & P.S. Hagan, 1994, Ohmic heating of foods during aseptic processing, *IMA Jl. Maths. Bus. Ind.*, **5**, 283-301.

[10] P. Skudder & S. Biss, 1987, Aseptic processing of food products using Ohmic heating, *The Chemical Engineer*, **2**, 26-28.

[11] R. Stirling, 1987, Ohmic heating - a new process for the food industry, *Power Eng. J.*, **6**, 365.

[12] L. Zhang & P. J. Fryer, 1993, Models for the electrical heating of solid-liquid food mixtures, *Chem. Eng. Sci.*, **48**, 633-642.

A.A. Lacey
Department of Mathematics, Heriot-Watt University,
Edinburgh, EH14 4AS, Scotland

D.E. Tzanetis & P.M. Vlamos
Department of Mathematics, National Technical University of Athens,
Zografou Campus, 157 80, Greece.

P A MARTIN AND J A DeSANTO

Some negative results on the use of Helmholtz integral equations for rough-surface scattering

1 Introduction

A plane time-harmonic sound wave is scattered by a bounded three-dimensional obstacle with surface S. The scattered field can be found by solving a Helmholtz integral equation. For example, if S is sound-hard, the scattered field on S satisfies

$$u(p) - \int_S u(q) \frac{\partial G}{\partial n_q}(p,q)\, dS_q = \int_S \frac{\partial u_{\text{inc}}}{\partial n} G(p,q)\, dS_q, \tag{1}$$

where p and q are points on S, G is the free-space Green's function, $\partial/\partial n_q$ denotes normal differentiation at q, and u_{inc} is the incident field. The Helmholtz integral equations are familiar boundary integral equations, for which there is a complete theoretical framework (Kleinman and Roach, 1974; Colton and Kress, 1983). Many examples of their numerical treatment, usually using boundary elements, can also be found.

Suppose now that S is unbounded. A typical problem is the reflection of a plane wave by an infinite two-dimensional rough surface. This is a classical problem, going back to Lord Rayleigh. Standard texts usually treat the problem using approximate techniques, such as perturbation theory or Kirchhoff theory. More recently, there has been an effort to validate these approximate techniques by comparing them with 'exact' methods. In particular, comparisons have been made with numerical solutions of Helmholtz integral equations, such as (1).

However, the derivation of a Helmholtz integral equation for an infinite rough surface is not straightforward. Indeed, we are not aware of any correct derivations in the literature (even though such equations have been the subject of extensive computational studies). In fact, we can show that such an integral equation is definitely not valid in certain cases! These will be illustrated using some explicit examples.

2 Bounded obstacles

Consider a bounded three-dimensional obstacle with a smooth surface S, insonified by a plane wave. The problem is to calculate the scattered field u. In order to have a well-posed boundary-value problem, one imposes the Sommerfeld radiation condition,

$$r\left(\frac{\partial u}{\partial r} - iku\right) \to 0 \qquad \text{as } r \to \infty, \tag{2}$$

uniformly in all directions. Here, r is a spherical polar coordinate, k is the wavenumber, and the time-dependence is $e^{-i\omega t}$.

Equation (1) is derived by applying Green's theorem to $u(Q)$ and $G(P, Q)$ in the region bounded internally by S and externally by C_r, a large sphere of radius r; the point P is in this region. The radiation condition implies that the integral

$$I(u; C_r) \equiv \int_{C_r} \left(u\,\frac{\partial G}{\partial r} - G\,\frac{\partial u}{\partial r}\right) dS \to 0 \qquad \text{as } r \to \infty, \tag{3}$$

and so only boundary integrals over S remain (Colton and Kress, 1983). Equation (1) follows by letting $P \to p \in S$.

3 Unbounded obstacles

The simplest scattering problem for unbounded obstacles is reflection of a plane wave by an infinite flat plane, S. It is well known that the incident wave is reflected specularly as a single propagating plane wave. More generally, if S is an infinite rough surface, an incident plane wave will be scattered into a spectrum of plane waves. For such problems, the Sommerfeld radiation condition is not appropriate as it is not satisfied by a plane wave. However, it is common to proceed, *assuming* that the scattered field can be represented in terms of plane waves, at least at some distance from S.

Typically, this requires the discarding of an integral such as (3), but with the large sphere C_r replaced by a large *hemisphere* H_r. Can this step be justified?

Assume that the two-dimensional rough surface, S, is sound-hard and that it is given by $z = s(x, y)$, $-\infty < x, y < \infty$, with $-h < s(x, y) \leq 0$ for some constant $h \geq 0$. We can write the total field as $u_{\text{tot}} = u_{\text{inc}} + u$, where u is the scattered field. The incident plane wave is

$$u_{\text{inc}}(r, \theta, \phi) = \exp\{\mathrm{i}\mathbf{k}_{\mathrm{i}} \cdot \mathbf{x}\}, \qquad 0 \leq \theta_{\mathrm{i}} \leq \tfrac{1}{2}\pi, \tag{4}$$

where $\mathbf{k}_{\mathrm{i}} = k(\sin\theta_{\mathrm{i}}, 0, -\cos\theta_{\mathrm{i}})$, θ_{i} is the angle of incidence, (r, θ, ϕ) are spherical polar coordinates, and $\mathbf{x} = r\hat{\mathbf{x}} = r(\sin\theta\, \cos\phi, \sin\theta\, \sin\phi, \cos\theta)$. All the fields u_{tot}, u_{inc} and u satisfy the Helmholtz equation, $(\nabla^2 + k^2)u = 0$, for $z > s$. The boundary condition is

$$\partial u_{\text{tot}}/\partial n = 0 \qquad \text{on } S, \tag{5}$$

where $\partial/\partial n$ denotes normal differentiation *out* of the acoustic medium.

4 Reflection by a flat surface

Let us return to the simplest problem, reflection by a flat surface, so that $s = 0$. The text-book solution for u is

$$u(r, \theta, \phi) = \exp\{\mathrm{i}\mathbf{k}_{\mathrm{s}} \cdot \mathbf{x}\} \qquad \text{for } 0 \leq \theta_{\mathrm{i}} < \tfrac{1}{2}\pi, \tag{6}$$

where $\mathbf{k}_{\mathrm{s}} = k(\sin\theta_{\mathrm{i}}, 0, \cos\theta_{\mathrm{i}})$. When $\theta_{\mathrm{i}} = \frac{1}{2}\pi$ ('grazing incidence'), we have $u \equiv 0$. So, for $0 \leq \theta_{\mathrm{i}} < \frac{1}{2}\pi$, $u_{\text{tot}} = 2\,\mathrm{e}^{\mathrm{i}kx\sin\theta_{\mathrm{i}}} \cos(kz\cos\theta_{\mathrm{i}})$ solves the problem. But consider

$$u'_{\text{tot}} \equiv u_{\text{tot}} + u_{\mathrm{g}} \tag{7}$$

with $u_{\mathrm{g}} = V(\beta)\,\mathrm{e}^{\mathrm{i}k(x\cos\beta + y\sin\beta)}$, where β and $V(\beta)$ are arbitrary, with $-\pi < \beta \leq \pi$. u'_{tot} also 'solves' the problem, in that it satisfies the Helmholtz equation and the boundary condition. Of course, we disallow this second solution, unless $V \equiv 0$: but why? The answer is: because of the radiation condition (which we have yet to specify). For example, take $\beta = 0$ and $V(0) = 1$, so that $u_{\rm g} = {\rm e}^{ikx}$; this gives an 'outgoing' grazing wave at $x = +\infty$ but it is an 'incoming' grazing wave at $x = -\infty$ — we must therefore exclude it. Indeed, we must exclude *all* contributions $u_{\rm g}$, for any β and V.

A similar condition is imposed on the two-dimensional problem (DeSanto and Martin, 1997). However, the three-dimensional problem has an extra feature: we could consider replacing $u_{\rm g}$ in (7) by

$$\frac{1}{2\pi}\int_{-\pi}^{\pi} V(\beta)\,{\rm e}^{{\rm i}k(x\cos\beta+y\sin\beta)}\,d\beta$$

where V is a continuous function; but, as $u_{\rm g}$ has been excluded, we must also exclude all linear combinations of such plane grazing waves. In particular, by taking $V(\beta) = (-{\rm i})^n\,{\rm e}^{{\rm i}n\beta}$, we see that we must exclude the cylindrical standing waves

$$J_n(kR)\,{\rm e}^{{\rm i}n\phi}, \tag{8}$$

where $R = r\sin\theta$ and (R, ϕ, z) are cylindrical polar coordinates of the point at $\mathbf{x}$. On the other hand, the exact scattered field, given by (6), when evaluated on any plane $z =$ constant, has an azimuthal Fourier component proportional to

$$J_n(k_{\rm i}R)\,{\rm e}^{{\rm i}n\phi}, \tag{9}$$

where $k_{\rm i} = k\sin\theta_{\rm i} \leq k$. Thus, if one wants to formulate a radiation condition, mathematically, it must be such that fields (8) are excluded but fields (9) are permitted. All this suggests that the specification of a *mathematical* radiation condition for the present class of problems (plane-wave scattering by an infinite two-dimensional rough surface) will not be straightforward. However, the *physical* purpose of a radiation condition is clear: it is to exclude all 'incoming' waves apart from the incident wave.

5 A radiation condition

The scattered field in the half-space $z > 0$ may be written using an angular-spectrum representation, as a superposition (integral) of propagating and evanescent plane waves

(DeSanto and Martin, 1996). A typical propagating plane wave is

$$v(r,\theta,\phi;\alpha,\beta) = \exp\{\mathrm{i}\mathbf{k}\cdot\mathbf{x}\}, \qquad 0 \le \alpha \le \tfrac{1}{2}\pi, \quad |\beta| \le \pi, \tag{10}$$

where $\mathbf{k} = k\hat{\mathbf{k}} = k(\sin\alpha\,\cos\beta, \sin\alpha\,\sin\beta, \cos\alpha)$. It turns out that the propagating waves give the dominant contribution to $I(u;H_r)$, where H_r is a large hemisphere of radius r and centre O.

We can now give our radiation condition. We require that all propagating plane-wave components $v(r,\theta,\phi;\alpha,\beta)$ in u propagate *outwards* through H_r, away from O. When imposing this, we have to be careful with grazing waves ($\alpha = \frac{1}{2}\pi$; see the discussion following (7)). One convenient way is to partition the half-space $z > 0$ and the hemisphere H_r into four parts. Thus, with

$$H_r^m = \{(r,\theta,\phi) : 0 \le \theta \le \tfrac{1}{2}\pi,\ \tfrac{1}{2}(m-3)\pi \le \phi < \tfrac{1}{2}(m-2)\pi\}, \quad m = 1,2,3,4,$$

being the surfaces of four octants of a sphere, we require the following conditions for the regions specified:

$$\begin{aligned} m=1: &\quad \text{in } x<0,\ y\le 0, \quad \text{use } -\pi \le \beta < -\tfrac{1}{2}\pi;\\ m=2: &\quad \text{in } x\ge 0,\ y<0, \quad \text{use } -\tfrac{1}{2}\pi \le \beta < 0.\\ m=3: &\quad \text{in } x>0,\ y\ge 0, \quad \text{use } 0 \le \beta < \tfrac{1}{2}\pi;\\ m=4: &\quad \text{in } x\le 0,\ y>0, \quad \text{use } \tfrac{1}{2}\pi \le \beta < \pi; \end{aligned} \tag{11}$$

This partitioning makes it easy to ensure that only plane waves propagating out through H_r^m are included. We have used this form of radiation condition to derive boundary integral equations.

6 Boundary integral equations

Apply Green's theorem to u and G in the region D_r with boundary $\partial D_r = H_r \cup S_r \cup T_r$, where $S_r = \{(x,y,z) : z = s(x,y),\ 0 \le x^2+y^2 < r^2\}$ is a truncated rough surface, and

$$T_r = \{(x,y,z) : x^2+y^2 = r^2,\ s(x,y) \le z \le 0\} \tag{12}$$

is the surface of a truncated circular cylinder joining H_r and S_r. After using the boundary condition (5), the result is

$$2u(P) = \int_{S_r} \left\{ u(q) \frac{\partial G}{\partial n_q}(P,q) + \frac{\partial u_{\text{inc}}}{\partial n} G(P,q) \right\} dS_q + I(u;H_r) + I(u;T_r), \tag{13}$$

where $P \in D_r$,

$$I(u;\mathcal{S}) = \int_{\mathcal{S}} \left\{ u(q) \frac{\partial G}{\partial n_q}(P,q) - \frac{\partial u}{\partial n} G(P,q) \right\} dS_q$$

and $\partial/\partial n_q$ denotes normal differentiation at q in a direction away from the origin (so that $\partial/\partial n = \partial/\partial r$ on H_r, consistent with (3)).

The next step is to estimate $I(u;H_r)$ and $I(u;T_r)$ for large r. This is described in detail by DeSanto and Martin (1998). Here, we consider a single propagating plane-wave component in the angular-spectrum representation of u, and focus on $I(v;H_r)$ as $r \to \infty$, where v is defined by (10). In fact, for this calculation, it is enough to set $\beta = 0$, and so we write

$$v(r,\theta,\phi;\alpha,0) = v(r,\theta,\phi;\alpha) = v(\mathbf{x};\alpha), \qquad 0 \le \alpha \le \tfrac{1}{2}\pi.$$

The simplest way to evaluate the large-r limit is to use the method of stationary phase. We then consider the contribution from $I(u;T_r)$ to (13). The results are surprising, and motivate some explicit examples.

7 The method of stationary phase

We use the method of stationary phase to estimate $I(v;H_r)$. We are interested in large values of $r = |\mathbf{x}|$, for fixed $\mathbf{y}$ and k. We have

$$G(P,q) \simeq (B/r) \exp\{ik(r - \mathbf{y}\cdot\hat{\mathbf{x}})\} \tag{14}$$

where P and q have position vectors $\mathbf{x}$ and $\mathbf{y}$, respectively, and $B = -\tfrac{1}{2}/\pi$. Hence, for large r, $I(v;H_r) \simeq \mathrm{i}B\,\mathrm{e}^{\mathrm{i}kr}\,L(kr)$, where

$$L(\lambda) = \lambda \int_{\mathcal{D}} g(\theta,\phi)\,\mathrm{e}^{\mathrm{i}\lambda F(\theta,\phi)}\,d\theta\,d\phi, \tag{15}$$

$$g(\theta,\phi) = (1 - \hat{\mathbf{k}}\cdot\hat{\mathbf{x}}) \exp\{-ik\mathbf{y}\cdot\hat{\mathbf{x}}\} \sin\theta, \tag{16}$$

$F(\theta,\phi) = \hat{\mathbf{k}}\cdot\hat{\mathbf{x}}$ and $\mathcal{D} = \{(\theta,\phi) : 0 \le \theta \le \frac{1}{2}\pi, -\pi \le \phi \le \pi\}$ is the rectangular domain of integration.

It turns out that there are three cases, depending on the value of α: $\alpha = 0$, $0 < \alpha < \frac{1}{2}\pi$ and $\alpha = \frac{1}{2}\pi$.

Case I: $\alpha = 0$. This corresponds to a plane wave v propagating along the z-axis. We have

$$L(\lambda) = \lambda \int_0^{\pi/2} (1 - \cos\theta)\, b(\theta;\mathbf{y})\, \mathrm{e}^{\mathrm{i}\lambda\cos\theta} \sin\theta\, d\theta \tag{17}$$

where

$$b(\theta;\mathbf{y}) = \int_{-\pi}^{\pi} \exp\{-ik\mathbf{y}\cdot\hat{\mathbf{x}}\}\, d\phi = 2\pi\, \mathrm{e}^{-ik\rho\cos\theta\,\cos\Theta}\, J_0(k\rho\sin\theta\,\sin\Theta)$$

and

$$\mathbf{y} = \rho\hat{\mathbf{y}} = \rho(\sin\Theta\,\cos\Phi, \sin\Theta\,\sin\Phi, \cos\Theta). \tag{18}$$

The integral (17) can be estimated for large λ using the (one-dimensional) method of stationary phase. The only stationary-phase point is at $\theta = 0$; as the integrand vanishes at $\theta = 0$, we deduce that $L(\lambda) = O(1)$ as $\lambda \to \infty$. In fact, an integration by parts shows that $L(\lambda) \sim \mathrm{i}\, b(\frac{1}{2}\pi;\mathbf{y})$ as $\lambda \to \infty$, whence

$$I(v;H_r) = \mathrm{e}^{ikr} J_0(k\rho\sin\Theta) + O((kr)^{-1/2}) \qquad \text{as } kr \to \infty, \text{ for } \alpha = 0. \tag{19}$$

Case II: $0 < \alpha < \frac{1}{2}\pi$. We can estimate $L(\lambda)$, defined by (15), for large λ, using the method of stationary phase for two-dimensional integrals. Thus, we look for stationary-phase points $\mathbf{c} = (\theta,\phi) \in \mathcal{D}$ at which grad $F = \mathbf{0}$; such points may be in the interior of $\mathcal{D}$ or on the boundary, $\partial\mathcal{D}$. Each $\mathbf{c}$ contributes a term to $L(\lambda)$ proportional to $g(\mathbf{c}) \exp\{\mathrm{i}\lambda F(\mathbf{c})\}$, the next term being $O(\lambda^{-1})$. The relevant stationary-phase points are at $\mathbf{c} = (0, \pm\frac{1}{2}\pi)$ and $(\alpha, 0)$. At these points, $g(\mathbf{c}) = 0$, whence $L(\lambda) = o(1)$ as $\lambda \to \infty$. In fact, we find that

$$I(v;H_r) = O((kr)^{-1/2}) \qquad \text{as } kr \to \infty, \text{ for } 0 < \alpha < \tfrac{1}{2}\pi.$$

Case III: $\alpha = \frac{1}{2}\pi$. Now, two more stationary-phase points appear, at $\mathbf{c} = (\frac{1}{2}\pi, \pm\pi)$. These give a non-trivial contribution; the result is

$$I(v; H_r) = 2\,\exp\{\mathrm{i}\mathbf{k}\cdot\mathbf{y}\} + O((kr)^{-1/2}) \qquad \text{as } kr \to \infty, \text{ for } \alpha = \tfrac{1}{2}\pi.$$

Let us summarise these results. The symmetry axis of the hemisphere H_r is the z-axis. We considered plane waves v propagating out of the hemisphere, at an angle α to the z-axis. We saw that $I(v; H_r) \to 0$ as $r \to \infty$, for $0 < \alpha < \frac{1}{2}\pi$. For $\alpha = \frac{1}{2}\pi$ ('grazing waves', with respect to the plane $z = 0$), $I(v; H_r) \to 2\,\exp\{\mathrm{i}\mathbf{k}\cdot\mathbf{y}\}$, a finite quantity, as $r \to \infty$. For $\alpha = 0$ ('normal waves', with respect to $z = 0$), $I(v; H_r) \sim \mathrm{e}^{ikr}\,J_0(k\rho\sin\Theta)$, which means that $I(v; H_r)$ does not have a limit (in this case) as $r \to \infty$. This is an unpleasant result!

8 Asymptotic behaviour of $I(u; T_r)$

The truncated cylindrical surface T_r is defined by (12). A point $q \in T_r$, with position vector $\mathbf{x}$, has cylindrical polar coordinates (r, ϕ, z). Then, for large r, $|\mathbf{x} - \mathbf{y}| \simeq r - \rho\sin\Theta\,\cos(\phi - \Phi)$, where the fixed point P has position vector $\mathbf{y}$ and spherical polar coordinates (ρ, Θ, Φ) defined by (18). It follows that

$$I(u; T_r) \sim \frac{1}{2\pi}\,\mathrm{e}^{ikr}\int_{-\pi}^{\pi} \mathcal{E}(r,\phi)\,\exp\{-ik\rho\sin\Theta\,\cos(\phi-\Phi)\}\,d\phi \qquad \text{as } r \to \infty,$$

where

$$\mathcal{E}(r,\phi) = \int_s^0 \left(\frac{\partial u}{\partial r} - iku\right) dz$$

and the lower limit is $s(r\cos\phi, r\sin\phi)$.

When is it true that $I(u; T_r) \to 0$ as $r \to \infty$? To answer this question, we need additional assumptions. For example, a sufficient condition is that $s \to 0$ as $r \to \infty$, for all ϕ, which means that the rough surface approaches the flat plane $z = 0$ at large distances, in all directions. We could also put some restriction on the behaviour of u near S.

However, there are situations in which $I(u; T_r)$ does not vanish as $r \to \infty$. An explicit example of this is given next.

9 An example

Recall the integral representation (13) for $u(P)$ when $P \in D_r$, the region bounded by the hemisphere H_r, the truncated rough surface S_r and the truncated circular cylinder T_r:

$$2u(P) = I(u; S_r) + I(u; H_r) + I(u; T_r). \tag{20}$$

We note that the left-hand side of (20) does not depend on r, so that the right-hand side must have a limit as $r \to \infty$. Nevertheless, it is instructive to see how the different surface pieces contribute. We do this by examining a very simple example, which is for a plane wave *normally* incident upon a *flat* surface at $z = -h$. Thus

$$u_{\mathrm{inc}} = A_0\,\mathrm{e}^{-\mathrm{i}kz} \text{ and } u = \mathrm{e}^{\mathrm{i}kz} \text{ with } A_0 = \mathrm{e}^{-2\mathrm{i}kh}.$$

This is the exact solution. Let us see how this solution is reconstructed by the representation (20). For simplicity, we take P at the origin; this will permit all the integrals to be evaluated exactly (without any asymptotic approximations).

On H_r, we have $u = \mathrm{e}^{\mathrm{i}kr\cos\theta}$, $\partial u/\partial n = \mathrm{i}ku\cos\theta$, $G = Br^{-1}\,\mathrm{e}^{\mathrm{i}kr}$ and $\partial G/\partial n = (\mathrm{i}k - r^{-1})G$. Integrating over θ gives

$$I(u; H_r) = \mathrm{e}^{\mathrm{i}kr}. \tag{21}$$

On S_r, we have $u = \mathrm{e}^{-\mathrm{i}kh}$, $\partial u/\partial n = -\mathrm{i}ku$,

$$G = BR_\sigma^{-1}\mathrm{e}^{\mathrm{i}kR_\sigma}, \qquad \partial G/\partial n = hR_\sigma^{-2}\left(\mathrm{i}k - R_\sigma^{-1}\right)G$$

and $R_\sigma = \sqrt{\sigma^2 + h^2}$, whence the integral over S_r is

$$\begin{aligned} I(u; S_r) &= 2\pi B\,\mathrm{e}^{-\mathrm{i}kh}\int_0^r \mathrm{e}^{\mathrm{i}kR_\sigma}\left\{\mathrm{i}k + hR_\sigma^{-1}\left(\mathrm{i}k - R_\sigma^{-1}\right)\right\}\frac{\sigma\,d\sigma}{R_\sigma} \\ &= -\mathrm{e}^{-\mathrm{i}kh}\int_h^R \mathrm{e}^{\mathrm{i}kt}\left\{\mathrm{i}k + ht^{-1}\left(\mathrm{i}k - t^{-1}\right)\right\}dt \\ &= -\mathrm{e}^{-\mathrm{i}kh}\left[\mathrm{e}^{\mathrm{i}kt} + ht^{-1}\,\mathrm{e}^{\mathrm{i}kt}\right]_{t=h}^R \\ &= 2 - \left(1 + \frac{h}{R}\right)\mathrm{e}^{\mathrm{i}k(R-h)} \end{aligned} \tag{22}$$

where $R = \sqrt{r^2 + h^2}$.

On T_r, we have $u = \mathrm{e}^{\mathrm{i}kz}$ and $\partial u/\partial n = \partial u/\partial r = 0$, whence

$$\begin{aligned} I(u; T_r) &= 2\pi B r \int_{-h}^{0} \mathrm{e}^{\mathrm{i}kz} \frac{\partial}{\partial r} \left(\frac{\mathrm{e}^{\mathrm{i}kR_z}}{R_z} \right) dz \\ &= -\left[(1 - zR_z^{-1})\, \mathrm{e}^{\mathrm{i}k(R_z+z)} \right]_{z=-h}^{0} \\ &= -\mathrm{e}^{\mathrm{i}kr} + \left(1 + \frac{h}{R} \right) \mathrm{e}^{\mathrm{i}k(R-h)} \end{aligned} \tag{23}$$

where $R_z = \sqrt{r^2 + z^2}$.

Adding equations (21), (22) and (23), we see that their sum is exactly 2, which is $2u(P)$ evaluated at the origin. Note that, as $r \to \infty$,

$$\begin{aligned} I(u; S_r) &\sim 2 - \mathrm{e}^{\mathrm{i}k(r-h)}, \\ I(u; T_r) &\sim -\mathrm{e}^{\mathrm{i}kr} + \mathrm{e}^{\mathrm{i}k(r-h)}, \end{aligned}$$

and $I(u; H_r)$ does not simplify further. Thus, the boundary integral over the truncated rough surface does not have a limit as $r \to \infty$. Moreover, the integral over the truncated cylinder T_r does not have a limit as $r \to \infty$, and it is not negligible. This is a genuine three-dimensional effect, which is not seen in the two-dimensional case (DeSanto and Martin, 1997).

10 Discussion

The results described above show that if the scattered field includes a plane wave propagating along the z-axis away from the rough surface ('normal waves'), then the usual Helmholtz integral equation is not valid: the boundary integral diverges. It is possible to devise a modified integral equation which reduces to the standard Helmholtz integral equation when normal waves are absent (DeSanto and Martin, 1998).

This is unsatisfactory, even though the mathematical difficulty may be overcome. Indeed, this difficulty is due entirely to the unphysical problem posed at the outset: *plane-wave* reflection by an *infinite* rough surface. Clearly, we can realise neither a plane wave nor an infinite rough surface. Moreover, the mathematical difficulty disappears if we consider either point-source insonification or a finite patch of roughness on an otherwise flat surface.

At the Metsovo conference, Ralph Kleinman asked if the results in section 9 could be extended to point-source insonification: they can. Thus, suppose that the incident field is generated by a point source at $(x,y,z)=(0,0,z_0)$ with $z_0>0$:

$$u_{\text{inc}} = AR^{-1}\,\mathrm{e}^{\mathrm{i}kR} \quad \text{with} \quad R=\sqrt{x^2+y^2+(z-z_0)^2}.$$

If we take $A = A_0 z_0\,\mathrm{e}^{-\mathrm{i}kz_0}$ with $A_0 = \mathrm{e}^{-2\mathrm{i}kh}$, then $u_{\text{inc}} \sim A_0\,\mathrm{e}^{-\mathrm{i}kz}$ as $z_0 \to \infty$, which means that the point-source incident field reduces to the incident plane wave considered in section 9.

We know the exact solution for the field scattered by an infinite plane surface at $z=-h$, using an image singularity. It is

$$u = AR_1^{-1}\,\mathrm{e}^{\mathrm{i}kR_1} \quad \text{with} \quad R=\sqrt{x^2+y^2+(z+z_0+2h)^2}.$$

If we take P at O, we can evaluate the integrals discussed in section 9. For example, we find that

$$I(u;H_r) = \frac{A}{z_0+2h}\left(1-\frac{r}{R_0}\right)\mathrm{e}^{\mathrm{i}k(R_0+r)},$$

where $R_0=\sqrt{r^2+(z_0+2h)^2}$. Substituting for A gives, exactly,

$$I(u;H_r) = \frac{z_0}{z_0+2h}\left(1-\frac{r}{R_0}\right)\mathrm{e}^{\mathrm{i}k(R_0+r-z_0-2h)}.$$

If we fix z_0 and h, we see that $I(u;H_r)\to 0$ as $r\to\infty$, because $R_0 \sim r$. On the other hand, if we fix r and h and let $z_0\to\infty$, we recover (21).

References

Colton, D. and Kress, R. (1983) *Integral Equation Methods in Scattering Theory*, Wiley, New York.

DeSanto, J. A. and Martin, P. A. (1996) On angular-spectrum representations for scattering by infinite rough surfaces, *Wave Motion* **24**, 421–433.

DeSanto, J. A. and Martin, P. A. (1997) On the derivation of boundary integral equations for scattering by an infinite one-dimensional rough surface, *J. Acoust. Soc. Amer.* **102**, 67–77.

DeSanto, J. A. and Martin, P. A. (1998) On the derivation of boundary integral equations for scattering by an infinite two-dimensional rough surface, *J. Math. Phys.*, to appear.

Kleinman, R. E. and Roach, G. F. (1974) Boundary integral equations for the three-dimensional Helmholtz equation, *SIAM Rev.* **16**, 214–236.

P. A. Martin
Department of Mathematics
University of Manchester
Manchester M13 9PL
U.K.

J. A. DeSanto
Department of Mathematical
and Computer Sciences
Colorado School of Mines
Golden, CO 80401, U.S.A.

B D SLEEMAN

Wave and scattering problems for fractal domains

1. The Weyl Conjecture

Let Ω be a non-empty bounded open set in R^n, $n \geq 1$ with boundary Γ and consider the variational eigenvalue problem

$$-\Delta u = \lambda u \text{ in } \Omega \qquad \text{(P)}$$
$$u = 0 \text{ on } \Gamma$$

where Δ is the Dirichlet Laplacian in Ω and the problem (P) is to be understood in the following sense; the scalar λ, is an eigenvalue of (P) if there exists $u \neq 0$ in $H_0^1(\Omega)$ satisfying $-\Delta u = \lambda u$ in the distributional sense.

The spectrum of (P) is discrete and consists of an infinite sequence of positive eigenvalues of finite multiplicity which may be ordered as

$$0 < \lambda_1 \leq \lambda_2 \leq .... \leq \lambda_\kappa \leq ... \qquad (1.1)$$

with $\lambda_\kappa \to +\infty$ as $\kappa \to +\infty$

Let

$$N(\lambda) \equiv \#\{\lambda_\kappa < \lambda\} \qquad (1.2)$$

be the counting function for the eigenvalues λ_κ.

Weyl's conjecture is

$$N(\lambda) = (2\pi)^{-n} \omega_n |\Omega|_n \lambda^{n/2} - C(n)|\Gamma|_{n-1} \lambda^{(n-1)/2} + O(\lambda^{(n-1)/2}) \qquad (1.3)$$

where ω_n is the volume of the unit ball in R^n.

For sufficiently smooth boundaries the formula (1.3) holds, with $C(n) = \left[4(4\pi)^{(n-1)/2}\Gamma\left(\frac{n+1}{2}\right)\right]^{-1}$, under the "billiard condition" (IV1, IV2). Ivrii's condition is generic and can be checked only in very simple cases (see Safarov- Vassiliev (SV)). If the boundary Γ is very irregular or even fractal, which is the case of interest here, then questions concerning the billiard condition cannot be posed since billiard trajectories cannot be determined. Nevertheless even for rough boundaries Melrose [M] has proved that the first term in (1.3) holds.

In a well known paper [B], Berry discussed diffraction and scattering of waves by fractal surfaces and formulated the conjecture:if Γ has Hausdorff dimension $h = h(\Gamma) < n$ and is Hausdorf f measurable then

$$N(\lambda) = (2\pi)^{-n} \omega_n |\Omega|_n \lambda^{n/2} - C(n,h)|\Gamma|_h \lambda^{h/2} + o(\lambda^{h/2}) \tag{1.4}$$

as $\lambda \to \infty$, where $|\Gamma|_h$ is the Hausdorf f measure of Γ. However Brossard and Carmona [BC] showed that (1.4) is false in general by constructing counter examples. Indeed their examples uggested that instead of using the Hausdorff dimension h (1.4) may hold if one employs the interior Minkowski dimension δ. We refer to (1.4) (with h replaced by δ) as the modified Weyl-Berry conjecture (MWB).From the work of Fleckinger, Lapidus and Vassiliev [LFl], [LF2], [F], [L], [V], it is known that if Γ has interior Minkowski dimension δ then

$$N(\lambda) = (2\pi)^{-n} \omega_n |\Omega|_n \lambda^{n/2} + O(\lambda^{\delta/2}), \tag{1.5}$$

as $\lambda \to \infty$

In the next section we discuss refinements of this.

2. The Modified Weyl-Berry Conjecture

We begin by defining Minkowski dimension and measure of Γ . Given $\varepsilon > 0$ define

$$\Omega_\varepsilon^i = \{\chi \in \Omega \mid d(\chi,\Gamma) < \varepsilon\} \tag{2.1}$$

where $d(\chi,\Gamma)$ denotes the Euclidean distance of χ to the boundary Γ. The set Ω_ε^i is called the interior ε - neighbourhood of Γ. For $l \geq 0$ let

$$\mu^*(l,\Gamma) = \limsup_{\varepsilon \to 0} \varepsilon^{-(n-1)} |\Omega_\varepsilon^i|_n \tag{2.2}$$

$$\mu_*(l,\Gamma) = \liminf_{\varepsilon \to 0^=} \varepsilon^{-(n-1)} |\Omega_\varepsilon^i|_n \tag{2.3}$$

The interior Minkowski dimension of Γ is defined as

$$\delta = \inf\{l \in R_+ \mid \mu^*(l,\Gamma)=0\} = \sup\{l \in R_+ \mid \mu^*(l,\Gamma)=+\infty\}. \tag{2.4}$$

Observe that $\delta \in [n-1,n]$ and $\mu^*(\delta,\Gamma) \in [0,\infty]$. On the other hand if H is the Hausdorff dimension of Γ then we know $H \leq \delta$. If δ is the interior Minkowski dimension of Γ and

$$0 < \mu_*(\delta,\Gamma) = \mu^*(\delta,\Gamma) < +\infty$$

we say that Γ is interior δ - Minkowski measurable and denote by

$$\mu(\delta,\Gamma) = \lim_{\varepsilon \to 0_+} \varepsilon^{-(n-\delta)} \left|\Omega_\varepsilon^i\right|_n \tag{2.5}$$

the interior δ - Minkowski measure of Γ.

With this definition the MWB conjecture can be restated as: If Γ is δ -Minkowski measurable then

$$\mathrm{N}(\lambda) = (2\pi)^{-n} \omega_n |\Omega|_n \lambda^{n/2} - C_{n,\delta}\, \mu(\delta,\Gamma) \lambda^{\delta/2} + \mathrm{O}(\lambda^{\delta/2}) \tag{2.6}$$

as $\lambda \to \infty$, where $C_{n,\delta}$, is a positive constant depending only on n and δ.

In the case n=1 (fractal strings) Lapidus and Pomerance $[LP]$ have given necessary and sufficient conditions for to be δ -Minkowski measurable and have shown that the MWB conjecture is true. For $n \geq 2$ Fleckinger and Vassiliev [FV] have constructed examples to show that MWB may be false in general. It is therefore natural to study (2.6) in a more general context, namely under the condition

$$0 < \mu_*(\delta,\Gamma) \leq \mu^*(\delta,\Gamma) < +\infty \tag{2.7}$$

and modify (2.6) to read

$$C_{n,\delta}\, \mu_*(\delta,\Gamma) \lambda^{\delta/2} \leq (2\pi)^{-n} \omega_n |\Omega|_n \lambda^{n/2} - \mathrm{N}(\lambda) \leq C_{n,\delta}\, \mu^*(\delta,\Gamma) \lambda^{\delta/2} + \mathrm{O}(\lambda^{\delta/2}) \tag{2.8}$$

We call (2.8) the weak modified Weyl-Berry conjecture (WMWB) .

In order to study (2.8) two basic "tools" are employed. The first is the idea of constructing a sequence of finer and finer tessellation of R" (i.e. the Whitney covering) by cubes $\{Q_\xi^\kappa\}_{\xi \in Z^n}$ .That is, for each *k*, $\{Q_\xi^\kappa\}_{\xi \in Z^n}$ is a tessellation of R^n into a countable family of congruent and non-overlapping open cubes with sides of length b_κ such that

$$R^n = \bigcup_{\xi \in Z^n} \overline{Q}_\kappa^\xi \tag{2.9}$$

By finer and finer tessellation we mean that $b_{\kappa+1} < b_{\kappa}$ and $b_{\kappa} \to 0$ as $\kappa \to \infty$. Note that the construction of the tessellation starts at the origin. This idea enables one to determine the influence of Ω and especially the irregular fractal nature of Γ. The second tool is substantial use of the Dirichlet-Neumann bracketing method as described in Courant and Hilbert [CH] or Edmunds and Evans [EE].

To illustrate the type of results obtained consider first the case n=1 (the fractal string). Here $\Omega \subset R$ is of finite length $|\Omega|_1$ with boundary Γ. Ω is the union of its connected components viz..

$$\Omega = \bigcup_{j=1}^{\infty} I_j \tag{2.10}$$

where I_j are open intervals, pairwise disjoint and of length $I_{j \cdot i} |\Omega|_1 = \sum_{j=1}^{\infty} I_j < \infty$.

Assume

$I_1 \geq I_2 \geq \ldots\ldots \geq I_j \geq \ldots\ldots$

For a vibrating string of length l the eigenvalues are given by $\lambda_k = (\pi/l)^2 \kappa^2, \kappa = 1,2,\ldots$ and so for such a string

$$\mathrm{N}(\lambda) = \#\{\kappa \geq 1, \kappa < l\sqrt{\lambda/\pi}\} \equiv \left[l\sqrt{\lambda/\pi}\right] \tag{2.11}$$

where $[x]$ is the integer part of x. So for our problem

$$\mathrm{N}(\lambda) = \sum_{j=1}^{\infty} \mathrm{N}(\lambda, I_j) = \sum_{j=1}^{\infty} [I_j, x] \tag{2.12}$$

where $x = \pi^{-1} \lambda^{1/2}$.

Suppose the component lengths I_j satisfy the asymptotic estimate

$$I_j \sim L j^{-1/D} \text{ as } j \to \infty \;, \tag{2.13}$$

then Lapidus and Pomerance [LP] have shown that Γ is measurable and that

$$\mathrm{N}(\lambda) = |\Omega|_1 x + \varsigma(D) L^D x^D + \mathrm{O}(x^D) \tag{2.14}$$

where $\varsigma(D)$ is the Riemann Zeta function. For this example MWB is true.

Now suppose I_j does not have the asymptotic form (2.13), i .e . suppose $I_j = \alpha^j$ for some $0 < \alpha < 1$ (as in the case of the "Cantor string")?. That is suppose we have 2^{j-1} copies of open intervals I_j of length 3^{-j}. This example has been studied by Lapidus and Pomerance [LP] and by Chen and Sleeman [CHS]. This example is particularly interesting in that we can use simple number theoretical arguments to obtain an informative exact expression for $\mathrm{N}(\lambda)$ as proved in [S]. Clearly

$$\mathrm{N}(\lambda) = \sum_{j=1}^{\infty} 2^{j-1}\left[3^{-j}\lambda^{1/2}/\pi\right] = \sum_{j=1}^{\infty} 2^{j-1}3^{-j}\lambda^{1/2}/\pi - S \tag{2.15}$$

where

$$S = \sum_{j=1}^{\infty} 2^{j-1}\left\{3^{-j}\lambda^{1/2}/\pi - \left[3^{-j}\lambda^{1/2}/\pi\right]\right\} \tag{2.16}$$

Let us fix $x = \lambda^{1/2}/\pi$ and let J be the maximum number for which $\lfloor x3^{-J} \rfloor \geq 1$. That is $J \leq [\ln x/\ln 3]$. We express x in ternary form;

$$x = \alpha_o + 3\alpha_1 + ... + 3^J \alpha_J \tag{2.17}$$

where $\alpha_o \in [0,3)$, $\alpha_j = 0,1,2;\ j = 1,..,J-1$ and $\alpha_j = 1$ or 2. It is simple to see that

$$S = \alpha_o + 2\alpha_1 + ... + 2^J \alpha_J$$

and that

$$\mathrm{N}(\lambda) = \frac{\lambda^{1/2}}{\pi} - \left(\frac{\lambda^{1/2}}{\pi}\right)^{\delta} \frac{(\alpha_0 + 2\alpha_1 + ... + 2^J \alpha_J)}{(\alpha_0 + 3\alpha_1 + ... + 3^J \alpha_J)^{\delta}} \tag{2.18}$$

where $\delta = \ln 2/\ln 3$ and the coefficient of $\left(\frac{\lambda^{1/2}}{\pi}\right)$ is periodic of period 1 in $\left\{\frac{\ln \lambda^{1/2}/\pi}{\ln 3}\right\}$. In other words the second term in the expansion of $N(\lambda)$ is oscillatory and so the MWB is not true in this case but WMWB is true.

Let us now turn to the case $n \geq 2$. In this case we have the following explicit results [CHS] for both fractal and non-fractal cases. Suppose $\delta = n-1$ and $\mu^*(\delta, \Gamma) < +\infty$ then

Theorem 2.1

Let $\Omega \subset R^n$ be an open bounded set with boundary Γ. If the Minkowski dimension of Γ is $\delta = n-1$ and $\mu^*(\delta,\Gamma) < +\infty$ then

$$\psi(\lambda) = |\phi(\lambda) - \mathrm{N}(\lambda)| \le C_n \mu^*(\delta,\Gamma) \log_\alpha \lambda \cdot \lambda^{\frac{n-1}{2}} + \mathrm{O}\left(\lambda^{\frac{n-1}{2}}\right),$$

as $\lambda \to \infty$, where $C_n = \left[4(4\pi)^{\frac{n-1}{2}} \Gamma\left(1+\frac{n-1}{2}\right)\right]^{-1}$ and $\alpha = \exp\left(\frac{1}{2e} n^{-3/2}\right) > 1$.

In the fractal case where $\delta \in (n-1, n]$, we have

Theorem 2.2

Let $\Omega \subset R^n (n \ge 1)$ be an open bounded set with boundary Γ. If the Minkowski dimension of Γ is $\delta \in (n-1, n)$ and $\mu^*(\delta,\Gamma) < +\infty$ then

$$\psi(\lambda) \le C_{n,\delta}(\alpha_0) \mu^*(\delta,\Gamma) \lambda^{d/2} + \mathrm{O}(\lambda^{\delta/2})$$

as $\lambda \to +\infty$, for any $\alpha_0 \in (0,1)$ and where

$$C_{n,d}(\alpha_0) = \left(1+\frac{1}{\pi}\right)^{\frac{n-1}{2}} \left(\frac{\sqrt{n}}{\alpha_0}\right)^{n-\delta} \left[1-\alpha_0^{\delta-n+1}\right]^{-1} + \left(1+\frac{1}{\pi}\right)^{\frac{n-1}{2}} n^{\frac{n-\delta}{2}} + (2\pi)^{-n} \omega_n n^{\frac{n-\delta}{2}}$$

In certain circumstances we can say more. For example if the sequence of tessellations of Γ by open cubes asymptotes at a certain rate (c.f. [CHS] for details) then as $\lambda \to \infty$ we have the refinement of Theorem 2.2 as

$$\psi(\lambda) = F_n(\lambda) \lambda^{\delta/2} + \mathrm{o}(\lambda^{\delta/2}) \tag{2.19}$$

where $F_n(\lambda)$ is a well defined, bounded and left continuous function of λ; furthermore the set of points of discontinuity is dense in R. This result covers a particularly important example of Fleckinger-Vassiliev [FV] in which $F_n(\lambda)$ is oscillatory and periodic. Further developments of the above results are to be found in [CHS1,2,3]. We close this section with the remark that there are similar results to those above for the Neumann problem. However for their validity one must assume a type of "segment property" to hold on Γ to ensure that the spectrum of the Neumann Laplacian is discrete. We shall return to this property later in addressing the scattering.

3. Fractals on Manifolds

We are concerned with the Dirichlet eigenvalue problem for the Laplace-Beltrami operator defined on a manifold *M*. Thus if the manifold is endowed with the metric tensor g_{ik} and the surface element $ds^2 = g_{ik} dx^i dx^\kappa$ then the eigenvalue problem is

$$\Delta u + \lambda u = \frac{1}{\sqrt{g}} \frac{\partial}{\partial x^i} \left(\sqrt{g} g^{ik} \frac{\partial u}{\partial x^\kappa} \right) + \lambda u = 0, \tag{3.1}$$

$u = 0$ on $\partial \mathrm{M}$,

where $g^{ik} g_{kJ} = \delta^i_j, g = \det g_{ij}$

In the particular case of interest here where M is a subset of the surface of the unit sphere we introduce the polar co-ordinates,

$$\begin{aligned} x_1 &= \sin \vartheta \cos \varphi, 0 \le \vartheta \le \pi, \\ x_2 &= \sin \vartheta \sin \varphi, 0 \le \varphi \le 2\pi, \\ x_3 &= \cos \vartheta, \end{aligned} \tag{3.2}$$

and let $t = \cos \vartheta$ Equation (3.1) then becomes

$$\left(1 - t^2\right) \frac{\partial^2 u}{\partial t^2} - 2t \frac{\partial u}{\partial t} + \frac{1}{\left(1 - t^2\right)} \frac{\partial^2 u}{\partial \vartheta^2} + \lambda u = 0. \tag{3.3}$$

The manifold *M* is taken to be the surface segment bounded by $\varphi = 0, \varphi = \beta$, for all $\vartheta \in [0, \pi]$. Further we assume that $\beta/\pi = b/\alpha$ is rational where $(\alpha, b) = 1$.

To construct a fractal surface we consider a triadic Cantor set of segments generated by the sequence $\beta_1 = 3^{-1}\pi$; two copies of $\beta_2 = 3^{-2}\pi$; four copies of $\beta_3 = 3^{-3}\pi$; etc . . . . , 2^{k-1} copies of $\beta_k = 3^{-k}\pi$. In order to develop counting function asymptotic for such a fractal surface we first consider the counting function for a particular segment. This problem was discussed by Gromes [G]. Our approach is different and is based on the simple number theoretic ideas above.

By the method of separation of variables we see that the eigenvalues of (3.3) are given by $\lambda = \nu(\nu + 1)$ where

$$\nu = \frac{m\pi}{\beta} + x, m = 1,2..., x = 0,1,.... \tag{3.4}$$

Hence the counting function is given by

$$\mathrm{N}(\lambda) = \sum_{m\geq 1, x\geq 0} 1 \tag{3.5}$$

subject to $\frac{m\pi}{\beta} + x \leq \sqrt{\lambda + 1/4} - 1/2 = \nu$.

In terms of ν we have

$$\mathrm{N}(\lambda) = \sum_{m\geq 1}^{[\nu\beta/\pi]} \sum_{x\geq 0}^{[\nu - m\pi/\beta]} 1 = \sum_{m\geq 1}^{[\nu\beta/\pi]} [\nu - m\pi/\beta] + [\nu\beta/\pi] \tag{3.6}$$

Suppose $\beta = 3^{-k}$ and suppose K is the maximum value of k for which $\nu 3^{-k} \geq 1$, and expand ν to the base 3 as

$$\nu = \alpha_0 + 3\alpha_1 + ... + 3^K \alpha_k \tag{3.7}$$

where $0 \leq \alpha_0 \leq 2, \alpha_i = 0,1,2, i = 1,....,\mathrm{K}$.

The arguments are now similar to those developed for the "string" problem of section 2 leading to the exact expression

$$\mathrm{N}(\nu) = \left(\nu + 1 - \frac{3^k}{2} - \{\alpha_0\}_F\right)[3^{-k}\nu] - \frac{3^k}{2}[3^{-k}\nu]^2, \tag{3.8}$$

where $\{\alpha_0\}_F = \alpha_0 - [a_0]$ is the fractional part of α_0. We should note that as $\lambda = \nu(\nu + 1) \to \infty$ then we can deduce f rom (3.8) that

$$\mathrm{N}(\lambda) = \frac{3^{-k}}{2}\lambda - \lambda^{1/2} + \delta(\lambda)\lambda^{1/2} + \frac{1}{4} + O(\lambda^{-1/2}) \tag{3.9}$$

where $\delta(\lambda) = -\frac{1}{3^k}\{(\sqrt{\lambda + 1/4} - 1/2)\} + O(\lambda^{-1/2})$ in which $\{x\} = x - [x] - 1/2$.

The result (3.9) is precisely that given by Gromes [G]. For the fractal problem it.is clear that

$$\mathrm{N}(\lambda) = \sum_{k=1}^{\infty} 2^{k-1} \mathrm{N}_k(\lambda) \tag{3.10}$$

where $N_k(\lambda)$ is the counting function for the segment of polar angle $3^{-k}\pi$. From (3.8) we can write

$$N(\lambda)=(\nu+1-\{\alpha_0\}_F)\sum_{k=1}^{\infty}2^{k-1}[3^{-k}\nu]-1/4\sum_{k=1}^{\infty}6^k[3^{-k}\nu]-1/4\sum_{k=1}^{\infty}6^k[3^{-k}\nu]^2. \quad (3.11)$$

$$=E-F-G. \quad (3.12)$$

Each of the terms E,F,G are evaluated again using the ideas above. For example

$$E=(\nu+1-\{\alpha_0\}_F)\left(\nu-\nu^{\delta}\frac{(\alpha_0+2\alpha_1+.....+2^K\alpha_K)}{(\alpha_0+3\alpha_1+.....+3^K\alpha_K)^{\delta}}\right),$$

$$F=-\frac{1}{2}\nu+\frac{1}{2}\nu^{\delta+1}\frac{(\alpha_0+6\alpha_1+...+6^K\alpha_K)}{(\alpha_0+3\alpha_1+...+3^K\alpha_K)^{\delta+1}}$$

and G is a structurally similar but much more complicated expression (see [SCH]) and $\delta=\ln 2/\ln 3$.

The important thing to deduce from these calculations is that once again MWB does not hold in the sense that subsequent terms in the expansion of $N(\lambda)$ oscillate.

The study of spectral asymptotics for manifolds is a subject of considerable interest to analysis and we cite Sakharov and Vassiliev [SV] for details.

4. Isospectal Fractal Domains

Let *(M,g)* be a compact Riemann manifold with boundary and let Δ be the Laplace operator defined on *M*. The spectrum of *M* is a sequence of eigenvalues of Δ. Two Riemann manifolds are isospectral if their spectra (counting multiplicities) coincide. A fundamental question concerning the relationship between spectral analysis and geometry raised by Kac[K] is: must two isospectral planar domains necessarily be isometric? Despite many results for non-planar domains being known for some time it was not until the ground breaking work of Gordon, Webb and Wolpert [GWW] in 1992 who gave a resounding NO to the Kac question. Furthermore their work allows one to construct a whole variety of pairs of isospectral non-isometric planar domains. Following Gordon et.al [GWW], Berard [Be] gave a simple proof showing that the eigenvalues of two planar isospectral, non isometric domains A, B say are identical by constructing a map which takes an eigenfunction of A say and maps it onto an eigenfunction for B with the same eigenvalue. The idea of transposition maps has been developed further by Buser et.al [Bu] to generate new examples of isospectral domains. In [SCHl,2] we have developed these ideas further to construct isospectral, non-isometric planar domains with fractal boundaries. The key to these constructions is a "paper folding" argument due to Chapman [C] which is a constructive method of realising the transposition maps of Berard [Be] and Buser et.al [Bu]. An example is given in Figures 1 and 2.

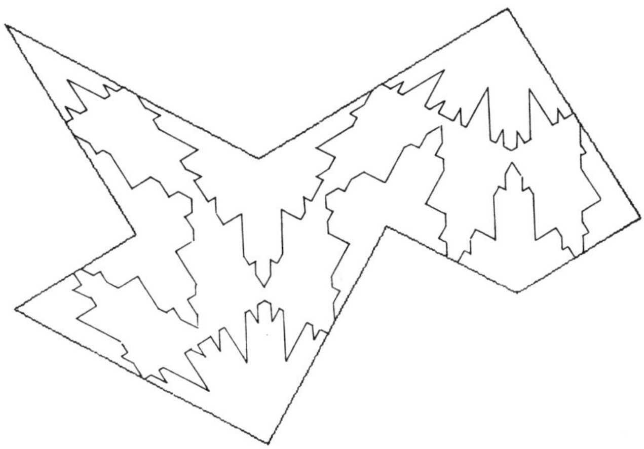

Figure 1

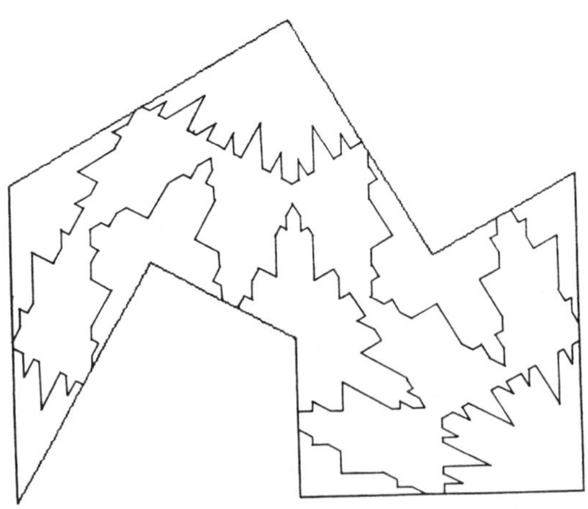

Figure 2

Figures 1 and 2 show a pair of (i) isospectral non-isometric domains with piecewise continuous boundaries represented by the outer curves and (ii) the first steps in constructing a pair of isospectral non-isometric connected fractal domains represented by the domains enclosed by the inner boundaries. The fractals are constructed following a similar procedure as for creating the Koch island curve based on the complement of the 1/3 Cantor set.

5. Acoustic Scattering by Rough Surfaces

In this section we formulate the problem of scattering of plane incident waves by an obstacle $\Gamma = R^n - \Omega$, where Ω is an exterior domain. Here the boundary $\partial\Gamma$ is assumed to be rough in the sense of being irregular and even fractal. In order to do this we must formulate the problem suitably. We follow in the spirit of Wilcox [W].

Let $w_0\,(p,x)$ represent an incident plane wave propagating in the direction p and a time harmonic dependence $\exp(-i\,|p\,|t)$ is suppressed. Let $w(p,x)$ be the wave scattered by the obstacle Γ.

Then $w(p,x)$ satisfies

$$\left(\Delta + |p|^2\right)w(p,x) = 0, x \in \Omega \tag{5.1}$$

where $w(p,x) = w_0\,(p,x) + w(p,x)$, $|x| \geq 0$,and satisfies a Neumann boundary condition on $\partial\Gamma$. Furthermore $w(p,x)$ is required to satisfy the Sommerfeld radiation condition. Since $\partial\Gamma$ is rough we must use a generalised form of the Neumann condition.

Definition 5.1

A function $u \in L_2^1(\Delta, \Omega)$ is said to satisfy the generalised Neumann condition for Ω if and only if

$$\int_\Omega \{(\Delta u)v + \nabla u \cdot \nabla v\}dx = 0 \forall v \in L_2^1(\Omega). \tag{5.2}$$

Note that (5.2) is meaningful for arbitrary domains Ω and defines a closed subspace

$$L_2^{\mathrm{N}}(\Delta, \Omega) = L_2^1(\Delta, \Omega) \cap \{u : u\,satisfies(5.2)\}$$

in the Hilbert space $L_2^1(\Delta, \Omega)$.

Remark: (5.2) is equivalent to the classical Neumann condition if $\partial\Gamma$ is sufficiently smooth. In this case functions $v \in L_2^1(\Delta, \Omega)$ will have a. trace on $\partial\Gamma$ and Δu of functions $u \in L_2^1(\Delta, \Omega)$ will have a trace on $\partial\Gamma$ by the Sobolev imbedding theorems.

Definition 5.2

Let Ω be an exterior domain. Then the outgoing scattered wave *w(p,x)* for Ω, and $p \in R^n$ is the function characterised by $w_0(p,\cdot) + w'(p,\cdot) \in L_2^{N,loc}$, satisfies (5.1) and *w'(p,x)* satisfies the radiation condition.

In order to establish existence and uniqueness of our problem we need a local compactness result for Ω as well as a corresponding limiting absorption principle for obstacles with rough boundaries. The tool needed for these considerations is an extension of the Rellich selection theorem. To this end we introduce

Definition 5.3

A domain $\Omega \subset R^n$ is said to have the local compactness property if and only if for each set of functions $S \subset L_2^{1,loc}(\overline{\Omega})$ and each $R > 0$ the condition

$$\|u\|_{L_2^1(\Omega_R)} \le C(R)$$

for all u = $v \mid_{\Omega R}$ with $v \in S$ implies that $\{u = v|_{\Omega_R} : v \in S\}$ is pre compact in $L_2(\Omega_R)$, where

$$\Omega_R = \Omega \cap \{x : |x| < R\}.$$

The class of domains with the local compactness property is denoted by L.C.

Remark: Rellich's original selection theorem states that bounded domains with smooth boundaries are in class L.C. AGM [A] generalised the result for domains having a "segment" property. While this covers a large class of domains there are exceptions, as pointed out by Wilcox [W]. To overcome these limitations Wilcox [W] introduced the set of domains having the "tiling" property. To do this assume there is an open set $O \subset R^n$, compact sets $K_1, K_2, \ldots, K_N \subset R^n$ and non-zero vectors $x^{(1)}, x^{(2)}, \ldots, x^{(N)}$ such that

$$\partial\Gamma \subset O, \tag{5.3}$$

$$O \cap \Omega \subset \bigcap_{j=1}^{N} K_j \, . \tag{5.4}$$

Each set K_j lies in a (curvilinear) co-ordinate patch O_j such that

$$\{x = x_0 + tx^{(j)} : 0 < t < 1\} \subset \Omega \tag{5.5}$$

for every $x_0 \in \Omega \cap K_j$, holds when $x=(x_1,.x_2 \cdot\cdot,x_n)$ are the corresponding co-ordinates.

Definition 5.4

An exterior domain Ω will be said to have the "finite tiling" property if (5.3), (5.4), (5.5) hold for a suitable set O, compact sets K_j and co-ordinate patches O_j.

Theorem 5.1

If Ω is an exterior domain having the finite tiling property then $\Omega \in LC$.

Theorem 5.2

Let Ω be an exterior domain such that $\Omega \in LC$. Then for each $p \in R^n$ there exists a unique outgoing distorted plane wave $w_+(p,x)$.

Corollary 5.1

Under the conditions of Theorem 5.2 there exists a function $\partial_+(\eta, p) \in C^\infty(S^{n-1} x R^n - \{0\})$ such that

$$w_+(p,x) = w_0(p,x) + \frac{e^{i|p||x|}}{|x|^{\frac{n-1}{2}}} \vartheta\left(p, \frac{x}{|x|}\right) + q_+(p,x)$$

where $q_+(p,x) = O(|x|^{-(n+1)/2}), |x| \to \infty$ uniformly for $\eta = \frac{x}{|x|} \in S^{n-1}$ and p in any compact subset of $R^n - \{0\}$.

$\vartheta_+(p,\eta)$ is called the far-field amplitude or radiation pattern.

Our purpose now is to discuss the asymptotic behaviour of $s(\lambda) = phase\,\vartheta\ \ \vartheta_+(p,\eta)$ as $\lambda \equiv |p|^2 \to \infty$.

In the case of the Dirichlet problem we already have the result (CHS3)

Theorem 5.3

Let K be an obstacle in R^n with boundary $\partial\mathrm{K}$. If $\partial\mathrm{K}$ is δ-Minkowski measurable with $\delta \in (n-1, n)$ and δ-Minkowski measure $\mu^*(\delta, \partial\mathrm{K})$, then there exists a constant $C_{n,d} > 0$ depending only on n and δ such that

$$\left|s(\lambda) - C_n |\mathrm{K}|_n \lambda^n\right| \leq C_{n,d}\, \mu^*(\delta, \partial\mathrm{K}) \lambda^\delta \tag{5.6}$$

where $C_n = (4\pi)^{-n/2} \left|\Gamma\left(1 + \frac{1}{2}n\right)\right.$.

This result which is proved using an extension of the abstract theory of Petkov and Popov [PP] for smooth and non-trapping obstacles and does not require the "tiling" property definecl above. In the case of the Neumann problem we have the following

Theorem 5.4

Let K be an obstacle (i.e. compact with connected complement $\Omega = R^n - \mathrm{K}$) in $R^n(n \geq 1)$ with boundary ∂K and which satisfies the "tiling" property.If δ is the Minkowski dimension of ∂K , $\delta \in [n-1, n)$ and $\mu^*(\delta, \partial K) < \infty$ then

1) If $\delta = n-1$, and $s_n(\lambda)$ is the Neumann scattering phase for the obstacle K, there is a constant $d_n > 0$ depending on n such that

$$\left|s_n(\lambda) - C_n \lambda^n\right| \leq d_n \mu^*(n-1, \partial K) \ln \lambda \cdot \lambda^n,$$

as $\lambda \to \infty$.

2) If $\delta \in (n-1, n)$ there is a constant $d_{n,\delta} > 0$, depending on n and δ such that

$$\left|s_n(\lambda) - C_n |\mathrm{K}| \lambda^n\right| \leq d_{n,\delta}\, \mu^*(\delta, \partial\mathrm{K}) \lambda^\delta,$$

as $\lambda \to \infty$.

This result recently obtained by Sleeman and Chen Hua [SCH3] depends on a tessellation of domains argument together with some new monotonicity theorems for the Neumann scattering phase. A result similar to Theorem 5.4 has been obtained by Christiansen [Chr] and in the case of the Dirichlet problem some very general results have been obtained by Robert [R1,2,3].

Acknowledgements

A number of the results reported in this paper have been obtained in collaboration with Chen Hua of Wuhan University and supported by grants from the Royal Society of London and the National Natural Sciences Foundation of China. Thanks are extended T. Christiansen and D Robert for bringing their work to the authors attention.

References

[A] Agmon,S: Elliptic Boundary Value Problems.Van Nostrand (1965)

[Be] Berard, P: Domaines plans isospectraux a la Gordon-Webb-Wolpert: une preure terra a terra (preprint) (see also Transplantation et Isospectralite I, II Math Ann (1992)

[B] Berry, M.V: Some geometric aspects of wave motion: Wave front dislocations, diffraction catastrophes, diffractals. In: Geometry of the Laplace Operator, Proc.Symp. Pure Math. Vol. **36**, Providence, RI: Am. Math. Soc; 1980, pp.13-38

[BC] Brossard, J; Carmona, R: Can one hear the dimension of a fractal. Commun. Math. Phys. **104**, 103-122 (1986)

[Bu] Buser, P; Conway, J; Doyle, P; Semmler, K-D: Some planar isospectral domains (preprint)

[C] Chapman, S. J: Drums that sound the same. Amer. Math. Monthly 124-138 (1995)

[CH] Courant, R: Hilbert, D: Methods of mathematical physics.Vol. **1**, Interscience, New York, London (1953)

[Chr] Christiansen,T: Spectral asymptotics for compactly supported perturbations of the Laplacian on R^n (pre-print) 1997

[CHS] Chen, Hua; Sleeman,B.D . Fractal Drums and the n-dimensional modified Weyl-Berry conjecture. Comm. Math. Phys; **168**, 581-607 (1995)

[CHS1] Chen, Hua; Sleeman, B. D: Counting function asymptotics and the weak Weyl-Berry conjecture for connected domains with fractal boundaries. Acta. Mathematica. Sinica. (to appear) (1998)

[CHS2] Chen, Hua; Sleeman, B.D: Estimates for the remainder term in the asymptotics of the counting function for domains with irregular boundaries. Rendiconti del Seminario Matematico, Universitae Politcnio Torino (to appear)(1998)

[CHS3] Chen, Hua; Sleeman, B.D: An analogue of Berry's conjecture for. the phase in fractal obstacle scattering.I.M.A.J. Appl. Math **49** 193-202 (1992)

[EE] Edmunds, D. E; Evans, W. D: Spectral Theory and Differential Operators, Oxford University Press, Oxford (1987)

[F] Fleckinger, J: On eigenvalue problems associated with fractal domains. Pitman Research Notes in Math.Series, **216,** Proceedings of the tenth Dundee Conference (Eds Sleeman, B. D; Jarvis, R. J) 1989, pp 60-72

[FV] Fleckinger, J; Vassiliev, D. G: An example of a two term asymptotics for the "counting function" of a fractal drum. Trans. Am. Math. Soc. **337** (1), 99-116 (1993)

[GWW] Gordon, C; Webb, D; Wolpert, S: Isospectral plane domains and surfaces via Riemannian orbifolds. Invent. Math; **110**, 1-22 (1992)

[G] Gromes, D: Úber die asymptotische Verteilung der Eigenwerte des Laplace-Operators fur Gebeite auf der Kugeloberflache. Math. Zeitschr. **94,** 110-121 (1966)

[IVl] Ivrii,V. Ja: Second term of the spectral asymptotic expansion of the Laplace-Beltrami operator on manifolds with boundary.Funct. Anal. Appl. **14**, 98-106 (1980)

[IVl] Ivrii, V. Ja: Precise Spectral asymptotics for elliptic operators acting on fiberings over manifolds with boundary. Lecture Notes in Math; Vol. **1100**, Berlin, Heidelberg, New York: Springer-Verlag, 1984

[K] Kac, M: Can one hear the shape of a drum?. Am. Monthly (Slaught Memorial paper, No. 11) (4) 73, 1-23 (1966)

[L] Lapidus, M.L: Fractal drum, inverse spectral problems for elliptic operators and a partial resolution of the Weyl-Berry conjecture. Trans. Am. Math. Soc. **325**, 465-529 (1991)

[LFl] Lapidus, M. L; Fleckinger, J: The vibrations of a fractal drum", Lectures in Pure and Appl. Math; Diff. Equa; N.Y.-Basel: Marcel Dekker, 1989, pp 423-436

[LF2] Lapidus, M. L; Fleckinger, J: Tambour fractal: Vers une resolution de la conjecture de Weyl- Berry pour les valeurs propres du Laplacian. C. R. Acad. Sci. Paris, Ser, 1 Math. **306**, serie 1, 171-175 (1988)

[LP] Lapidus, M.L; Pomerance, C: The Riemann zeta-function and the one-dimensional Weyl-Berry conjecture for fractal drums. Proc. London. Math. Soc. (3) **66**, 41-69 (1993)

[M] Melrose, R: Weyl's conjecture for manifolds with concave boundary. Geometry of the Laplacian Operator. Proc. Symp. Pure. Math; Vol. **36**, Providence, RI: Am. Math. Soc.1980

[PP] Petkov, V; Popov, G: Asymptotic behaviour of the scattering phase for non-trapping obstacles. Ann. Inst. Four. **32**, 111-149 (1982)

[R1] Robert, D: Sur la formule de Weyl pour des ouverts non borne's. C. R. Acad. Sci. Paris, t. **319**, Se'rie 1, 29-34 (1994)

[R2] Robert, D: A trace formula for obstacles problems and applications. Operator Theory; Advances and Applications, Vo1. **70**, 283-292 (1994)

[R3] Robert, D: On the Weyl formula for obstacles, Rapport de Recherche, Universite' de Nantes, 95/10-1

[SV] Safarov,Yu: Vassiliev, D: The Asymptotic distribution of Evgenvalues of Partial Differential Operators. Translations of Mathematical Monographs Vol **155**, Providence, RI, Am. Math. Soc. 1997

[S] Sleeman, B.D: Some contributions to thc Wcyl-Berry conjecture for fractal domains. Intl. J. Appl. Sci & Comp.**2** 344-361 (1995)

[SCH] Sleeman, B. D; Chen, Hua: Counting function asymptotics on fractal manifolds (preprint) (1998)

[SCH1] Sleeman, B. D; Chen, Hua: On the construction of non-isometric isospectral domains and Weyl-Berry conjecture. J. Sem. Ceremath-UTi, University of Toulousel (1996)

[SCH2] Sleeman, B. D; Chen, Hua: On non-isometric isospectral connected fractal domains (in preparation) (1998)

[SCH3] Sleeman, B. D; Chen, Hua: Asymptotic estimates for the Neumann scattering phase in non-smooth obstacle scattering, (in preparation) 1998.

[V] Vassiliev, D: One can hear the dimension of a connected fractal in R^2. In: Petkov and Lazarov-Integral Equations and Inverse Problems, London: Longman Academic, Scientific and Technical, 1991, pp 270-273

[W] Wilcox, C: Scattering theory for the d'Alembert Equation in Exterior Domains, Lecture notes in Mathematics series vol. **442**, Springer-Verlag, Heidelberg (1975)

Address for correspondence

School of Mathematics,

University of Leeds,

Leeds, LS2-9JT.

England,UK.

e-mail bds@amsta.leeds.ac.uk

SECTION B:
Mathematical Methods in Biomedical Technology

G CONSTANTINIDES, D GINTIDES, S KATTIS, K KIRIAKI, C PARASKEVA, A PAYATAKES, D POLYZOS, S TSINOPOULOS AND S YANNOPOULOS

Particle shape and size analyzer

1 Introduction

As it is well known, the problem of recovering the shape or/and the physical characteristics of a body from the particular way it disturbs the propagation of a time-harmonic incident wave, constitutes the inverse scattering problem, which is of great interest in applications. Inversion techniques usually utilize results of the study of the direct scattering problem, that is, the determination of the scattered field caused by a body, which is excited by an incident harmonic wave. The investigation of the direct and inverse electromagnetic scattering problem is very important because in many applications results of this scattering process are exploited.

The direct electromagnetic scattering problem has been studied extensively by many researchers. Several methods have been developed to solve it. These can be classified in three classes, namely, analytical, semi-analytical and numerical methods. Most analytical methods are based on the theories of Mie, Rayleigh, Fraunhofer, Rayleigh-Gans, anomalous diffraction, etc. Analytical and semi-analytical methods are used in the work of Tranquilla and Al-Rizzo [1], [15] and Barber and Hill [4] to handle bodies with special geometry. Numerical methods constitute a broad class and have a great interest, especially those which are based on integral equations. Among them, the boundary element method [9] is a powerful tool to face numerically the direct scattering problem for scatterers with general shape. In [16] this method is used to examine a three dimensional dielectric body.

In the examination of the inverse scattering problems the algorithms which are proposed, are based on the continuity dependence of the far-field data on the characteristics of the scatterer. This is necessary in order to stabilize the inverse problem [3],[6]. In applications a priori informations about the characteristics of the scatterers, which are always available, are incorporated in the inversion schemes to simplify the whole procedure and to produce fast algorithms. In many commercial applications, the size of particles in liquid suspensions (or aerosols) is determined by measuring the light scattered by particles, based on the principle that the light scattering pattern depends on the geometrical characteristics of the scatterer [11]. The main difficulty in this technique is the inversion of the data, in order to extract the desired information for the shape and the size of the particles. For this reason, most commercial instruments treat even non-spherical particles as if they were spherical, and assign to each measured particle an equivalent diameter, using empirical calibration functions.

In the present work a new instrument is developed for the accurate and fast determination of the shape, the size and the refractive index of particles in liquid suspensions, based on the scattering of a monochromatic electromagnetic (EM) wave by the particles. In Section 2 the mathematical formulation of the direct and inverse electromagnetic scattering problem is presented. A uniquely solvable boundary integral equations system is used in order to reformulate the boundary value problem which describes the scattering process in integral form. Based on it the continuity dependence of the far-field data upon the shape and the refractive index of the particles is established. In Section 3 the experimental instrument, which is developed, is presented. In Section 4 a fast inversion algorithm for the determination of the shape, the size and the refractive index is constructed. A database for spheroids and spheres is produced by solving the direct scattering problem with a Boundary Element Method. A parametric study of the database, produces an important class of light scattering parameters which are connected with the shape, the size and the refractive index of the scatterers. An inversion algorithm which incorporates a priori data about the shape and the refractive index of the scatterers in a pattern recognition program is constructed. This program finds which light scattering pattern from the database best fits to the measured data. The computation time for the recognition is less that 2 sec in a typical Pentium PC. The performance of the new apparatus is demonstrated in Section 6, using first spherical latex and glass prototypes and in the sequel non-spherical biological particles, namely, red blood cells and bacteria .

2 Mathematical formulation of the direct and the inverse scattering problem

We assume a bounded, convex and closed subset of R^3, let Ω_2, with smooth boundary S. The complement of $\overline{\Omega}_2$ will be denoted by Ω_1 and it will be assumed to be simply connected. The dielectric constant in Ω_2 is ε_2, the magnetic permeability μ_2 and the electric conductivity σ_2. The surrounding medium has the parameters ε_1, μ_1, σ_1 respectively.

We consider in Ω_1 the propagation of a time harmonic electromagnetic wave of the form

$$\mathbf{E}_{in}(\mathbf{r}) = \widehat{\mathbf{b}}\, e^{i\,k_1 \widehat{\mathbf{k}}\cdot\mathbf{r}}, \;\; \mathbf{H}_{in}(\mathbf{r}) = (\widehat{\mathbf{k}}\times \widehat{\mathbf{b}})\, e^{i\,k_1 \widehat{\mathbf{k}}\cdot\mathbf{r}} \tag{1}$$

where $\widehat{\mathbf{k}}$ is the unit propagation vector and $\widehat{\mathbf{b}}$ is the unit polarization vector of the electric field $\widehat{\mathbf{k}}\cdot\widehat{\mathbf{b}} = 0$, k_1 is the wavenumber expressed by the relation

$$k_1^2 = (\varepsilon_1 + \frac{i\,\sigma_1}{\omega})\,\mu_1\omega^2 \tag{2}$$

where $Im\,k_1 \geq 0$. We define by $\mathbf{E}_i$ and $\mathbf{B}_i$ the fields in $\Omega_i, i = 1, 2$. The total fields in Ω_1 are the superposition of the incident and the scattered fields and are given by the relations

$$\mathbf{E}_1(\mathbf{r}) = \mathbf{E}_{in}(\mathbf{r}) + \mathbf{E}_{sc}(\mathbf{r}), \;\; \mathbf{H}_1(\mathbf{r}) = \mathbf{H}_{in}(\mathbf{r}) + \mathbf{H}_{sc}(\mathbf{r}) \tag{3}$$

The electric and magnetic fields satisfy the time reduced Maxwell's equations

$$curl\,\mathbf{E}_i\,(\mathbf{r}) - ik_i\mathbf{H}_i\,(\mathbf{r}) = \mathbf{0}, \;\; curl\,\mathbf{H}_i\,(\mathbf{r}) + ik_i\mathbf{E}_i\,(\mathbf{r}) = \mathbf{0} \tag{4}$$

in Ω_i for $i = 1, 2$, k_2 is given by a similar relation as (2) . The dielectric boundary conditions are expressing the continuity of the tangential components of the electric and magnetic field on the boundary of the scatterer and are given by the relations

$$\widehat{\eta} \times \mathbf{E}_1(\mathbf{r}') = \widehat{\eta} \times \mathbf{E}_2(\mathbf{r}'), \;\; \widehat{\eta} \times \mathbf{H}_1(\mathbf{r}') = \widehat{\eta} \times \mathbf{H}_2(\mathbf{r}') \quad , \mathbf{r}' \in S \tag{5}$$

For the scattered fields we suppose that the Silver-Müller radiation conditions hold

$$\mathbf{H}_{sc}(\mathbf{r})\times\frac{\mathbf{r}}{r}-\mathbf{E}_{sc}(\mathbf{r})=o\,(\frac{1}{r}),\ r\to\infty \tag{6}$$

$$\mathbf{E}_{sc}(\mathbf{r})\times\frac{\mathbf{r}}{r}+\mathbf{H}_{sc}(\mathbf{r})=o\,(\frac{1}{r}),\ r\to\infty \tag{7}$$

where $r=|\mathbf{r}|$, uniformly over all directions.

In [2] the conductivity problem for the Maxwell's equations using the method of integral equations is considered. The derived system of equations is uniquely solvable for all frequencies. This is very important for the examination of the inverse problem because in shape recognition the eigenfrequencies of the adjoint interior boundary value problem, as it is obvious are unknown.

In the sequel we will use the layer derived system of equations proposed in [2]. We express the scattered electric field as a combination of magnetic and electric dipole distributions of the form

$$\mathbf{E}_{sc}(\mathbf{r})=k_1\,curl\int_S \mathbf{a}\,(\mathbf{r}')\ \Phi_1(\mathbf{r},\mathbf{r}')\ \ dS(\mathbf{r}')+curlcurl\int_S \mathbf{b}(\mathbf{r}')\ \Phi_1(\mathbf{r},\mathbf{r}')dS(\mathbf{r}') \tag{8}$$

where $\Phi_j(\mathbf{r},\mathbf{r}')=\frac{e^{i\,k_j|\mathbf{r}-\mathbf{r}'|}}{4\pi|\mathbf{r}-\mathbf{r}|},\mathbf{r}\neq\mathbf{r}',\ j=1,2$ is the free-space Green's function for the Helmholtz equation. The tangential density vectors $\mathbf{a}$ and $\mathbf{b}$ are defined on the set $C_D^{0,a}(S)=\{\mathbf{h}\in C_T^{0,a}(S):Div\mathbf{h}\in C_T^{0,a}(S)\}$ where $Div\mathbf{h}$ is the surface divergence of $\mathbf{h}$ and $C_T^{0,a}(S)=\{\mathbf{h}\in C^{0,a}(S):\widehat{\boldsymbol{\eta}}\cdot\mathbf{h}=0 \text{ on } S\}$, $C^{0,a}(S)$ is the space of continuous functions satisfying a Hölder condition with $0<a\leq 1$. The interior electric field is similarly expressed in the form

$$\mathbf{E}_2(\mathbf{r})=k_2\,curl\int_S \mathbf{a}\,(\mathbf{r}')\ \Phi_2(\mathbf{r},\mathbf{r}')\,dS(\mathbf{r}')+curlcurl\int_S \mathbf{b}(\mathbf{r}')\ \Phi_2(\mathbf{r},\mathbf{r}')dS(\mathbf{r}') \tag{9}$$

Proceeding in the usual layer approach, from the continuity and jump relations of the potentials on the boundary, we infer that the vector densities have to satisfy the integral equations system

$$(I+K)\begin{pmatrix}\mathbf{a}\\ \mathbf{b}\end{pmatrix}=\frac{1}{k_1+k_2}\begin{pmatrix}-2\widehat{\boldsymbol{\eta}}\times\mathbf{E}_{in}\\ -2\,i\,\widehat{\boldsymbol{\eta}}\times\mathbf{H}_{in}\end{pmatrix} \tag{10}$$

where K is a compact operator defined with the space definition

$$K : C_D^{0,a}(S) \times C_D^{0,a}(S) \longmapsto C_D^{0,a}(S) \times C_T^{0,a}(S)$$

and given by the relation

$$K = \frac{1}{k_1 + k_2} \begin{pmatrix} k_1 M_1 - k_2 M_2 & P_1 - P_2 \\ P_1 - P_2 & M_1 - M_2 \end{pmatrix} \tag{11}$$

and the boundary operators in (11) are defined as

$$(M_j \mathbf{h})(\mathbf{r}) = \widehat{\boldsymbol{\eta}} \times curl \int_S \mathbf{h}(\mathbf{r}')\ \Phi_j(\mathbf{r}, \mathbf{r}')\, dS(\mathbf{r}'),\ j = 1, 2 \tag{12}$$

$$(P_j \mathbf{h})(\mathbf{r}) = \widehat{\boldsymbol{\eta}} \times curl\, curl \int_S \mathbf{h}(\mathbf{r}')\ \Phi_j(\mathbf{r}, \mathbf{r}')\, dS(\mathbf{r}'),\ j = 1, 2 \tag{13}$$

According to [2] this system is uniquely solvable for arbitrary values of k_1 and k_2.

The behavior of the scattered electric field $\mathbf{E}_{sc}$ is characterized from the far field amplitude $\mathbf{E}^\infty$ since the following asymptotic relation holds [6]

$$\mathbf{E}_{sc}(\mathbf{r}) = \frac{e^{i\,k_1 r}}{r} \mathbf{E}^\infty(\widehat{\mathbf{r}}) + O(\frac{1}{r^2}),\ \ r \to \infty \tag{14}$$

which is established from the Silver-Müller conditions.

The inverse problem that we examine in this work is the determination of the shape and the refractive index of the scatterer from far field measurements of the vector field $\mathbf{E}^\infty(\widehat{\mathbf{r}})$ over the unit sphere S_0 or in a compact subset of S_0. We can assume that the unknown body is star-shaped. Then continuity dependence of the far-field data on the shape of the scatterer can be established if we take into account the uniquely solvable integral equations system [7]. We also can establish continuity dependence of the far-field data upon the refractive index again exploiting the unique solvability of (10). Based on this result an algorithm can be implemented using a non-linear minimization procedure on a penalized functional [7]. The disadvantage of minimizing this functional is that an extremely great amount of computational time is needed. So, for applications this procedure is not practically useful. A priori assumptions about some characteristics of the scatterers are usually taken into consideration and in this way the whole procedure is simplified.

3 Description of the developed apparatus

In what follows we will describe the experimental instrument which we have developed.

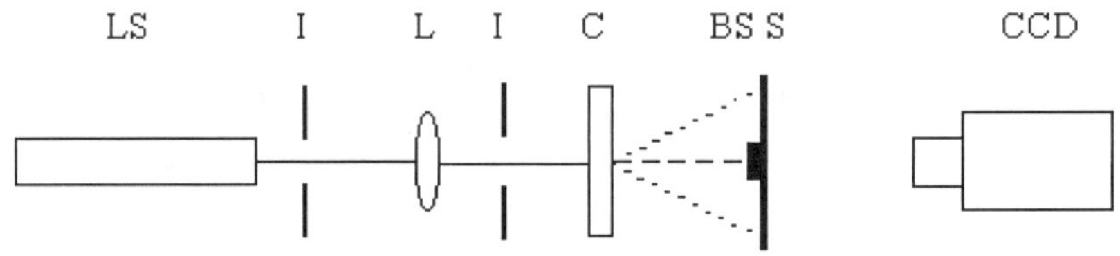

Fig.1 Schematic representation of the apparatus for particle analysis

A schematic diagram of the apparatus is given in Fig. 1. It consists of a monochromatic light source (LS), two iris diaphragms (I), a Fourier convex lens (L), the flow chamber (C), a semi-transparent screen (S), a beam stopper (BS), and a CCD camera. The light source is a 5 mW He-Ne laser ($\lambda = 632.8nm$), and the beam diameter is $0.8mm$. The Fourier lens is an achromatic doublet to correct for spherical aberration. The flow chamber is an optical cuvette with an observation window with dimensions $8 \times 35mm^2$.The optical cuvette consists of two parallel quartz plates leaving a gap of 100 μm in between them. The suspension is fed continuously into the flow chamber at constant flowrate, using a syringe pump. For a sufficiently large mean velocity in the flow chamber ($v = 1m/s$), relatively large shear stresses are developed, which force axisymmetric particles to become aligned with respect to the macroscopic flow direction. Particles in the flow chamber are excited by the laser beam, and the forward light scattering pattern appears on the semi-transparent screen (S). The beam stopper (BS) is used to prevent the light of the central beam from "blinding" the CCD camera. Using the CCD camera, an image analysis board and the appropriate software, the scattering pattern is transmitted to a PC for analysis. Special non-linear image filters are used to reduce the noise of the scattering pattern. For each scattering pattern the light intensity is measured along specific directions. For spheroidal particles, the light intensity is measured along the major and the minor semi-axes of the elliptic scattering pattern, starting from the center of the image.

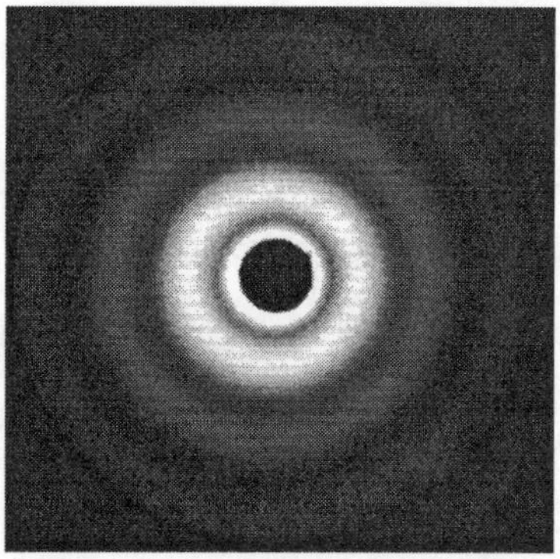

Fig.2 Latex particles diameter $a = 5\mu m$ **Fig.3** Bacteria (bacilus-subtilis)

A first test of the performance of the apparatus was done using spherical particles. In Fig. 2 the spot which is taken on the CCD camera is presented. The black spot at the center of the image is caused by the beam stopper, which is used to avoid saturation of the CCD camera. The scattering pattern consists of light and dark fringes, which are the result of the interference of the waves scattered from different parts of the particles. As one can see, spherical particles produce circular fringes. Secondly, we use the apparatus to examine prolate spheroidal particles. Then the spot is elliptic as it is presented in Fig. 3.

4 A fast inversion algorithm for the determination of the shape and the refractive index. The database construction

From the above discussion the apparatus which we have developed permits the measurement of the scattering intensity of the individual particles carried by a flow. As we have already seen the mathematical theory of the inverse scattering problem is established on the knowledge of the far-field data. But, as it is well known, the intensity is the square of the electric field

$$I(\mathbf{r}) = |\mathbf{E}(\mathbf{r})|^2 = \frac{|\mathbf{E}^{\infty}(\widehat{\mathbf{r}})|^2}{r^2} \tag{15}$$

So, we can propose an inversion algorithm from the knowledge of the intensity data. For (15) it is obvious that the intensity far - field data are over the unit sphere. So, the intensity data are measured as a function of the scattering angles θ and ϕ.

In order to propose a simple algorithm for the applications we will do some a priori assumptions concerning:

A. The shape of the scatterer. We consider scatterers with spheroidal shape. As it is well known the red blood cells under hydrodynamic alignment in the flow cell of the aforementioned design take a shape too closed to the spheroidal one with the axis of symmetry parallel to the walls and normal to the flow direction. Similarly, rod-like cells (bacteria) become aligned with their axis of symmetry parallel to the flow cell walls and normal to the flow direction which is very convenient for signal analysis and interpretation. In view of all the above mentioned, our a priori assumption about the spheroidal shape does not constitute a restriction for applications.

B. The size and the refractive index of the particles. Our purpose is to recognize particles with characteristic dimensions in the range of $1\mu m$ to $8\mu m$, that is we work in the resonance region. We consider the relative refractive index m, defined as the ratio $m = \frac{k_1}{k_2}$, in the range of 1.08 to 1.2.

C. The observation angles. The measurements are performed in the range of scattering angles between 5^0 and 25^0. The limitations to the angles of observation are posed from the experimental device As it is obvious the existence of the beam stopper and the dimensions of the CCD impose the limits to the range of the observation angles. The choice of 5^0 to 25^0 was found to be optimal in the sense that the inversion algorithm which we will present in the sequel provides optimal results.

A database was produced numerically by a computer code which was developed using the Boundary Element Method [16]. In this paper the authors have proposed the B. E. M. to solve the electromagnetic scattering by penetrable bodies. They exploit a computationally attractive surface integral equation containing only weakly and strongly singular integrals in the contrast to most formulations involving not only strongly singular but hypersingular integrals as well. They use successfully advanced methods for the integration in order to compute this kind of integrals.

The database with B.E.M. was produced for spheroids and spheres for recorded normalized intensities in the interval of angles from 5^0 to 25^0. The relative refractive index m was in the range $1.08 \leq m \leq 1.2$, the length of the major axis a of the particle was in the range $1\mu m$ to $8\mu m$ and the ellipticity ratio τ, that is the ratio of the length of the minor axis to the length of the major axis, $\tau = \frac{b}{a}$.

4.1 Relations between light scattering parameters and shape, size and refractive index

In the sequel the relation of m, a and τ with a class of light scattering parameters is examined. More precisely three parameters were adopted, as a sufficient set of light scattering parameters to provide m, a and τ from the measurements of the normalized intensity. Recent results [5] using approximate solutions based on Rayleigh-Debye-Gans approximation or anomalous diffraction establish a relation between the shape, size and refractive index of the particles with certain light scattering parameters. The three types of parameters are the following :

A. The visibility factor $V(\theta_0)$ after a certain angle θ_0, defined as

$$V(\theta_0) = \frac{I_{max} - I_{min}}{I_{max} + I_{min}} \tag{16}$$

where I_{max} is the first local maximum and I_{min} is the next local minimum after θ_0. We have found by numerical testing, that $\theta_0 = 10^0$ gives best inversion results.

B. The difference $\Delta\theta(\theta_0)$ between the angles at which we measure two successive minima and two successive maxima after the same angle θ_0.

C. The ellipticity ratio of the elliptical isointensity curves of the elliptical projected pattern (by trial and error it was found that it is better to consider the isointensity curve corresponding to 50% of the normalized intensity).

We have found numerically from the database, that the visibility factor V depends on the refractive index m and that the fringe pitch $\Delta\theta$ depends monotonically on the size. In Figures 4 and 5 we present this relation. In Fig. 6 we see that the ellipticity ratio depends on the ellipticity of the pattern almost linearly.

The fast inversion algorithm does not use all the information received by the CCD. It is sufficient to get data only on the major and the minor axes of the elliptic spot because these measurements give enough information for the computation of these aforementioned desired parameters. We apply a procedure of smoothing these data by a polynomial fitting technique.

Using these facts the algorithm lists as follows:

(i). Smooth the measured data.

(ii). Extract the light scattering parameters.

(iii). Enter all a priori information about size and relative refractive index of the scatterer.

(iv). Use a pattern recognition program to find which pattern best fits to the measured light scattering parameters.

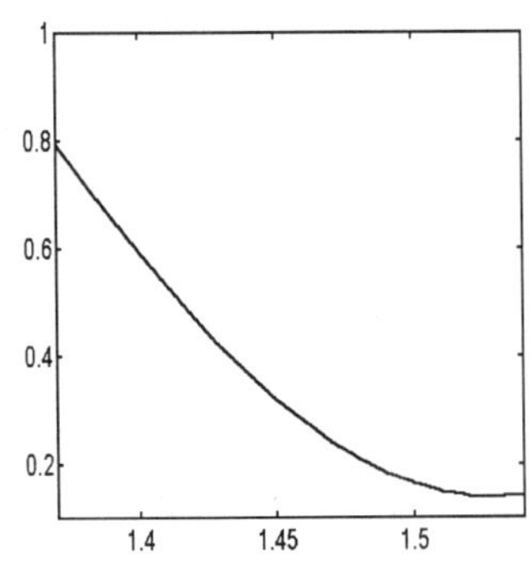

Fig.4 Visibility V versus relative refractive index m

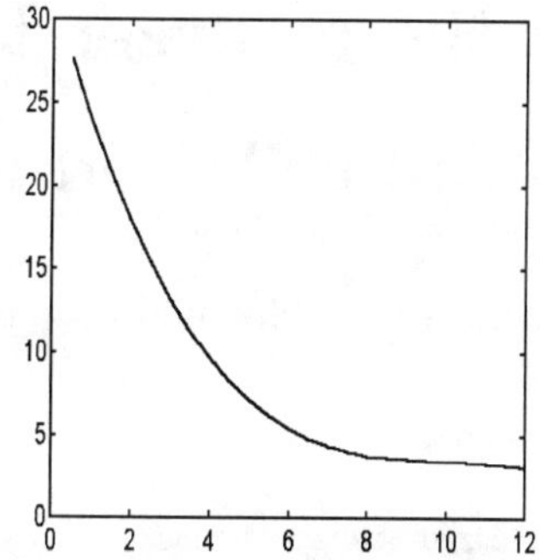

Fig.5 Fringe pitch $\Delta\theta$ versus radius a

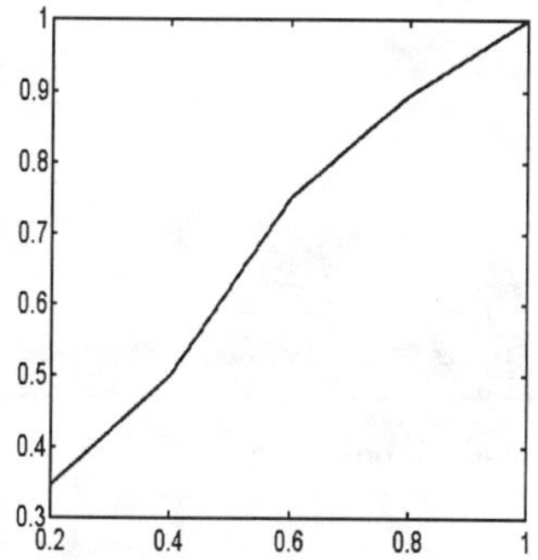

Fig.6 Particle ellipticity versus pattern ellipticity.

A numerical testing procedure was applied to this scheme. We produced numerical results, for the direct problem, from the range of the parameters determining the database with noise similar distributed as in the experimental data. We found that, except few regions , the algorithm produces a unique description of the scatterer with an error which is a function of the added noise.

As a second step in the testing procedure we have used real experimental data taken from known scatterers. The inversion results were absolutely satisfying. So, the simple inversion algorithm with the a priori information that the particles are spheroids is established. As it is obvious the sphere can be considered as a degenerate case of the spheroid.

5 Application of the analyzer for latex and glass particles, red blood cells and bacteria

A first test of the inversion scheme was performed in glass and latex particles in water and ethanol suspensions.

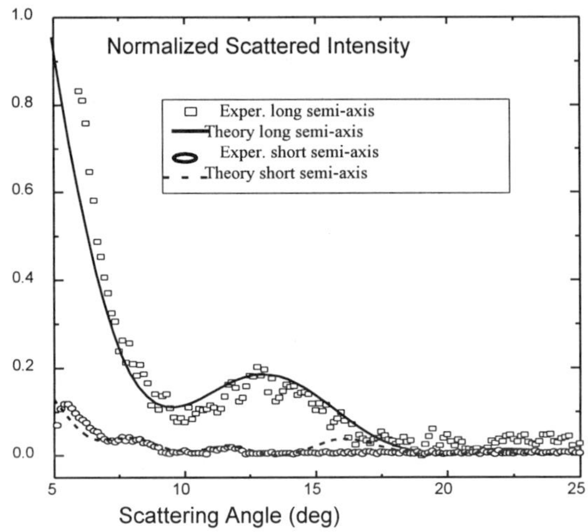

Fig.7 R.B.C.: theoretical and experimental results

The samples were firstly analyzed using a commercial particle analyzer (Mastersizer S, Marvern). As we have tested the results are the same.Fig. 7 shows the light scattering pattern of normal human red blood cells (RBC), the experimentally measured values of the light intensity along the major and the minor semi-axes and the respective theoretical predictions. Normal human RBC, in a hydrodynamic alignment, are non-spherical, virtually axisymmetric particles. Their shape can be described as that of an oblate spheroid that has been squeezed in the middle, forming symmetrical depressions on both sizes. Their equatorial diameter is $7.82\mu m$ and their thickness along the axis of symmetry (short axis) is $0.81\mu m$. The scattered light pattern is clear and strongly asymmetric due to the hydrodynamic alignment, which orients every RBC with its axis parallel to the flow cell walls and normal to the flow direction (see above). The exact refractive index for the RBC and the size is deduced. This capability may become very important for the diagnosis of diseases which change the refractive index of RBC, for example, by changing the concentration of the hemoglobin in RBC or the shape of the RBC.Fig. 8 shows the results from the study of a culture of bacteria bacilus-subtitlis. The culture contains bacteria which have shapes very close to prolate spheroids with the same ellipticity ratio. The bacteria were examined carefully under the microscope to measure the aspect ratio, τ, which was found to be $\tau = 0.6$ with a small deviation. Based on these results and the measured aspect ratio, we determined the composition of the culture. It was found that the culture contains bacteria with length of their major axis $a = 1.0\mu m$. Note, here, that hydrodynamic alignment orients the bacteria with their major axis parallel to the flow cell walls and normal to the direction of the flow. The shape ($\tau = 0.6$) and the relative refractive index ($m = 1.09$) were calculated correctly using the fast inversion algorithm described above.

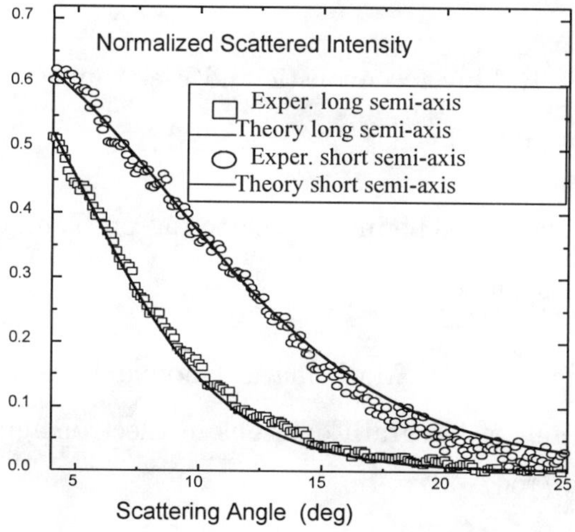

Fig.8 Bacteria bacilus-subtitlis: experimental and theoretical results

References

[1] Al-Rizzo, H.M. and Tranquilla, J.M., "Electromagnetic Wave Scattering by

Highly Elongated and Geometrically Composite Objects of Large Size Parameters: The Generalized Multipole Technique", Applied Optics, 34, 3502 (1995)

[2] Angell T. S. and Kirsh A. "The conductive boundary condition for Maxwell's Equations" Siam J. Appl. Math., V. 52, No 6, (1992), pp. 1597-1610.

[3] Angell T. S., Kleinman R. E. and Roach G. F. "An inverse transmission problem for the Helhmoltz equation", Inverse Problems, 3, (1986), pp.149-180.

[4] Barber, P.W. and Hill, S.C., "Light Scattering by Particles: Computational Methods", Word Scientific, Singapore (1990)

[5] Chernyshev, A.V., Prots, V.I., Doroshkin, A.A., and Maltsev, V.P., "Measurement of Scattering Properties of Individual Particles with a Scanning Flow Cytometer", Applied Optics, 34, 6301 (1995).

[6] Colton D. and Kress R. "Inverse Acoustic and Electromagnetic Scattering Theory" Springer - Verlag (1992).

[7] Gintides D. and Kiriaki K."The inverse scattering problem for dielectric bodies" submitted for publication(1997).

[8] Hsiao G.C., Kleinman R.E., "Mathematical Foundations for Error Estimation in Numerical Solutions of Integral Equations in Electromagnetics", IEEE Trans. Ant. Prop., 45, 316 (1997).

[9] Manolis G.D. and Beskos D.E., "Boundary Element Methods in Elastodynamics", Unwin-Hyman, London (1988).

[10] "Scattering by Three Dimensional Objects", Special Edition of J. Opt. Soc. Am. A, 11(4) (1994).

[11] Shapiro H. M., "Practical Flow Cytometry", Alan R. Liss, Inc., New York,1985.

[12] Stamatakos S. G., Yowa D. and Uzunoglu N. K."Integral equation model of light scattering by an oriented monodisperse system of triaxial dielectric ellipsoids: application in ectacytometry", Applied Optics,Vol.38, No 22 (1997).

[13] Streekstra G. L., Hoekstra A. G. and Heethaar R. M., "Anomalous Diffraction by arbitrary oriented ellipsoids: application in ectacytometry", Appl. Opt. 33, 7288-7296 (1994).

[14] Torres R. H. "A transmission problem in the scattering of electromagnetic waves by a penetrable object", Siam J. Math. Anal., Vol. 27, No. 5, (1996), pp. 1406-1423.

[15] Tranquilla, J.M. and Al-Rizzo, H.M., "Electromagnetic Scattering from Dielectric-Coated Axisymmetric Objects Using the Generalized Point-Matching Technique (GPMT)", IEEE Trans. Ant. Prop., 43, 63 (1995).

[16] Tsinopoulos, S.V., Kattis, S.E. and Polyzos, D., "Three Dimensional Boundary Element Analysis of Electromagnetic Wave Scattering by Penetrable Bodies", Computational Mechanics (To appear).

G. N. Constantinides[(4)], D. Gintides[(3)], S. E. Kattis[(2),(4)], K. Kiriaki[(3)], C. A. Paraskeva[(1),(4)], A. C. Payatakes[(1),(4)], D. Polyzos[(2),(4)], S. Tsinopoulos[(2),(4)], S. N. Yannopoulos[(4)]

[(1)]Chem. Engin. Dept., Univ. of Patras, 265 00 Patras, Greece

[(2)]Mech. Engin. Dept., Univ. of Patras, 265 00 Patras, Greece

[(3)]Dept. of Math. Nat. Tech. Univ. of Athens Zografou Campus, 15780 Athens, Greece

[(4)]Inst. of Chem. Engin. and High Temp. Chem. Proc.- FORTH, P.O. BOX 1414, 265 00 Patras, Greece

A CHARALAMBOPOULOS, D I FOTIADIS AND C V MASSALAS

Biomechanics of the human cranial system

1. Introduction

Several researchers have proposed models to study brain injuries due to an external cause such as those resulting from car accidents. The human head - neck system is very complicated structure and various geometrical and material behavior approximations have been used in the modeling studies.

Various approaches are given to the problem both from the engineering and from the medical point of view [1-4]. Among the most important contributions Landkof and Goldsmith [5] performed an analytical and experimental study involving non-destructive, axisymmetric impact on a fluid - filled shell constrained by a viscoelastic artificial neck, and Hickling and Wenner [6] developed a mathematical model using three - dimensional equations of linear viscoelasticity for the brain and the skull to predict the response of a human head to axisymmetric impact.

In previous communications we have presented an extensive investigation of the dynamic characteristics of the human head - neck system [7-9]. The mathematical analysis was based on the expansion of the solution in terms of the Navier eigenvectors and determination of the eigenfrequencies from the frequency equation defined by the existing boundary conditions. In some cases the solution of the problem required the use of arguments from complex analysis and complicated numerical schemes.

In this work we present the analysis concerning the mathematical formulation of the general problem, in which the system under consideration is assumed to constitute a stratified spherical medium. In particular we study the elastic, isotropic and homogeneous human skull under an external stimulus. Our results can be extended to other systems as they are presented in [10]. The methodology we follow includes determination of the eigenvectors of the homogeneous system and computation of the displacement fields when a Dirac force in space and time is applied on the external surface of the human skull.

2. Problem Formulation

The system under consideration is shown in Fig. 1. There exist n elastic spherical layers simulating the several regions of the human skull. Every spherical layer of the model is assumed to be filled with an isotropic, homogeneous, elastic material. The well - posedness of the problem requires the satisfaction of suitable boundary conditions on the discontinuity surfaces of the system.

Our goal is the determination of the response of the system when a Dirac force, in space and time, is applied on it. More precisely, we suppose that the system is subjected to an external force $\boldsymbol{F}(\boldsymbol{r},t) = \boldsymbol{F}_o\delta(\boldsymbol{r}-\boldsymbol{r}_f)\delta(t)$, per unit mass, where $\boldsymbol{F}_o$ is an arbitrary vector and $\boldsymbol{r}_f$ is an arbitrary point of region V_j. It is noticed here that response of the system to more general external forces can be deduced easily from the response to the Dirac force, which constitutes the Green function of the stimulus procedure. In addition, the position of $\boldsymbol{r}_f$ can be arbitrary but its selection depends on the nature of the external force. For the case of an external stimulus, $\boldsymbol{r}_f$ is placed very close to the external skull surface.

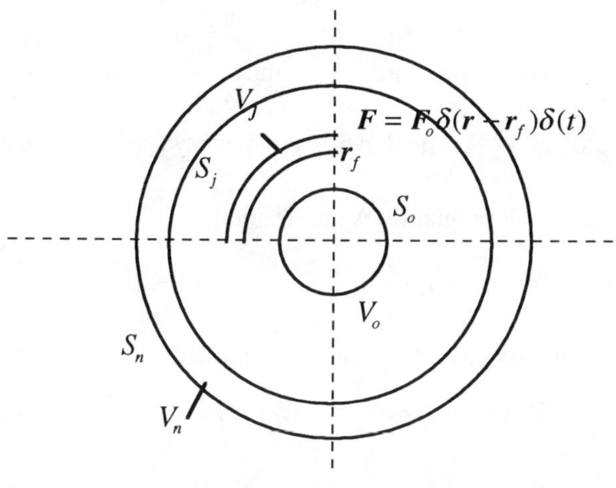

Figure 1: Geometry of the Human Elastic Skull

In order to study the kinematic behavior of the system, we will determine the displacement field of every particular region. Since the region V_i is an elastic one then the motion of this region is characterized by the displacement field $\boldsymbol{u}_i(\boldsymbol{r},t)$ satisfying the time - dependent non - homogeneous equation of elasticity.

$$\mu_i\nabla^2\boldsymbol{u}_i(\boldsymbol{r},t)+(\lambda_i+\mu_i)\nabla(\nabla\cdot\boldsymbol{u}_i(\boldsymbol{r},t))+\rho_i\delta_i^{(j)}\boldsymbol{F}_o\delta(\boldsymbol{r}-\boldsymbol{r}_f)\delta(t)=\rho_i\frac{\partial^2\boldsymbol{u}_i(\boldsymbol{r},t)}{\partial t^2} \quad (1)$$

where μ_i, λ_i are Lamè's constants, ρ_i is the density of region V_i , ∇ is the del operator, $\delta_i^{(j)}=1$ only if $i=j$ which means that $\boldsymbol{r}_f$ belongs to the V_i region, and repetition of a subscript does not mean addition with respect to it

We apply Fourier transform analysis to the problem defining

$$\hat{\boldsymbol{u}}_i(\boldsymbol{r},\omega)=\int_{-\infty}^{+\infty}\boldsymbol{u}_i(\boldsymbol{r},t)e^{i\omega t}dt. \tag{2}$$

We suppress the dependence of previous transformed functions on their argument ω and taking advantage of Fourier transform properties we get equations involving the transformed functions from the equations governing the initial fields.

For every region V_i we obtain

$$\mu_i\nabla^2\hat{\mathbf{u}}_i(\boldsymbol{r})+(\lambda_i+\mu_i)\nabla(\nabla\cdot\hat{\mathbf{u}}_i(\boldsymbol{r}))+\rho_i\omega^2\hat{\mathbf{u}}_i(\boldsymbol{r})=-\rho_i\delta_i^{(j)}\boldsymbol{F}_o\delta(\boldsymbol{r}-\boldsymbol{r}_f). \tag{3}$$

Introducing the velocities $c_{p,i}=\sqrt{\dfrac{\lambda_i+2\mu_i}{\rho_i}}$, $c_{s,i}=\sqrt{\dfrac{\mu_i}{\rho_i}}$ which characterize completely the elastic properties of elastic medium V_i, they give place to the dimensionless quantities $c'_{p,i}=\dfrac{c_{p,i}}{c_{p,n}}$, $c'_{s,i}=\dfrac{c_{s,i}}{c_{p,n}}$. The density ρ_i in every region V_i is replaced by $\rho'_i=\dfrac{\rho_i}{\rho_n}$, and the following dimensionless quantities are defined

$$\boldsymbol{r}'=\frac{\boldsymbol{r}}{\alpha},\quad \Omega=\frac{\omega\alpha}{c_{p,n}},\quad \nabla'=\alpha\nabla,\quad (\alpha=r_n).$$

The differential equations governing materials motion take the dimensionless form

$$c'^2_{s,i}\,\nabla'^2\,\hat{\mathbf{u}}_i(\boldsymbol{r}')+(c'^2_{p,i}-c'^2_{s,i})\nabla'(\nabla'\cdot\hat{\boldsymbol{u}}_i(\boldsymbol{r}'))+\Omega^2\hat{\mathbf{u}}_i(\boldsymbol{r}')=-\delta_i^{(j)}\frac{\boldsymbol{F}_o\alpha}{c^2_{p,n}}\delta(\boldsymbol{r}'-\boldsymbol{r}'_f). \tag{4}$$

$\forall\ i=0,1,2,\ldots,n.$

The boundary conditions for equations (4) are: the exterior surface S_n is stress free, ($\boldsymbol{T}\hat{\boldsymbol{u}}_n(\boldsymbol{r}')\big|_{r'=\alpha}=\boldsymbol{0}$, where $\boldsymbol{T}$ is the surface traction operator [7]). Every surface S_i separating two elastic media must support equal displacement and stress fields from the two sides. Denoting uniformly

$$\left.\begin{aligned}\hat{\boldsymbol{u}}(\boldsymbol{r}')&=\hat{\boldsymbol{u}}_i(\boldsymbol{r}'),\ \text{for } \boldsymbol{r}'\in V_i\\ c'_s(\boldsymbol{r}')&=c'_{s,i},\ \text{for } \boldsymbol{r}'\in V_i\\ c'_p(\boldsymbol{r}')&=c'_{p,i},\ \text{for } \boldsymbol{r}'\in V_i\end{aligned}\right\} \tag{5}$$

the equation governing the displacement fields can be written through only one equation with non-constant coefficients

$$c'_s(\mathbf{r}')^2\nabla'^2\hat{\mathbf{u}}(\mathbf{r}')+\left[c'_p(\mathbf{r}')^2-c'_s(\mathbf{r}')^2\right]\nabla'(\nabla'\cdot\hat{\mathbf{u}}(\mathbf{r}'))+\Omega^2\hat{\mathbf{u}}(\mathbf{r}')=-\frac{\mathbf{F}_o\alpha}{c_{p,n}^2}\delta(\mathbf{r}'-\mathbf{r}'_f),\ \ \mathbf{r}'\in V. \tag{6}$$

Equation (6) with the above mentioned boundary conditions is a well -posed non-homogeneous boundary value problem.

Its solvability reduces to the corresponding homogeneous problem. More precisely, let us consider the boundary value problem consisting of the equation

$$c'_s(\mathbf{r}')^2\nabla'^2\hat{\mathbf{u}}(\mathbf{r}')+\left[c'_p(\mathbf{r}')^2-c'_s(\mathbf{r}')^2\right]\nabla'(\nabla'\cdot\hat{\mathbf{u}}(\mathbf{r}'))+\lambda^2\hat{\mathbf{u}}(\mathbf{r}')=0,\ \ \mathbf{r}'\in V \tag{7}$$

and the same set of the boundary conditions satisfied by the solution of the inhomogeneous problem. But this is exactly the problem studied in [7]. The solution of this problem is based on the representation of the displacement field of the skull in terms of the Navier eigenvectors [11]

$$\hat{\mathbf{u}}(\mathbf{r}')=\sum_{l=1}^{2}\left\{\alpha_n^{m,l}\mathbf{L}_n^{m,l}(\mathbf{r}')+\beta_n^{m,l}\mathbf{M}_n^{m,l}(\mathbf{r}')+\gamma_n^{m,l}\mathbf{N}_n^{m,l}(\mathbf{r}')\right\} \tag{8}$$

Where $\mathbf{L},\mathbf{M},\mathbf{N}$ stand for the Navier eigenvectors, given in [7].

The frequency equation is constructed by imposing the satisfaction of the boundary conditions. This, in matrix form, is written as

$$\mathbf{Dx}=\mathbf{0} \tag{9}$$

or

$$\begin{bmatrix} A_n^1(r'_1) & A_n^2(r'_1) & 0 & 0 & D_n^1(r'_1) & D_n^2(r'_1) \\ B_n^1(r'_1) & B_n^2(r'_1) & 0 & 0 & E_n^1(r'_1) & E_n^2(r'_1) \\ 0 & 0 & C_n^1(r'_1) & C_n^2(r'_1) & 0 & 0 \\ A_n^1(r'_0) & A_n^2(r'_0) & 0 & 0 & D_n^1(r'_0) & D_n^2(r'_0) \\ B_n^1(r'_0) & B_n^2(r'_0) & 0 & 0 & E_n^1(r'_0) & E_n^2(r'_0) \\ 0 & 0 & C_n^1(r'_0) & C_n^2(r'_0) & 0 & 0 \end{bmatrix}\begin{bmatrix}\alpha_n^{m,1}\\ \alpha_n^{m,2}\\ \beta_n^{m,1}\\ \beta_n^{m,2}\\ \gamma_n^{m,1}\\ \gamma_n^{m,2}\end{bmatrix}=\mathbf{0}. \tag{10}$$

The elements of the above matrix are given in [7]. The existence of a non trivial solution for (10) imposes that

$$\det(\mathbf{D})=0. \tag{11}$$

The problem (11) can be solved numerically and this leads to a sequence of eigenvalues λ (Ω_n^k, $n=1,2,3,...;\ k=1,2,3,...$) and a sequence of the corresponding eigenvectors $\hat{\boldsymbol{u}}_n^{m,k}(r')$, $\boldsymbol{r}'\in V$, $n=1,2,...;\ |\mathrm{m}|\leq n;\ k=1,2,3,...$

Generalized Sturm - Liouville theory guarantees that the set of functions $\hat{\boldsymbol{u}}_n^m(\boldsymbol{r}')$ constitutes a complete orthogonal set of functions in the space of square integrable functions in V. Orthogonality can be deduced by suitable application of Green's type theorem in space V using Betti's formulae and exploiting the boundary conditions satisfied by these solutions.

Orthogonality is expressed through the equation

$$\int \hat{\boldsymbol{u}}_n^{m,k}(\boldsymbol{r}')\cdot\hat{\boldsymbol{u}}_{n'}^{m',k'^*}(\boldsymbol{r}')d\boldsymbol{r}'=0,\quad n\neq n' \ \text{ or } m\neq m' \ \text{ or } k\neq k'. \tag{12}$$

Completeness of $\hat{\boldsymbol{u}}_n^{m,k}(\boldsymbol{r}')$ permits the following representation of the solution $\hat{\boldsymbol{u}}(\boldsymbol{r}')$ of the non-homogeneous problem

$$\hat{\boldsymbol{u}}(\boldsymbol{r}')=\sum_{n,m,k}\delta_n^{m,k}(\boldsymbol{r}'_f)\hat{\boldsymbol{u}}_n^{m,k}(\boldsymbol{r}') \tag{13}$$

where summation extends over all possible values of indices n,m,k, $\delta_n^{m,k}(\boldsymbol{r}'_f)$ are the coefficients to be determined in order for $\hat{\boldsymbol{u}}(\boldsymbol{r}')$ to be found which depend on the location of the impact.

Introducing (13) in (8), we find that

$$\sum_{n,m,k}\delta_n^{m,k}(\boldsymbol{r}'_f)(\Omega^2-\Omega_n^{k2})\hat{\boldsymbol{u}}_n^{m,k}(\boldsymbol{r}')=-\frac{\boldsymbol{F}_o\alpha}{c_{p,n}^2}\delta(\boldsymbol{r}'-\boldsymbol{r}'_f). \tag{14}$$

Using the orthogonality of $\hat{\boldsymbol{u}}_n^{m,k}(\boldsymbol{r}')$ we obtain after projecting (14) on $\hat{\boldsymbol{u}}_n^{m,k}(\boldsymbol{r}')$ and integrating over V

$$\delta_n^{m,k}(\boldsymbol{r}'_f)=-\frac{\alpha\boldsymbol{F}_o\cdot\hat{\boldsymbol{u}}_n^{m,k^*}(\boldsymbol{r}'_f)}{c_{p,n}^2(\Omega^2-\Omega_n^{k2})}\frac{1}{\int_V\left|\hat{\boldsymbol{u}}_n^{m,k}(\boldsymbol{r}')\right|^2 d\boldsymbol{r}'}. \tag{15}$$

Denoting $\left\|\hat{\boldsymbol{u}}_n^{m,k}\right\|=\left(\int_V\left|\boldsymbol{u}_n^{m,k}(\boldsymbol{r}')\right|^2 d\boldsymbol{r}'\right)^{1/2}$, the norm of $\hat{\boldsymbol{u}}_n^{m,k}(\boldsymbol{r}')$ in the space of square integrable functions, equation (13) can be written as

$$\hat{\boldsymbol{u}}(\boldsymbol{r}')=-\frac{\alpha\boldsymbol{F}_o}{c_{p,n}^2}\cdot\sum_{n,m,k}\frac{\hat{\boldsymbol{u}}_n^{m,k}(\boldsymbol{r}'_f)\hat{\boldsymbol{u}}_n^{m,k}(\boldsymbol{r}')}{(\Omega^2-\Omega_n^{k2})\left\|\hat{\boldsymbol{u}}_n^{m,k}\right\|^2}. \tag{16}$$

We can now determine $\boldsymbol{u}(\boldsymbol{r}',t)$ by taking the inverse Fourier transform of (16).

Consequently

$$\boldsymbol{u}(\boldsymbol{r}',t)=\frac{1}{2\pi}\int_{-\infty}^{+\infty}\hat{\boldsymbol{u}}(\boldsymbol{r}',\omega)e^{-i\omega t}d\omega=$$
$$=-\frac{1}{2\pi}\boldsymbol{F}_o\cdot\sum_{n,m,k}\hat{\boldsymbol{u}}_n^{m,k^*}(\boldsymbol{r}'_f)\hat{\boldsymbol{u}}_n^{m,k}(\boldsymbol{r}')\frac{1}{\left\|\hat{\boldsymbol{u}}_n^{m,k}\right\|^2}\int_{-\infty}^{+\infty}\frac{e^{-i\omega t}}{\omega^2-\omega_n^{k2}}d\omega \tag{17}$$

where $\omega_n^k=\frac{c_{p,n}}{\alpha}\Omega_n^k$.

Using complex analysis integration arguments, we finally obtain

$$\boldsymbol{u}(\boldsymbol{r}',t)=-\frac{i}{2}\boldsymbol{F}_o\cdot\sum_{n,m,k}\hat{\boldsymbol{u}}_n^{m,k^*}(\boldsymbol{r}'_f)\hat{\boldsymbol{u}}_n^{m,k}(\boldsymbol{r}')\frac{e^{i\omega_n^k t}}{\left\|\hat{\boldsymbol{u}}_n^{m,k}\right\|^2\omega_n^k}. \tag{18}$$

The explicit form of $\hat{\boldsymbol{u}}_n^{m,k}(\boldsymbol{r}')$ in equation (18) is given by the solution of (10). As a matter of fact

$$\hat{\boldsymbol{u}}_n^{m,k}(\boldsymbol{r}')=\hat{\boldsymbol{u}}_{n,i}^{m,k}(\boldsymbol{r}')\text{ for }\boldsymbol{r}'\in V_i \tag{19}$$

with

$$\hat{\boldsymbol{u}}_{n,i}^{m,k}(\boldsymbol{r}')=\sum_{l=1}^{2}\left\{\alpha_{n,i}^{m,l}\boldsymbol{L}_{n,i}^{m,l}(\boldsymbol{r}')+\beta_{n,i}^{m,l}\boldsymbol{M}_{n,i}^{m,l}(\boldsymbol{r}')+\gamma_{n,i}^{m,l}\boldsymbol{N}_{n,i}^{m,l}(\boldsymbol{r}')\right\}\Big|_{\Omega=\Omega_n^k}.$$

The norm $\left\|\hat{\boldsymbol{u}}_n^{m,k}\right\|$ appearing in (18) can be calculated as it is shown in [10]

$$\int_{V_i}\hat{\boldsymbol{u}}_{n,i}^{m,k}(\boldsymbol{r}')\cdot\hat{\boldsymbol{u}}_{n,i}^{m,k^*}(\boldsymbol{r}')d\boldsymbol{r}'=\sum_{l=1}^{2}\left\{\begin{matrix}\left|\alpha_{n,i}^{m,l}\right|^2\left\|\boldsymbol{L}_{n,i}^{m,l}(\boldsymbol{r}')\right\|^2+\left|\beta_{n,i}^{m,l}\right|^2\left\|\boldsymbol{M}_{n,i}^{m,l}(\boldsymbol{r}')\right\|^2+\left|\gamma_{n,i}^{m,l}\right|^2\left\|\boldsymbol{N}_{n,i}^{m,l}(\boldsymbol{r}')\right\|^2\\+2\,\mathrm{Re}\left\{\alpha_{n,i}^{m,l}\gamma_{n,i}^{m,l^*}\int_{V_i}\boldsymbol{L}_{n,i}^{m,l}(\boldsymbol{r}')\cdot\boldsymbol{N}_{n,i}^{m,l^*}(\boldsymbol{r}')d\boldsymbol{r}'\right\}\end{matrix}\right\}\Bigg|_{\Omega=\Omega_n^k} \tag{33}$$

where $\left\|\boldsymbol{L}_{n,i}^{m,l}(\boldsymbol{r}')\right\|$, $\left\|\boldsymbol{M}_{n,i}^{m,l}(\boldsymbol{r}')\right\|$, $\left\|\boldsymbol{N}_{n,i}^{m,l}(\boldsymbol{r}')\right\|$ stand for the L^2 - norms of the corresponding functions in space V_i. Details on the expressions in (20) are given in [10].

3. Numerical Results

The homogeneous problem is solved and the eigenfrequencies Ω_n^k and the corresponding vectors of coefficients $\boldsymbol{x}$ from the frequency equation can be found. Then the eigenvector can be easily determined from the expansion in terms of Navier eigenvectors. A quicksort [12] algorithm is used to order Ω_n^k in ascending order. Equation (18) is used to compute the displacement fields under the state of an external Dirac force $\boldsymbol{F}_0$.

The parameters for the human skull used in our model are

$$E = 1.379 \times 10^9 N/m^2,\ \nu = 0.25,\ \rho = 2.132 \times 10^3 kg/m^3,$$
$$r_1 = 0.082m,\ r_0 = 0.076m$$

and the parameters for the external force are

$$\boldsymbol{r}_f = (r_1, 0, 0),\ \ \boldsymbol{F}_0 = (1, 0, 0).$$

The results obtained for the displacement fields are shown in Fig. 2 and 3 as a function of time for $\theta = 0$ and $\theta = \pi/2$ respectively. The time scale displayed corresponds to $0 \le t \le 1000$ μsec and in the enclosed framed figure this is extended up to 500 μsec. The corresponding displacement fields as function of $0 \le \varphi \le 2\pi$ for discrete time steps are shown in Fig. 4. The number of eigenfrequencies (k') used in (18) strongly influences the convergence of the computed displacement fields as it is shown in Fig. 5. In our computations we have used the first 44 eigenfrequencies and we have computed 45 eigenfrequencies for $0 \le n \le 20$ and $0.0 \le \Omega_n^k \le 10.0$.

4. Conclusions

We have presented a method for the computation of the response of the human skull to an external stimulus. The method is based on the fact that the elastic medium is stratified and can be extended to any morphology of the human head - neck system. However, the complexity of its geometry and material behavior lead to much more difficult mathematical and numerical manipulations which are cited in [10].

Acknowledgment

The present work forms part of the project "New Systems for Early Medical Diagnosis and Biotechnological Applications" which is supported by the Greek General Secretariat for Research and Technology through the EU funded R & D Program EPET II.

References

[1] Goldsmith, W. (1970) Biomechanics of Head Injury. *In Biomechanics: Its Foundations and Objectives. (Edited by Fung, Y.C., Perrone N., and Anliker, M.), Prentice Hall, Englewood Cliffs, NJ.*

[2] Liu, Y.K., (1970) *The Biomechanics of Spinal and Head Impact: Problems of Mathematical Simulation. Proc. Symp. Biodynamic Modeling and its Applications*, 701-736, Dayton, OH.

[3] Advani, S.H. and Owings, R.P. (1975) *Structural Modeling of Human Head.* ASCE, J. Engng. Mech. Div. **3** 257-266.

[4] Khalil, T.B. and Hubbard R.P. (1977) *Parametric Study of Head Response by Finite Element Modeling*. J. Biomechanics **10** 119-132.

[5] Landkof, B., Goldsmith W., and Sackman J.L. (1976) *Impact on a Head - Neck Structure*. J. Biomechanics **9** 141-151.

[6] Hickling, R. and Wenner, M.L. (1973) *Mathematical Model of a Head Subjected to an Axisymmetric Impact.* J. Biomechanics **6** 115-132.

[7] Charalambopoulos, A., Dassios, G., Fotiadis, D.I., Kostopoulos, V., and Massalas, C.V. (1996) *On the Dynamic Characteristics of the Human Skull.* Int. J. Engng. Sci. **34**(12) 1339-1348.

[8] Charalambopoulos, A., Dassios, G., Fotiadis, D.I., and Massalas, C.V. (1998) *On the Dynamic Characteristics of the Human Skull - Brain System. Computer and Mathematical Modelling*, **27** 81-101.

[9] Charalambopoulos, A., Dassios, G., Fotiadis, D.I., and Massalas, C.V. (1997) *Frequency Spectrum of the Human Head - Neck System*, Int. J. Engng. Sci. **35**(8)753-768.

[10] Charalambopoulos, A., Fotiadis, D.I., and Massalas, C.V. (1998) *Dynamic Response of the Human Head to An External Stimulus* (in print).

[11] Press, W.H., Teukolsky, S.A., Vetterling, W.T., and Flannery, B.P. (1992) *Numerical Recipes in FORTRAN: The Art of Scientific Computing, 2nd Edition.* Cambridge University Press, Cambridge.

[12] Hansen W.W. (1935) *A New Type of Expansion in Radiation Problems*, Phys. Rev. **47** 139-143.

Figure 2: Displacements u_r, u_φ, u_θ as a function of time ($r = r_1$, $\varphi = \pi/2$, $\theta = 0$).

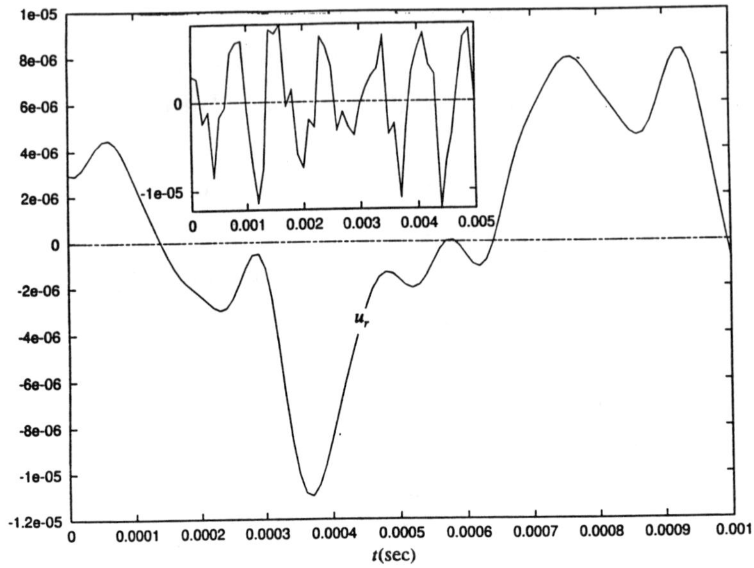

Figure 3: Displacements u_r, u_φ, u_θ as a function of time ($r = r_1$, $\varphi = \pi/2$, $\theta = \pi/2$).

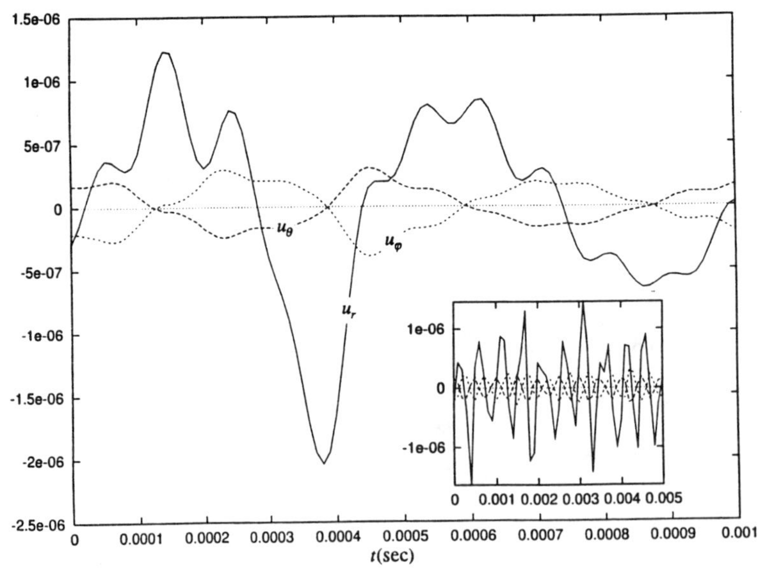

Figure 4: Displacements u_r, u_φ, u_θ for $r = r_1$
(A: $\theta = \pi$, $t = 0.0004$ sec, B: $\theta = \pi/2$, $t = 0.0003$ sec)

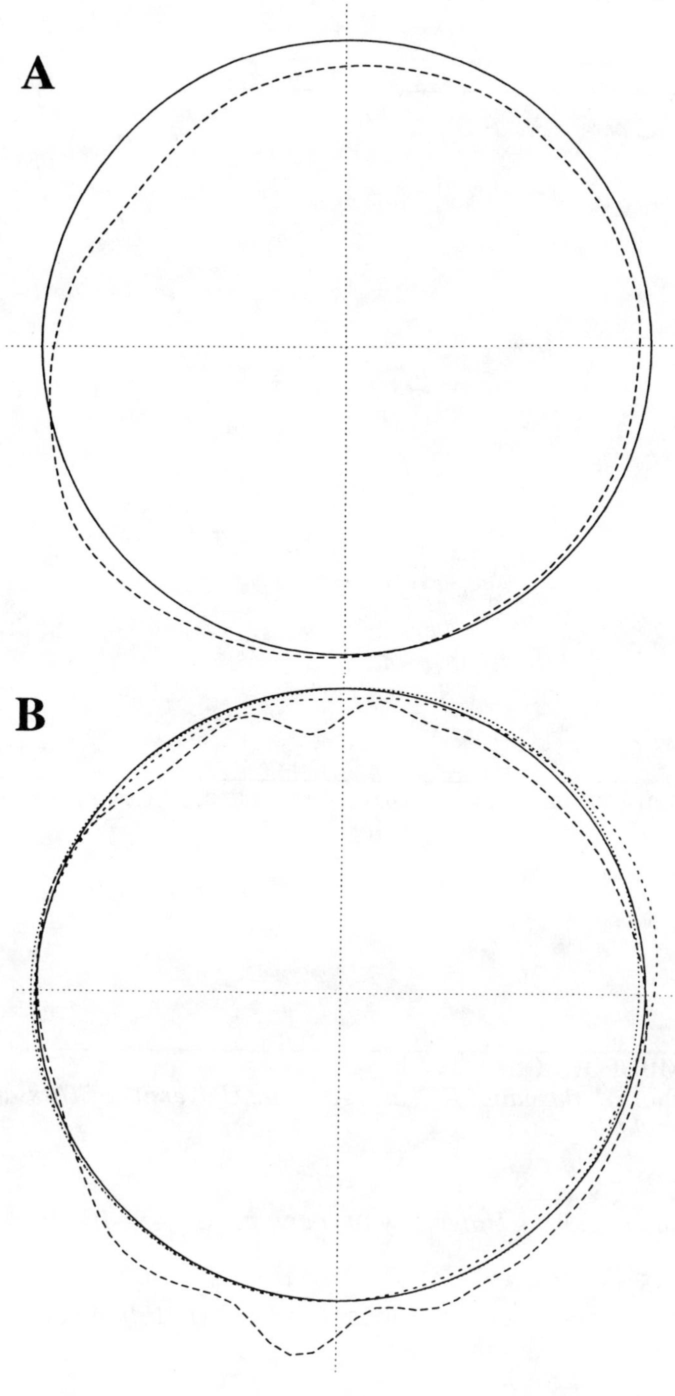

Figure 5: Convergence of u_r as a function of time t for $k' = 14, 24, 44$.
$(r = r_1,\ \theta = \pi/2,\ \varphi = \pi/2)$

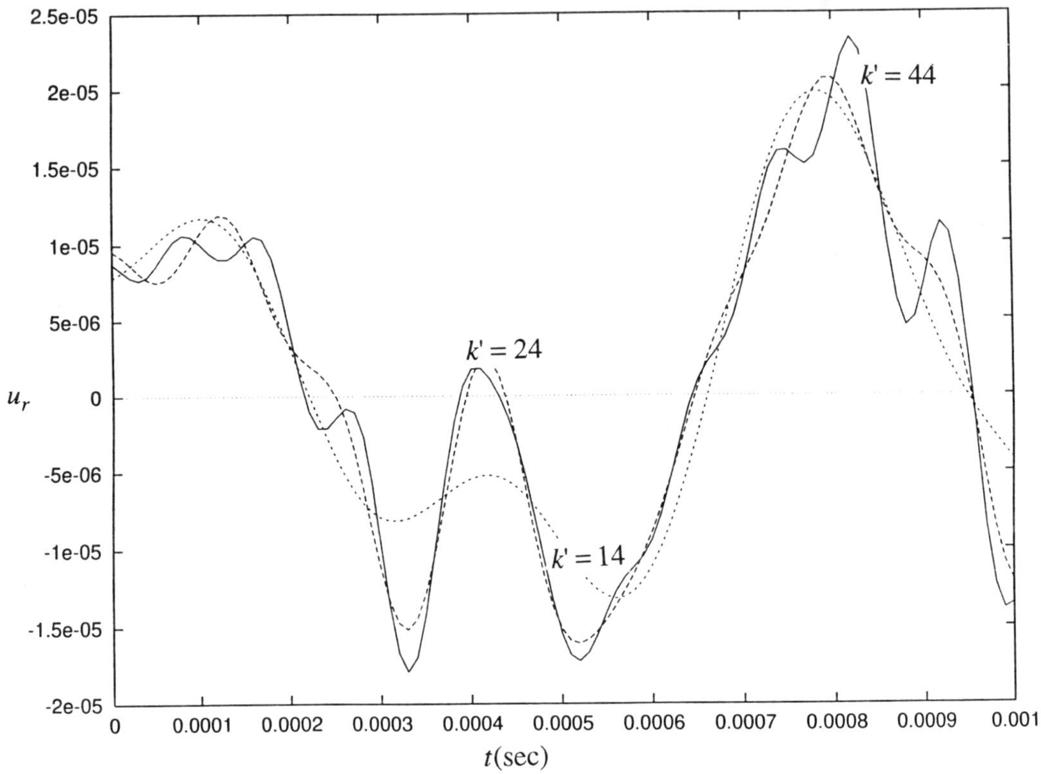

A. CHARALAMBOPOULOS
Polytechnic School, Mathematics Division, Aristotle University of Thessaloniki, GR 540 06 Thessaloniki, Greece

D.I. FOTIADIS
Dept. of Computer Science, University of Ioannina, GR 451 10 Ioannina, Greece

C.V. MASSALAS
Dept. of Mathematics, University of Ioannina, GR 451 10 Ioannina, Greece

DIMITRIOS I FOTIADIS, GEORGIA FOUTSITZI
AND CHRISTOS V MASSALAS

Wave propagation in human long bones

1. Introduction

Many researchers have worked on the modeling of the wave propagation problem in long bones. Wave propagation and vibration of bone can be used to determine the properties of bone *in vivo* which is of great importance in the examination of growth mechanisms. Vayo and Chista [1] modeled the wave propagation problem in long bones assuming that the bone is composed of two-layered elastic transversely isotropic cylinders corresponding to the cortical and spongy bone. Nowinski and Davis [2] have investigated theoretically the wave propagation in wet bone. They assumed that long bone is a solid poroelastic cylinder composed of linear perfect elastic solid and perfect fluid. Paul and Murali [3] analyzed the propagation of flexural waves in bone, treating bone as a hollow poroelastic cylinder. The problem of free vibrations of a double layered elastic isotropic cylinder was studied by Charalambopoulos et al. [4], by simulating the outer layer with the space of cortical bone and the inner one with the medullary space.

The bone tissue has been studied as a piezoelectric material of crystal class 6. Ambardar and Ferris [5] investigated the wave propagation problem in long bone considering it as a two-layered piezoelectric cylindrical shell of crystal class 6mm. Güzelsu and Saha [6] studied electromechanical wave propagation in dry long bone modeled as a piezoelectric transverse isotropic hollow cylinder. The wave propagation in a piezoelectric bone of arbitrary cross section with a cylindrical cavity of arbitrary shape was studied by Paul and Vankatensan [7], [8]. In Ref. [9] the human long bone was simulated with a hollow piezoelectric cylinder of crystal class 6 and the propagation of harmonic waves was studied.

In the present work an attempt has been made to study the wave propagation problem in a piezoelectric cylinder of crystal class 6, filled with an incompressible viscous fluid. The system under consideration simulates the human long bones where the piezoelectric solid cylinder corresponds to the cortical bone and the viscous fluid to the bone marrow. The mathematical modeling of the solid cylinder is based on the three-dimensional theory of piezoelectricity while the linearized Navier-Stokes equations and the continuity equation are employed to describe the dynamic behavior of the fluid medium. The present study is based on the analysis proposed in Ref. [9] and that of Ref. [10]. The frequency equation is obtained for the case where the exterior lateral surface of the cylinder is stress free and coated with electrodes that are shorted and the conditions of continuity of the fields entering the problem in the interior lateral surface.

2. Mathematical Model

Let us consider an infinite hollow cylinder filled with an incompressible viscous fluid (Fig. 1).

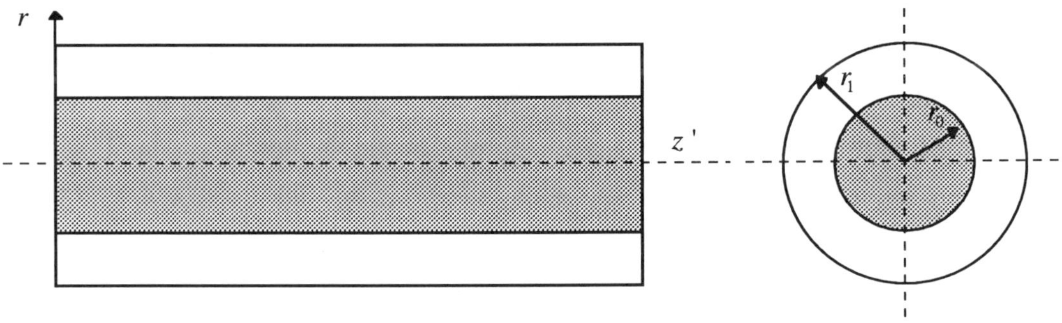

Figure 1: Bone Geometry

Piezoelectric Cylinder

The equations of motion and the Gaussian equation for a piezoelectric cylinder of crystal class 6 are [9]:

$$c_{11}\left(u'_{r,rr} + r^{-1}u'_{r,r} - r^{-2}u'_r\right) + c_{66}r^{-2}u'_{r,\theta\theta} + c_{44}u'_{r,z'z'} + \left(c_{66} + c_{12}\right)r^{-1}u'_{\theta,r\theta}$$

$$-\left(c_{66} + c_{11}\right)r^{-2}u'_{\theta,\theta} + \left(c_{44} + c_{13}\right)u'_{z,rz'} + \left(e_{15} + e_{31}\right)V'_{,rz'} - e_{14}r^{-1}V'_{,\theta z'} = \rho_s \frac{\partial^2 u'_r}{\partial t'^2}, \qquad (1)$$

$$\left(c_{66} + c_{12}\right)r^{-1}u'_{r,r\theta} + \left(c_{66} + c_{11}\right)r^{-2}u'_{r,\theta} + c_{66}\left(u'_{\theta,rr} + r^{-1}u'_{\theta,r} - r^{-2}u'_\theta\right) + c_{11}r^{-2}u'_{\theta,\theta\theta}$$

$$+c_{44}u'_{\theta,z'z'}+(c_{44}+c_{13})r^{-1}u'_{z,\theta z'}+e_{14}V'_{,rz'}+(e_{15}+e_{31})r^{-1}V'_{,\theta z'}=\rho_s\frac{\partial^2 u'_\theta}{\partial t'^2}, \quad (2)$$

$$(c_{44}+c_{13})\left(u'_{r,rz'}+r^{-1}u'_{r,z'}+r^{-1}u'_{\theta,\theta z'}\right)+c_{44}\left(u'_{z,rr}+r^{-1}u'_{z,r}+r^{-2}u'_{z,\theta\theta}\right)+c_{33}u'_{z,z'z'}$$

$$+e_{15}\left(V'_{,rr}+r^{-1}V'_{,r}+r^{-2}V'_{,\theta\theta}\right)+e_{33}V'_{,z'z'}=\rho_s\frac{\partial^2 u'_z}{\partial t'^2}, \quad (3)$$

$$\in_{11}\left(V'_{,rr}+r^{-1}V'_{,r}+r^{-2}V'_{,\theta\theta}\right)+\in_{33}V'_{,z'z'}-(e_{15}+e_{31})\left(u'_{r,rz'}+r^{-1}u'_{r,z'}+r^{-1}u'_{\theta,\theta z'}\right)$$

$$+e_{14}\left(r^{-1}u'_{r,\theta z'}-u'_{\theta,rz'}-r^{-1}u'_{\theta,z'}\right)-e_{15}\left(u'_{z,rr}+r^{-1}u'_{z,r}+r^{-1}u'_{z,\theta\theta}\right)-e_{33}u'_{z,z'z'}=0. \quad (4)$$

where u'_r, u'_θ and u'_z are the components of the displacement vector, V' is the electrostatic potential, c_{ij} are the elastic constants with $c_{66}=\frac{1}{2}(c_{11}-c_{12})$, e_{ij} are the piezoelectric constants, $\in_{ij}$ are the dielectric constants, ρ_s is the density of the solid cylinder and $(\ )_{,\xi}\equiv\partial(\ )/\partial\xi$.

We introduce the following dimensionless variables:

$$x=\frac{r}{R},\qquad z=\frac{z'}{R},\qquad u_x=\frac{1}{R}u'_r,\qquad u_\theta=\frac{1}{R}u'_\theta,\qquad u_z=\frac{1}{R}u'_z$$

$$V=\frac{e_{33}}{Rc_{44}}V'\ \ \tilde{c}_{ij}=\frac{c_{ij}}{c_{44}},\qquad \tilde{e}_{ij}=\frac{e_{ij}}{e_{33}},\qquad \varepsilon_{i3}^2=\frac{e_{33}^2}{c_{44}\in_{ii}},\qquad t=\frac{c_s}{R}\ t',c_s^2=\frac{c_{44}}{\rho_s}$$

where $R=r_1-r_0$.

We seek a solution of the following form:

$$u_x=\left(G_{,x}+\frac{1}{x}\psi_{,\theta}\right)e^{i(\lambda z-\Omega t)},\quad u_\theta=\left(\frac{1}{x}G_{,\theta}-\psi_{,x}\right)e^{i(\lambda z-\Omega t)},$$

$$u_z=iwe^{i(\lambda z-\Omega t)},\ V=i\phi\ e^{i(\lambda z-\Omega t)}, \quad (5)$$

where G, ψ, w and ϕ are functions of x and θ, $\lambda=R\gamma$, γ is the wavenumber, $\Omega^2=\frac{(R\omega)^2\rho_s}{c_{44}}$, ω is the angular frequency and $i=\sqrt{-1}$.

Following the method introduced in Ref. 9, the elastic displacements and the electric potential can be written as:

$$u_x = \sum_{j=1}^{4}\sum_{l=1}^{2}\left\{\left[a_j^{m,l}\delta_j^{p1}\frac{\partial\zeta^{m,l}(k_jx)}{\partial x}+\beta_j^{m,l}\delta_j^{p2}\frac{m}{x}\zeta^{m,l}(k_jx)\right]\cos(m\theta)\right.$$

$$\left.+\left[-a_j^{m,l}\delta_j^{p2}\frac{m}{x}\zeta^{m,l}(k_jx)+\beta_j^{m,l}\delta_j^{p1}\frac{\partial\zeta^{m,l}(k_jx)}{\partial x}\right]\sin(m\theta)\right\}e^{i(\lambda z-\Omega t)},$$

$$u_\theta = \sum_{j=1}^{4}\sum_{l=1}^{2}\left\{\left[-a_j^{m,l}\delta_j^{p2}\frac{\partial\zeta^{m,l}(k_jx)}{\partial x}+\beta_j^{m,l}\delta_j^{p1}\frac{m}{x}\zeta^{m,l}(k_jx)\right]\cos(m\theta)\right.$$

$$\left.-\left[a_j^{m,l}\delta_j^{p1}\frac{m}{x}\zeta^{m,l}(k_jx)+\beta_j^{m,l}\delta_j^{p2}\frac{\partial\zeta^{m,l}(k_jx)}{\partial x}\right]\sin(m\theta)\right\}e^{i(\lambda z-\Omega t)},$$

$$u_z = i\sum_{j=1}^{4}\sum_{l=1}^{2}\left\{\left[a_j^{m,l}\delta_j^{p3}\zeta^{m,l}(k_jx)\right]\cos(m\theta)+\left[\beta_j^{m,l}\delta_j^{p3}\zeta^{m,l}(k_jx)\right]\sin(m\theta)\right\}e^{i(\lambda z-\Omega t)},$$

$$V = i\sum_{j=1}^{4}\sum_{l=1}^{2}\left\{\left[a_j^{m,l}\delta_j^{p4}\zeta^{m,l}(k_jx)\right]\cos(m\theta)+\left[\beta_j^{m,l}\delta_j^{p4}\zeta^{m,l}(k_jx)\right]\sin(m\theta)\right\}e^{i(\lambda z-\Omega t)}, \qquad (6)$$

where $a_j^{m,l}$ and $\beta_j^{m,l}$ are arbitrary constants,

$\delta_j^{pq} = -d_1^{pq}k_j^6 + d_2^{pq}k_j^4 - d_3^{pq}k_j^2 + d_4^{pq}, \qquad p,q,j = 1,2,3,4,$

$$\zeta^{m,l}(k_jx) = \begin{cases}\begin{cases} J^m(k_jx), & l=1, \quad (Bessel\ of\ 1st\ kind)\\ Y^m(k_jx), & l=2, \quad (Bessel\ of\ 2nd\ kind)\end{cases} & if\ k_j^2>0,\\ \begin{cases} I^m(k_jx), & l=1, \quad (mod.\,Bessel\ of\ 1nd\ kind)\\ K^m(k_jx), & l=2, \quad (mod.\,Bessel\ of\ 2nd\ kind)\end{cases} & if\ k_j^2<0,\end{cases}$$

$k_j = |k_j^2|^{1/2}$ and k_j^2, $j = 1,2,3,4$ are the roots of the equation:

$ak_j^8 - bk_j^6 + ck_j^4 - dk_j^2 + e = 0,$

and the coefficients d_s^{pq}, $p,q,s = 1,2,3,4$ and a, b, c, d, e are given in Appendices A and B, respectively.

Using (6) the stresses are expressed as follows:

$$T_{xx} = \sum_{j=1}^{4}\sum_{l=1}^{2}\{\left[a_j^{m,l}P_{p,j}^{m,l}(k_jx)+\beta_j^{m,l}Q_{p,j}^{m,l}(k_jx)\right]\cos(m\theta)$$

$$+\left[-a_j^{m,l}Q_{p,j}^{m,l}(k_jx)+\beta_j^{m,l}P_{p,j}^{m,l}(k_jx)\right]\sin(m\theta)\}e^{i(\lambda z-\Omega t)},$$

$$T_{x\theta} = \tilde{c}_{66}\sum_{j=1}^{4}\sum_{l=1}^{2}\{\left[a_j^{m,l}R_{p,j}^{m,l}(k_jx)+\beta_j^{m,l}S_{p,j}^{m,l}(k_jx)\right]\cos(m\theta)$$

$$+\left[-a_j^{m,l}S_{p,j}^{m,l}(k_jx)+\beta_j^{m,l}R_{p,j}^{m,l}(k_jx)\right]\sin(m\theta)\}e^{i(\lambda z-\Omega t)},$$

$$T_{xz} = i\sum_{j=1}^{4}\sum_{l=1}^{2}\{\left[a_j^{m,l}T_{p,j}^{m,l}(k_jx)+\beta_j^{m,l}U_{p,j}^{m,l}(k_jx)\right]\cos(m\theta)$$

$$+\left[-a_j^{m,l}S_{p,j}^{m,l}(k_jx)+\beta_j^{m,l}R_{p,j}^{m,l}(k_jx)\right]\sin(m\theta)\}e^{i(\lambda z-\Omega t)}, \tag{7}$$

where $P_{p,j}^{m,l}$, $Q_{p,j}^{m,l}$, $R_{p,j}^{m,l}$, $S_{p,j}^{m,l}$, $T_{p,j}^{m,l}$ and $U_{p,j}^{m,l}$ are given Appendix C.

Incompressible Viscous Fluid

We consider the following dimensionless variables:

$$x=\frac{r}{R},\qquad z=\frac{z'}{R},\qquad t=\frac{c_s}{R}t',\qquad c_s^2=\frac{c_{44}}{\rho_s},$$

$$v_x=\frac{1}{c_s}v_r',\qquad v_\theta=\frac{1}{c_s}v_\theta',\qquad v_z=\frac{1}{c_s}v_z',\qquad p=\frac{1}{\rho_f c_s^2}p',\qquad R_e=\frac{\rho_f c_s R}{\eta},$$

where v_x, v_θ and v_z are the components of the velocity of the fluid, p is the pressure, ρ_f is the density of the fluid, η is the viscosity and R_e is the Reynolds number.

The equations governing the motion of the fluid are the linearized Navier-Stokes equations [11]:

$$\dot{v}_x=-p_{,x}+\frac{1}{R_e}\left(v_{x,xx}+\frac{1}{x}v_{x,x}+\frac{1}{x^2}v_{x,\theta\theta}+v_{x,zz}-\frac{2}{x^2}v_{\theta,\theta}-\frac{1}{x^2}v_x\right), \tag{8}$$

$$\dot{v}_\theta=-\frac{1}{x}p_{,\theta}+\frac{1}{R_e}\left(v_{\theta,xx}+\frac{1}{x}v_{\theta,x}+\frac{1}{x^2}v_{\theta,\theta\theta}+v_{\theta,zz}+\frac{2}{x^2}v_{x,\theta}-\frac{1}{x^2}v_\theta\right), \tag{9}$$

$$\dot{v}_z=-p_{,z}+\frac{1}{R_e}\left(v_{z,xx}+\frac{1}{x}v_{z,x}+\frac{1}{x^2}v_{z,\theta\theta}+v_{z,zz}\right), \tag{10}$$

and the continuity equation:

$$v_{x,x} + \frac{1}{x} v_x + \frac{1}{x} v_{\theta,\theta} + v_{z,z} = 0. \tag{11}$$

We seek a solution of the form:

$$v_x = \left(U_{,x} + \frac{1}{x} \Psi_{,\theta} \right) e^{i(\lambda z - \Omega t)}, \quad v_\theta = \left(\frac{1}{x} U_{,\theta} - \Psi_{,x} \right) e^{i(\lambda z - \Omega t)},$$

$$v_z = W \; e^{i(\lambda z - \Omega t)}, \quad p = P \; e^{i(\lambda z - \Omega t)}, \tag{12}$$

where U, Ψ, W and P are functions of x and θ.

The system of equations (8) - (11) is simplified as follows:

$$\left[\nabla^2 + iR_e\Omega - \lambda^2 \right] \Psi = 0, \tag{13}$$

$$\mathbf{C} \begin{bmatrix} U \\ W \\ P \end{bmatrix} = \begin{bmatrix} 0 \\ 0 \\ 0 \end{bmatrix}, \tag{14}$$

where

$$\mathbf{C} = \begin{bmatrix} \nabla^2 + iR_e\Omega - \lambda^2 & 0 & -R_e \\ 0 & \nabla^2 + iR_e\Omega - \lambda^2 & -i\lambda R_e \\ \nabla^2 & i\lambda & 0 \end{bmatrix},$$

and $\nabla^2 \equiv \frac{\partial^2}{\partial x^2} + \frac{1}{x}\frac{\partial}{\partial x} + \frac{1}{x^2}\frac{\partial^2}{\partial \theta^2}$.

Following Ref. 10, we introduce a new function $g = g(x,\theta)$ such that $\{\det \mathbf{C}\} g = 0$ or equivalently,

$$\left(\nabla^2 - \mu_1^2 \right)\left(\nabla^2 - \mu_2^2 \right) g = 0. \tag{15}$$

where $\mu_1^2 = \lambda^2 - i\Omega R_e$ and $\mu_2^2 = \lambda^2$.

Then the solution of the system (14) is given as:

$$U = \mathbf{C}_{i1} g, \qquad W = \mathbf{C}_{i2} g, \qquad P = \mathbf{C}_{i3} g, \qquad i = 1,2,3, \tag{16}$$

where $\mathbf{C}_{ij}$ are the algebraic components of the matrix $\mathbf{C}$ satisfying the relations:

$$\mathbf{C}_{ij} = c_1^{ij} \nabla^4 + c_2^{ij} \nabla^2 + c_3^{ij},$$

and the coefficients c_k^{ij}, $i,j,k=1,2,3$ are given in the Appendix D.

The solution of (14) is equivalent of solving equation(15). This equation admits a solution of the form:

$$g=\sum_{l=1}^{2}\left\{A_l^m\cos(m\theta)+B_l^m\sin(m\theta)\right\}I^m(\mu_l x),\qquad m\in \mathrm{N} \tag{17}$$

where A_l^m, B_l^m are arbitrary constants and $I^m(\mu_l x)$ are the modified Bessel functions.

Solving equation (13) we have:

$$\Psi=\left\{A_3^m\sin(m\theta)+B_3^m\cos(m\theta)\right\}I^m(\mu_3 x),\qquad m\in \mathrm{N} \tag{18}$$

where A_3^m, B_3^m are arbitrary constants and $\mu_3=\mu_1$.

Using the relations (12), (16) -(18), the solution for the system (8)-(11) is expressed as:

$$v_x=\sum_{l=1}^{2}\left\{\left[A_l^m\sigma_l^{i1}\frac{\partial I^m(\mu_l x)}{\partial x}+A_3^m\frac{m}{x}I^m(\mu_3 x)\right]\cos(m\theta)\right.$$

$$\left.+\left[B_l^m\sigma_l^{i1}\frac{\partial I^m(\mu_l x)}{\partial x}-B_3^m\frac{m}{x}I^m(\mu_3 x)\right]\sin(m\theta)\right\}e^{i(\lambda z-\Omega t)},$$

$$v_\theta=\sum_{l=1}^{2}\left\{\left[B_l^m\sigma_l^{i1}\frac{m}{x}I^m(\mu_l x)-B_3^m\frac{\partial I^m(\mu_3 x)}{\partial x}\right]\cos(m\theta)\right.$$

$$\left.-\left[A_l^m\sigma_l^{i1}\frac{m}{x}I^m(\mu_l x)+A_3^m\frac{\partial I^m(\mu_3 x)}{\partial x}\right]\sin(m\theta)\right\}e^{i(\lambda z-\Omega t)},$$

$$v_z=\sum_{l=1}^{2}\left\{\sigma_l^{i2}\left[A_l^m\cos(m\theta)+B_l^m\sin(m\theta)\right]I^m(\mu_l x)\right\}e^{i(\lambda z-\Omega t)},$$

$$p=\sum_{l=1}^{2}\left\{\sigma_l^{i3}\left[A_l^m\cos(m\theta)+B_l^m\sin(m\theta)\right]I^m(\mu_l x)\right\}e^{i(\lambda z-\Omega t)}, \tag{19}$$

where $\sigma_l^{ij}=c_1^{ij}\mu_l^4+c_2^{ij}\mu_l^2+c_3^{ij}$.

For an incompressible viscous fluid the stresses τ_{xx}, $\tau_{x\theta}$, τ_{xz} are given by

$$\tau_{xx} = -p + \frac{2}{R_e} v_{x,x}, \qquad \tau_{x\theta} = \frac{1}{R_e}\left(\frac{1}{x} v_{x,\theta} + v_{\theta,x} - \frac{1}{x} v_\theta\right), \qquad \tau_{xz} = \frac{1}{R_e}\left(v_{z,x} + v_{x,z}\right).$$

Replacing into the above expressions the relations (19) we obtain

$$\tau_{xx} = \sum_{j=1}^{3}\left[A_j^m q_j^m(\mu_j x)\cos(m\theta) + B_j^m r_j^m(\mu_j x)\sin(m\theta)\right]e^{i(\lambda z - \Omega t)},$$

$$\tau_{x\theta} = \sum_{j=1}^{3}\left[B_j^m s_j^m(\mu_j x)\cos(m\theta) + A_j^m t_j^m(\mu_j x)\sin(m\theta)\right]e^{i(\lambda z - \Omega t)},$$

$$\tau_{xz} = \sum_{j=1}^{3}\left[A_j^m w_j^m(\mu_j x)\cos(m\theta) + B_j^m z_j^m(\mu_j x)\sin(m\theta)\right]e^{i(\lambda z - \Omega t)}, \tag{20}$$

where $q_j^m(\mu_j x)$, $r_j^m(\mu_j x)$, $s_j^m(\mu_j x)$, $t_j^m(\mu_j x)$, $w_j^m(\mu_j x)$ and $z_j^m(\mu_j x)$ are given in Appendix E.

3. Frequency Equation

For the system under discussion we consider the following boundary conditions:

$$\left.\begin{array}{l} T_{xx} = -\tau_{xx},\ T_{x\theta} = -\tau_{x\theta},\ T_{xz} = -\tau_{xz} \\ \dot{u}_x = v_x,\ \dot{u}_\theta = v_\theta,\ \dot{u}_z = v_z,\ V = 0 \end{array}\right\} \text{for } x = \frac{r_0}{R}, \tag{21}$$

$$T_{xx} = 0,\ T_{x\theta} = 0,\ T_{xz} = 0,\ V = 0 \text{ for } x = \frac{r_1}{R}. \tag{22}$$

The conditions (22) express the fact that the lateral surface of the piezoelectric cylinder is free of mechanical traction and coated with electrodes which are shorted, while (21) indicates that conditions of continuity of the dynamic, kinetic and electrical quantities are met on the internal lateral surface of the cylinder.

Satisfaction of the boundary conditions leads to a 22×22 linear system of the form:

$$\mathcal{M}\mathbf{x} = \mathbf{0}$$

where

$$\mathbf{x} = \left(\alpha_j^{m,1},\ \alpha_j^{m,2},\ \beta_j^{m,1},\ \beta_j^{m,2},\ A_\alpha^m,\ B_\alpha^m\right), \qquad j = 1,2,3,4, \quad \alpha = 1,2,3.$$

In order for a nontrivial solution to exist, the determinant of the coefficients matrix must vanish:

$$\det(\mathcal{M}_{rs}) = 0, \qquad r,s = 1,2,\ldots,22, \qquad (23)$$

(the elements of the matrix $\mathcal{M}_{rs}$ are given in Appendix F).

Equation (23) relates the wavenumber λ with the frequency Ω. The frequency is generally complex and can be written as the sum of real and imaginary parts, e.g. $\Omega = \Omega_1 + i\Omega_2$. For various values of λ the angular frequency Ω_1 and the attenuation Ω_2 can be evaluated numerically [12]

References

[1] Vayo, H. W. & Ghista, D. N. (1971) *Wave propagation in bone media,* Bull. Math. Biophys. **33**, 463.

[2] Nowinski, J. & Davis, C.(1971) *Propagation of longitudinal waves in circular cylindrical bone elements,* J. Appl. Mech. **38**, 578.

[3] Paul, H. S. & Murali, V. M. (1992) *Wave propagation in a cylindrical poroelastic bone with cavity,* Int. J. Engng. Sci. **30**, 1629.

[4] Charalambopoulos, A., Fotiadis, D. I. & Massalas, C.V. (1998) *Free vibration in a double layered elastic isotropic cylindrical rod*', Int. J. Engng. Sci., (to appear).

[5] Ambardar, A. & Ferris C. D. (1978), *Wave propagation in a piezoelectric two-layered cylindrical shell with hexagonal symmetry: Some implications for long bone,* J. Acoust. Soc Am. **63**(3), 1781.

[6] Güzelsu, N. & Saha, S. (1981) *Electro-mechanical wave propagation in long bones,* J. Biomechanics, **14** 19.

[7] Paul, H. S. & Venkatesan, M. (1991) *Wave propagation in a piezoelectrical human bone of arbitrary cross section with a circular cylindrical cavity,* J. Acoust. Am. **89**, 196.

[8] Paul, H. S. & Venkatesan, M. (1991) *Wave propagation in a piezoelectrical bone with a cylindrical cavity of arbitrary shape,* Int. J. Engng. Sci. **29**, 1601.

[9] Foutsitzi, G., Fotiadis, D.I. & Massalas, C.V. (1997) *Wave propagation in human long bones,* in proc. of 15th IMACS World Congress on Scientific Computation, Modeling and Applied Mathematics, Vol. 3, p.785.

[10] Ding, H. & Chenbuo, L. (1996) *General solutions for coupled equations for*

piezoelectric media, Int. J. Solids Structures **16**, 2283.

[11] Landau, L. & Lifshitz, E. M.(1959) Fluid Mechanics, Addison-Wesley.

[12] Foutsitzi, G., Fotiadis, D.I. & Massalas, C.V. (1998) *Wave Propagation Modeling in a Piezoelectrical Cavity Filled with Fluid,* in preparation.

Appendix A

$$d_1^{11} = \tilde{c}_{66}\left(\tilde{e}_{15}^2 + \varepsilon_{13}^{-2}\right),$$

$$d_2^{11} = -\left\{\tilde{e}_{14}^2 + \tilde{e}_{15}^2 + \varepsilon_{13}^{-2} + \tilde{c}_{66}\left[2\tilde{e}_{15} + \tilde{c}_{33}\varepsilon_{13}^{-2} + \varepsilon_{33}^{-2}\right]\right\}\lambda^2 + \left\{\tilde{e}_{15}^2 + \varepsilon_{13}^{-2}\left[1 + \tilde{c}_{66}\right]\right\}\Omega^2,$$

$$d_3^{11} = \left\{\tilde{c}_{66} + 2\tilde{e}_{15} + \varepsilon_{33}^{-2} + \tilde{c}_{33}\left(\tilde{e}_{14}^2 + \varepsilon_{13}^{-2} + \tilde{c}_{66}\varepsilon_{33}^{-2}\right)\right\}\lambda^4$$
$$-\left\{\left(\tilde{e}_{14}^2 + 2\tilde{e}_{15}\right) + \varepsilon_{13}^{-2}(1 + \tilde{c}_{33}) + \varepsilon_{33}^{-2}(1 + \tilde{c}_{66})\right\}\lambda^2\Omega^2 + \varepsilon_{13}^{-2}\Omega^4,$$

$$d_4^{11} = -\left\{1 + \tilde{c}_{33}\varepsilon_{33}^{-2}\right\}\lambda^6 + \left\{1 + \varepsilon_{33}^{-2}(1 + \tilde{c}_{33})\right\}\lambda^4\Omega^2 - \varepsilon_{33}^{-2}\lambda^2\Omega^4,$$

$$d_1^{12} = 0,$$

$$d_2^{12} = -\tilde{e}_{14}\left\{\tilde{e}_{31} - \tilde{c}_{13}\tilde{e}_{15}\right\}\lambda^2,$$

$$d_3^{12} = -\tilde{e}_{14}\left\{\left[(1 + \tilde{c}_{13}) - \tilde{c}_{33}(\tilde{e}_{15} + \tilde{e}_{31})\right]\lambda^4 + (\tilde{e}_{15} + \tilde{e}_{31})\lambda^2\Omega^2\right\},$$

$$d_4^{12} = 0,$$

$$d_1^{13} = -\tilde{c}_{66}\left\{\tilde{e}_{15}(\tilde{e}_{15} + \tilde{e}_{31}) + \varepsilon_{13}^{-2}(1 + \tilde{c}_{13})\right\}\lambda,$$

$$d_2^{13} = \left\{(1 + \tilde{c}_{13})\left(\tilde{e}_{14}^2 + \varepsilon_{13}^{-2} + \tilde{c}_{66}\varepsilon_{33}^{-2}\right) + (\tilde{c}_{66} + \tilde{e}_{15})(\tilde{e}_{15} + \tilde{e}_{31})\right\}\lambda^3$$
$$-\left\{\tilde{e}_{15}(\tilde{e}_{15} + \tilde{e}_{31}) + \varepsilon_{13}^{-2}(1 + \tilde{c}_{13})\right\}\lambda\Omega^2,$$

$$d_3^{13} = \left\{\varepsilon_{33}^{-2}(1 + \tilde{c}_{13}) + (\tilde{e}_{15} + \tilde{e}_{31})\right\}\left(\Omega^2 - \lambda^2\right)\lambda^3,$$

$$d_4^{13} = 0,$$

$$d_1^{14} = -\tilde{c}_{66}(\tilde{c}_{13}\tilde{e}_{15} - \tilde{e}_{31})\lambda,$$

$$d_2^{14} = \left\{\tilde{c}_{66}\left(1+\tilde{c}_{13}\right)+\left(\tilde{c}_{13}\tilde{e}_{15}-\tilde{e}_{31}\right)-\tilde{c}_{66}\tilde{c}_{33}\left(\tilde{e}_{15}+\tilde{e}_{31}\right)\right\}\lambda^3$$
$$-\left\{\left(\tilde{c}_{13}\tilde{e}_{15}-\tilde{e}_{31}\right)-\tilde{c}_{66}\left(\tilde{e}_{15}+\tilde{e}_{31}\right)\right\}\lambda\Omega^2,$$

$$d_3^{14} = -\left\{\left(1+\tilde{c}_{13}\right)-\tilde{c}_{33}\left(\tilde{e}_{15}+\tilde{e}_{31}\right)\right\}\lambda^5$$
$$-\left\{-\left(1+\tilde{c}_{13}\right)+\left(1+\tilde{c}_{33}\right)\left(\tilde{e}_{15}+\tilde{e}_{31}\right)\right\}\lambda^3\Omega^2+\left(\tilde{e}_{15}+\tilde{e}_{31}\right)\lambda\Omega^4,$$

$$d_4^{14} = 0,$$

$$d_1^{21} = 0,$$

$$d_2^{21} = -\tilde{e}_{14}\left(\tilde{e}_{31}-\tilde{c}_{13}\tilde{e}_{15}\right)\lambda^2,$$

$$d_3^{21} = -\tilde{e}_{14}\left\{\left[\left(1+\tilde{c}_{13}\right)-\tilde{c}_{33}\left(\tilde{e}_{15}+\tilde{e}_{31}\right)\right]\lambda^4+\left(\tilde{e}_{15}+\tilde{e}_{31}\right)\lambda^2\Omega^2\right\},$$

$$d_4^{21} = 0,$$

$$d_1^{22} = \tilde{c}_{11}\left(\tilde{e}_{15}^2+\varepsilon_{13}^{-2}\right),$$

$$d_2^{22} = \left\{-\left(\tilde{e}_{15}+\tilde{e}_{31}\right)^2+2\left(1+\tilde{c}_{13}\right)\left(\tilde{e}_{15}+\tilde{e}_{31}\right)\bar{e}_{15}-\tilde{e}_{15}\left(2\tilde{c}_{11}+\tilde{e}_{15}\right)\right.$$
$$\left.+\varepsilon_{13}^{-2}\tilde{c}_{13}\left(2+\tilde{c}_{13}\right)-\tilde{c}_{11}\left(\tilde{c}_{33}\varepsilon_{13}^{-2}+\varepsilon_{33}^{-2}\right)\right\}\lambda^2+\left\{\tilde{e}_{15}^2+\varepsilon_{13}^{-2}\left(1+\tilde{c}_{11}\right)\right\}\Omega^2,$$

$$d_3^{22} = \left\{-2\left(1+\tilde{c}_{13}\right)\left(\tilde{e}_{15}+\tilde{e}_{31}\right)+\tilde{c}_{11}\left(1+\tilde{c}_{33}\varepsilon_{13}^{-2}\right)+\tilde{c}_{33}\left(\tilde{e}_{15}+\tilde{e}_{31}\right)^2+\right.$$
$$\left.+2\tilde{e}_{15}+\tilde{c}_{33}\varepsilon_{13}^{-2}-\tilde{c}_{13}\varepsilon_{33}^{-2}\left(2+\tilde{c}_{13}\right)\right\}\lambda^4$$
$$+\left\{-\left(\tilde{e}_{15}+\tilde{e}_{31}\right)^2-2\tilde{e}_{15}-\varepsilon_{13}^{-2}\left(1+\tilde{c}_{33}\right)-\varepsilon_{33}^{-2}\left(1+\tilde{c}_{11}\right)\right\}\lambda^2\Omega^2+\varepsilon_{13}^{-2}\Omega^4,$$

$$d_4^{22} = -\left\{1+\tilde{c}_{13}\varepsilon_{33}^{-2}\right\}\lambda^6+\left\{1+\left(1+\tilde{c}_{33}\right)\varepsilon_{33}^{-2}\right\}\lambda^4\Omega^2-\varepsilon_{33}^{-2}\lambda^2\Omega^4,$$

$$d_1^{23} = \tilde{c}_{11}\tilde{e}_{14}\tilde{e}_{15}\lambda,$$

$$d_2^{23} = -\tilde{e}_{14}\left\{\left[\tilde{c}_{11}-\tilde{e}_{31}-\tilde{c}_{13}\left(\tilde{e}_{15}+\tilde{e}_{31}\right)\right]\lambda^3-\tilde{e}_{15}\lambda\Omega^2\right\},$$

$$d_3^{23} = -\tilde{e}_{14}\left\{-\lambda^5+\lambda^3\Omega^2\right\},$$

$$d_4^{23} = 0,$$

$$d_1^{24} = -\tilde{e}_{14}\tilde{c}_{11}\lambda,$$

$$d_2^{24} = \tilde{e}_{14}\left\{\left[-\tilde{c}_{13}(2+\tilde{c}_{13}) + \tilde{c}_{11}\tilde{c}_{33}\right]\lambda^3 - (1+\tilde{c}_{11})\lambda\Omega^2\right\},$$

$$d_3^{24} = \tilde{e}_{14}\left\{-\tilde{c}_{33}\lambda^5 + (1+\tilde{c}_{33})\lambda^3\Omega^2 - \lambda\Omega^4\right\},$$

$$d_4^{24} = 0,$$

$$d_1^{31} = 0,$$

$$d_2^{31} = \tilde{c}_{66}\left\{\tilde{e}_{15}(\tilde{e}_{15} + \tilde{e}_{31}) + \varepsilon_{13}^{-2}(1+\tilde{c}_{13})\right\}\lambda,$$

$$d_3^{31} = \left\{-\tilde{e}_{14}^2(1+\tilde{c}_{13}) - (\tilde{e}_{15} + \tilde{e}_{31})(\tilde{c}_{66} + \tilde{e}_{15}) - (1+\tilde{c}_{13})\left(\varepsilon_{13}^{-2} + \tilde{c}_{66}\varepsilon_{33}^{-2}\right)\right\}\lambda^3$$
$$+\left\{\tilde{e}_{15}(\tilde{e}_{15} + \tilde{e}_{31}) + \varepsilon_{13}^{-2}(1+\tilde{c}_{13})\right\}\lambda\Omega^2,$$

$$d_4^{31} = \left\{(\tilde{e}_{15} + \tilde{e}_{31}) + \varepsilon_{33}^{-2}(1+\tilde{c}_{13})\right\}\left(\lambda^5 - \lambda^3\Omega^2\right),$$

$$d_1^{32} = 0,$$

$$d_2^{32} = -\tilde{e}_{14}\tilde{c}_{11}\tilde{e}_{15}\lambda,$$

$$d_3^{32} = -\tilde{e}_{14}\left[\left\{-\tilde{c}_{11} + \tilde{c}_{13}(\tilde{e}_{15} + \tilde{e}_{31}) + \tilde{e}_{31}\right\}\lambda^3 + \tilde{e}_{15}\lambda\Omega^2\right],$$

$$d_4^{32} = -\tilde{e}_{14}\left\{\lambda^5 - \lambda^3\Omega^2\right\},$$

$$d_1^{33} = \tilde{c}_{11}\tilde{c}_{66}\varepsilon_{13}^{-2},$$

$$d_2^{33} = -\left\{\tilde{c}_{11}\tilde{e}_{14}^2 + \tilde{c}_{66}(\tilde{e}_{15} + \tilde{e}_{31})^2 + \varepsilon_{13}^{-2}(\tilde{c}_{11} + \tilde{c}_{66}) + \tilde{c}_{11}\tilde{c}_{66}\varepsilon_{33}^{-2}\right\}\lambda^2$$
$$-\varepsilon_{13}^{-2}(\tilde{c}_{11} + \tilde{c}_{66})\Omega^2,$$

$$d_3^{33} = \left\{\tilde{e}_{14}^2 + (\tilde{e}_{15} + \tilde{e}_{31})^2 + \varepsilon_{13}^{-2} + \varepsilon_{33}^{-2}(\tilde{c}_{11} + \tilde{c}_{66})\right\}\lambda^4$$
$$-\left\{\tilde{e}_{14}^2 + (\tilde{e}_{15} + \tilde{e}_{31})^2 + 2\varepsilon_{13}^{-2} + \varepsilon_{33}^{-2}(\tilde{c}_{11} + \tilde{c}_{66})\right\}\lambda^2\Omega^2 + \varepsilon_{13}^{-2}\Omega^4,$$

$$d_4^{33} = -\varepsilon_{33}^{-2}\left\{\lambda^6 - 2\lambda^4\Omega^2 + \lambda^2\Omega^4\right\},$$

$$d_1^{34} = \tilde{c}_{11}\tilde{c}_{66}\tilde{e}_{15},$$

$$d_2^{34} = -\left\{\tilde{c}_{11}\tilde{c}_{66} + \tilde{c}_{11}\tilde{e}_{15} - \tilde{c}_{66}\tilde{e}_{31} - \tilde{c}_{13}\tilde{c}_{66}(\tilde{e}_{15} + \tilde{e}_{31})\right\}\lambda^2 + \tilde{e}_{15}(\tilde{c}_{11} + \tilde{c}_{66})\Omega^2,$$

$$d_3^{34} = -\left\{-\tilde{c}_{11} - \tilde{c}_{66} + \tilde{c}_{13}\tilde{e}_{15} + \tilde{e}_{31}(1+\tilde{c}_{13})\right\}\lambda^4$$
$$-\left\{\tilde{c}_{11} + \tilde{c}_{66} + \tilde{e}_{15}(1-\tilde{c}_{13}) - \tilde{e}_{31}(1+\tilde{c}_{13})\right\}\lambda^2\Omega^2 + \tilde{e}_{15}\Omega^4,$$

$$d_4^{34} = -\left\{\lambda^6 - 2\lambda^4\Omega^2 + \lambda^2\Omega^4\right\},$$

$$d_1^{41} = 0,$$

$$d_2^{41} = -\tilde{c}_{66}\left\{\tilde{c}_{13}\tilde{e}_{15} - \tilde{e}_{31}\right\}\lambda,$$

$$d_3^{41} = -\left\{-\tilde{c}_{66}(1+\tilde{c}_{13}) - \tilde{e}_{15}(\tilde{c}_{13} - \tilde{c}_{33}\tilde{c}_{66}) + \tilde{e}_{31}(1+\tilde{c}_{33}\tilde{c}_{66})\right\}\lambda^3$$
$$-\left\{\tilde{e}_{15}(\tilde{c}_{13} - \tilde{c}_{66}) - \tilde{e}_{31}(1+\tilde{c}_{66})\right\}\lambda\Omega^2,$$

$$d_4^{41} = -\left\{(1+\tilde{c}_{13}) - \tilde{c}_{33}(\tilde{e}_{15} + \tilde{e}_{31})\right\}\lambda^5 + \left\{(1+\tilde{c}_{13}) - (\tilde{e}_{15} + \tilde{e}_{31})(1+\tilde{c}_{33})\right\}\lambda^3\Omega^2 + (\tilde{e}_{15} + \tilde{e}_{31})\lambda\Omega^4,$$

$$d_1^{42} = 0,$$

$$d_2^{42} = -\tilde{e}_{14}\tilde{c}_{11}\lambda,$$

$$d_3^{42} = -\tilde{e}_{14}\left\{\left[\tilde{c}_{13}(2+\tilde{c}_{13}) - \tilde{c}_{11}\tilde{c}_{33}\right]\lambda^3 + (1+\tilde{c}_{11})\lambda\Omega^2\right\},$$

$$d_4^{42} = \tilde{e}_{14}\left\{-\tilde{c}_{33}\lambda^5 + (1+\tilde{c}_{33})\lambda^3\Omega^2 - \lambda\Omega^4\right\},$$

$$d_1^{43} = -\tilde{c}_{11}\tilde{c}_{66}\tilde{e}_{15},$$

$$d_2^{43} = \left\{\tilde{c}_{66}\left[\tilde{e}_{15} + \tilde{c}_{11} - (1+\tilde{c}_{13})(\tilde{e}_{15} + \tilde{e}_{31})\right] + \tilde{c}_{11}\tilde{e}_{15}\right\}\lambda^2 - \tilde{e}_{15}(\tilde{c}_{11} + \tilde{c}_{66})\Omega^2,$$

$$d_3^{43} = -\left\{\tilde{c}_{66} + \tilde{e}_{15} + \tilde{c}_{11} - (1+\tilde{c}_{13})(\tilde{e}_{15} + \tilde{e}_{31})\right\}\lambda^4$$
$$+\left\{\tilde{c}_{66} + 2\tilde{e}_{15} + \tilde{c}_{11} - (1+\tilde{c}_{13})(\tilde{e}_{15} + \tilde{e}_{31})\right\}\lambda^2\Omega^2 - \tilde{e}_{15}\Omega^4,$$

$$d_4^{43} = \lambda^6 - 2\lambda^4\Omega^2 + \Omega^4\lambda^2,$$

$$d_1^{44} = \tilde{c}_{11}\tilde{c}_{66},$$

$$d_2^{44} = \left\{-\tilde{c}_{11}(1+\tilde{c}_{33}\tilde{c}_{66}) + \tilde{c}_{13}\tilde{c}_{66}(2+\tilde{c}_{13})\right\}\lambda^2 + \left\{\tilde{c}_{11} + \tilde{c}_{66} + \tilde{c}_{11}\tilde{c}_{66}\right\}\Omega^2,$$

$$d_3^{44} = \left\{-\tilde{c}_{13}(2+\tilde{c}_{13}) + \tilde{c}_{33}(\tilde{c}_{66} + \tilde{c}_{11})\right\}\lambda^4 +$$
$$\left\{-1 + \tilde{c}_{13}(2+\tilde{c}_{13}) - (1+\tilde{c}_{33})(\tilde{c}_{66} + \tilde{c}_{11})\right\}\Omega^2\lambda^2 + (1+\tilde{c}_{66} + \tilde{c}_{11})\Omega^4,$$

$$d_4^{44} = -\tilde{c}_{33}\lambda^3 + (1 + 2\tilde{c}_{33})\Omega^2\lambda^4 - (2 + \tilde{c}_{33})\Omega^4\lambda^2 + \Omega^6.$$

Appendix B

$$a = \tilde{c}_{11}\tilde{c}_{66}\left(\tilde{e}_{15}^2 + \varepsilon_{13}^{-2}\right),$$

$$\begin{aligned} b = -\Big\{&\tilde{c}_{66}\left[\tilde{e}_{31}^2 - \tilde{e}_{15}^2 - 2\tilde{c}_{13}\tilde{e}_{15}\left(\tilde{e}_{15} + \tilde{e}_{31}\right) + \varepsilon_{13}^{-2}(1 + \tilde{c}_{13})^2\right] \\ &+\tilde{c}_{11}\tilde{e}_{14}^2 + \tilde{c}_{11}\tilde{c}_{66}\left[2\tilde{e}_{15} + \tilde{c}_{33}\varepsilon_{13}^{-2} + \varepsilon_{33}^{-2}\right] - (\tilde{c}_{11} + \tilde{c}_{66})\left(\tilde{e}_{15}^2 + \varepsilon_{13}^{-2}\right)\Big\}\lambda^2 \\ &+\Big\{(\tilde{c}_{11} + \tilde{c}_{66})\left(\tilde{e}_{15}^2 + \varepsilon_{13}^{-2}\right) + \tilde{c}_{11}\tilde{c}_{66}\varepsilon_{13}^{-2}\Big\}\Omega^2, \end{aligned}$$

$$\begin{aligned} c = \Big\{&(\tilde{e}_{15} + \tilde{e}_{31})^2(1 + \tilde{c}_{33}\tilde{c}_{66}) - (1 + \tilde{c}_{13})^2\left(\tilde{e}_{14}^2 + \varepsilon_{13}^{-2} + \tilde{c}_{66}\varepsilon_{33}^{-2}\right) \\ &-2(\tilde{e}_{15} + \tilde{e}_{31})(1 + \tilde{c}_{13})(\tilde{c}_{66} + \tilde{e}_{15}) + \tilde{e}_{14}^2(1 + \tilde{c}_{11}\tilde{c}_{33}) \\ &+(\tilde{c}_{11} + \tilde{c}_{66})\left(\tilde{c}_{33}\varepsilon_{13}^{-2} + \varepsilon_{33}^{-2} + 2\tilde{e}_{15}\right) + \tilde{e}_{15}^2 + \varepsilon_{13}^{-2} + \tilde{c}_{11}\tilde{c}_{66}\left(1 + \tilde{c}_{33}\varepsilon_{33}^{-2}\right)\Big\}\lambda^4 \\ &-\Big\{(\tilde{e}_{15} + \tilde{e}_{31})^2(1 + \tilde{c}_{66}) + \tilde{e}_{14}^2(1 + \tilde{c}_{11}) + (\tilde{c}_{11} + \tilde{c}_{66})\left(\tilde{c}_{33}\varepsilon_{13}^{-2} + \varepsilon_{13}^{-2} + \varepsilon_{33}^{-2} + 2\tilde{e}_{15}\right) \\ &-(1 + \tilde{c}_{13})\left[2(\tilde{e}_{15} + \tilde{e}_{31})\tilde{e}_{15} + (1 + \tilde{c}_{13})\varepsilon_{13}^{-2}\right] + 2\tilde{e}_{15}^2 + 2\varepsilon_{13}^{-2} + \tilde{c}_{11}\tilde{c}_{66}\varepsilon_{33}^{-2}\Big\}\lambda^2\Omega^2 \\ &+\Big\{\tilde{e}_{15}^2 + \varepsilon_{13}^{-2} + (1 + \tilde{c}_{11} + \tilde{c}_{66})\Big\}\Omega^4, \end{aligned}$$

$$\begin{aligned} d = \left(\Omega^2 - \lambda^2\right)\Big\{&\Big[-(\tilde{e}_{15} + \tilde{e}_{31})[2(1 + \tilde{c}_{13}) - (\tilde{e}_{15} + \tilde{e}_{31})\tilde{c}_{33}] + (\tilde{c}_{11} + \tilde{c}_{66})\left(1 + \tilde{c}_{33}\varepsilon_{33}^{-2}\right) \\ &+\tilde{c}_{33}\left(\tilde{e}_{14}^2 + \varepsilon_{13}^{-2}\right) + 2\tilde{e}_{15} - \tilde{c}_{13}\varepsilon_{33}^{-2}(2 + \tilde{c}_{13})\Big]\lambda^4 \\ &-\Big[(\tilde{e}_{15} + \tilde{e}_{31})^2 + \tilde{e}_{14}^2 + 2\tilde{e}_{15} + \varepsilon_{13}^{-2}(1 + \tilde{c}_{33}) + \varepsilon_{33}^{-2}(1 + \tilde{c}_{11}) + \tilde{c}_{66}\varepsilon_{33}^{-2}\Big]\lambda^2\Omega^2 + \varepsilon_{13}^{-2}\Omega^4\Big\}, \end{aligned}$$

$$e = \left\{1 + \tilde{c}_{33}\varepsilon_{33}^{-2}\right\}\lambda^8 - \left\{2 + \varepsilon_{33}^{-2}(1 + 2\tilde{c}_{33})\right\}\lambda^6\Omega^2 + \left\{1 + \varepsilon_{33}^{-2}(2 + \tilde{c}_{33})\right\}\lambda^4\Omega^4 - \varepsilon_{33}^{-2}\lambda^2\Omega^6.$$

Appendix C

$$\begin{aligned} P_{p,j}^{m,l}(k_j x) = &\left\{\delta_j^{p1}\left[2\tilde{c}_{66}\frac{m(m-1)}{x^2} + (2\tilde{c}_{66} - \tilde{c}_{12})k_j^2\right] - \lambda\left[\tilde{c}_{13}\delta_j^{p3} + \tilde{e}_{31}\delta_j^{p4}\right]\right\}\zeta^{m,l}(k_j x) \\ &+2\tilde{c}_{66}\delta_j^{p1}\frac{k_j}{x}\zeta^{m+,l}(k_j x), \end{aligned}$$

$$Q_{p,j}^{m,l}(k_j x) = 2\tilde{c}_{66}\delta_j^{p2} m\left[\frac{m-1}{x^2}\zeta^{m,l}(k_j x) - \frac{k_j}{x}\zeta^{m+1,l}(k_j x)\right],$$

$$R_{p,j}^{m,l}\left(k_j x\right) = -\delta_j^{p2}\left[2\frac{k_j}{x}\zeta^{m+1,l}\left(k_j x\right) + \left(2\frac{m(m-1)}{x^2} - k_j^2\right)\zeta^{m,l}\left(k_j x\right)\right],$$

$$S_{p,j}^{m,l}(k_j x) = 2\delta_j^{p1} m\left[\frac{m-1}{x^2}\zeta^{m,l}(k_j x) - \frac{k_j}{x}\zeta^{m+1,l}(k_j x)\right],$$

$$T_{p,j}^{m,l}(k_j x) = \left(\lambda\delta_j^{p1} + \delta_j^{p3} + \tilde{e}_{15}\delta_j^{p4}\right)\left[\frac{m}{x}\zeta^{m,l}(k_j x) - k_j\zeta^{m+1,l}(k_j x)\right],$$

$$U_{p,j}^{m,l}(k_j x) = \frac{m}{x}\left(\lambda\delta_j^{p2} - \tilde{e}_{14}\delta_j^{p4}\right)\zeta^{m,l}(k_j x).$$

Appendix D

$$c_1^{11} = c_2^{11} = 0, \qquad c_3^{11} = -\lambda^2 R_e,$$

$$c_1^{12} = 0, \quad c_2^{12} = -i\lambda R_e, \quad c_3^{12} = 0,$$

$$c_1^{13} = -1, \quad c_2^{12} = -\left(i\Omega R_e - \lambda^2\right), \quad c_3^{12} = 0,$$

$$c_1^{21} = c_2^{21} = 0, \qquad c_3^{21} = -i\lambda R_e,$$

$$c_1^{22} = 0, \quad c_2^{22} = R_e, \quad c_3^{22} = 0,$$

$$c_1^{23} = 0, \quad c_2^{23} = -i\lambda, \quad c_3^{23} = -i\lambda\left(i\Omega R_e - \lambda^2\right),$$

$$c_1^{31} = 0, \quad c_2^{31} = R_e, \quad c_3^{31} = R_e\left(i\Omega R_e - \lambda^2\right),$$

$$c_1^{32} = 0, \quad c_2^{32} = i\lambda R_e, \quad c_3^{32} = i\lambda R_e\left(i\Omega R_e - \lambda^2\right),$$

$$c_1^{33} = 1, \quad c_2^{33} = 2\left(i\Omega R_e - \lambda^2\right), \quad c_3^{33} = \left(i\Omega R_e - \lambda^2\right)^2.$$

Appendix E

$$q_l^m(\mu_l x) = \frac{2}{R_e}\sigma_l^{i1}\frac{\partial^2 I^m(\mu_l x)}{\partial x^2} - \sigma_l^{i3} I^m(\mu_l x), \quad l = 1,2$$

$$q_3^m(\mu_3 x) = \frac{2}{R_e}\frac{m}{x}\left(\frac{\partial I^m(\mu_3 x)}{\partial x} - \frac{1}{x}I^m(\mu_3 x)\right),$$

$$r_l^m(\mu_l x) = q_l^m(\mu_l x), \qquad l = 1,2$$

$$r_3^m(\mu_3 x) = -q_3^m(\mu_3 x),$$

$$s_l^m(\mu_l x) = \frac{2}{R_e}\frac{m}{x}\sigma_l^{i1}\left(\frac{\partial I^m(\mu_l x)}{\partial x} - \frac{1}{x}I^m(\mu_l x)\right), \qquad l = 1,2$$

$$s_3^m(\mu_3 x) = -\frac{1}{R_e}\left(\frac{\partial^2 I^m(\mu_3 x)}{\partial x^2} - \frac{1}{x}\frac{\partial I^m(\mu_3 x)}{\partial x} + \frac{m^2}{x^2}I^m(\mu_3 x)\right),$$

$$t_l^m(\mu_l x) = -s_l^m(\mu_l x), \qquad l = 1,2$$

$$t_3^m(\mu_3 x) = s_3^m(\mu_3 x),$$

$$w_l^m(\mu_l x) = \frac{1}{R_e}\left(i\lambda\sigma_l^{i1} + \sigma_l^{i2}\right)\frac{\partial I^m(\mu_l x)}{\partial x}, \qquad l = 1,2$$

$$w_3^m(\mu_3 x) = \frac{1}{R_e}\frac{m}{x}i\lambda\, I^m(\mu_3 x),$$

$$z_l^m(\mu_l x) = w_l^m(\mu_l x), \qquad l = 1,2$$

$$z_3^m(\mu_3 x) = -w_3^m(\mu_3 x),$$

Appendix F

$$\mathcal{M}_{1,j} = P_{p,j}^{m,1}\left(k_j\bar{r}_0\right), \quad \mathcal{M}_{1,j+4} = P_{p,j}^{m,2}\left(k_j\bar{r}_0\right), \quad \mathcal{M}_{1,j+8} = Q_{p,j}^{m,1}\left(k_j\bar{r}_0\right), \quad \mathcal{M}_{1,j+12} = Q_{p,j}^{m,2}\left(k_j\bar{r}_0\right),$$

$$\mathcal{M}_{1,17} = q_1^m(\mu_1\bar{r}_0), \quad \mathcal{M}_{1,18} = q_2^m(\mu_2\bar{r}_0), \quad \mathcal{M}_{1,19} = q_3^m(\mu_3\bar{r}_0), \quad \mathcal{M}_{1,20} = \mathcal{M}_{1,21} = \mathcal{M}_{1,22} = 0,$$

$$\mathcal{M}_{2,j} = -\mathcal{M}_{1,j+8}, \quad \mathcal{M}_{2,j+4} = -\mathcal{M}_{1,j+12}, \quad \mathcal{M}_{2,j+8} = \mathcal{M}_{1,j}, \quad \mathcal{M}_{2,j+12} = \mathcal{M}_{1,j+4},$$

$$\mathcal{M}_{2,17} = \mathcal{M}_{2,18} = \mathcal{M}_{2,19} = 0, \quad \mathcal{M}_{2,20} = r_1^m(\mu_1\bar{r}_0), \quad \mathcal{M}_{2,21} = r_2^m(\mu_2\bar{r}_0), \quad \mathcal{M}_{2,22} = r_3^m(\mu_3\bar{r}_0),$$

$$\mathcal{M}_{3,j} = \tilde{c}_{66} R_{p,j}^{m,1}\left(k_j \bar{r}_0\right), \qquad \mathcal{M}_{3,j+4} = \tilde{c}_{66} R_{p,j}^{m,2}\left(k_j \bar{r}_0\right), \qquad \mathcal{M}_{3,j+8} = \tilde{c}_{66} S_{p,j}^{m,1}\left(k_j \bar{r}_0\right),$$

$$\mathcal{M}_{3,j+12} = \tilde{c}_{66} S_{p,j}^{m,2}\left(k_j \bar{r}_0\right), \qquad \mathcal{M}_{3,17} = \mathcal{M}_{3,18} = \mathcal{M}_{3,19} = 0, \quad \mathcal{M}_{3,20} = s_1^m(\mu_1 \bar{r}_0),$$

$$\mathcal{M}_{3,21} = s_2^m(\mu_2 \bar{r}_0), \quad \mathcal{M}_{3,22} = s_3^m(\mu_3 \bar{r}_0),$$

$$\mathcal{M}_{4,j} = -\mathcal{M}_{3,j+8}, \qquad \mathcal{M}_{4,j+4} = -\mathcal{M}_{3,j+12}, \qquad \mathcal{M}_{4,j+8} = \mathcal{M}_{3,j}, \qquad \mathcal{M}_{4,j+12} = \mathcal{M}_{3,j+4},$$

$$\mathcal{M}_{4,17} = t_1^m(\mu_1 \bar{r}_0), \qquad \mathcal{M}_{4,18} = t_2^m(\mu_2 \bar{r}_0), \qquad \mathcal{M}_{4,19} = t_3^m(\mu_3 \bar{r}_0), \qquad \mathcal{M}_{4,20} = \mathcal{M}_{4,21} = \mathcal{M}_{4,22} = 0$$

$$\mathcal{M}_{5,j} = iT_{p,j}^{m,1}\left(k_j \bar{r}_0\right), \quad \mathcal{M}_{5,j+4} = iT_{p,j}^{m,2}\left(k_j \bar{r}_0\right), \quad \mathcal{M}_{5,j+8} = iU_{p,j}^{m,1}\left(k_j \bar{r}_0\right), \quad \mathcal{M}_{5,j+12} = iU_{p,j}^{m,2}\left(k_j \bar{r}_0\right),$$

$$\mathcal{M}_{5,17} = w_1^m(\mu_1 \bar{r}_0), \qquad \mathcal{M}_{5,18} = w_2^m(\mu_2 \bar{r}_0), \qquad \mathcal{M}_{5,19} = w_3^m(\mu_3 \bar{r}_0), \quad \mathcal{M}_{5,20} = \mathcal{M}_{5,21} = \mathcal{M}_{5,22} = 0,$$

$$\mathcal{M}_{6,j} = -\mathcal{M}_{5,j+8}, \qquad \mathcal{M}_{6,j+4} = -\mathcal{M}_{5,j+12}, \qquad \mathcal{M}_{6,j+8} = \mathcal{M}_{5,j}, \qquad \mathcal{M}_{6,j+12} = \mathcal{M}_{5,j+4},$$

$$\mathcal{M}_{6,17} = \mathcal{M}_{6,18} = \mathcal{M}_{6,19} = 0, \quad \mathcal{M}_{6,20} = z_1^m(\mu_1 \bar{r}_0), \quad \mathcal{M}_{6,21} = z_2^m(\mu_2 \bar{r}_0), \quad \mathcal{M}_{6,22} = z_3^m(\mu_3 \bar{r}_0),$$

$$\mathcal{M}_{7,j} = i\delta_j^{p1} F_{p,j}^{m,1}\left(k_j \bar{r}_0\right), \quad \mathcal{M}_{7,j+4} = i\delta_j^{p1} F_{p,j}^{m,2}\left(k_j \bar{r}_0\right), \quad \mathcal{M}_{7,j+8} = i\delta_j^{p2} G_{p,j}^{m,1}\left(k_j \bar{r}_0\right),$$

$$\mathcal{M}_{7,j+12} = i\delta_j^{p2} G_{p,j}^{m,2}\left(k_j \bar{r}_0\right), \quad \mathcal{M}_{7,17} = f_1^m(\mu_1 \bar{r}_0), \qquad \mathcal{M}_{7,18} = f_2^m(\mu_2 \bar{r}_0),$$

$$\mathcal{M}_{7,19} = f_3^m(\mu_3 \bar{r}_0), \qquad \mathcal{M}_{7,20} = \mathcal{M}_{7,21} = \mathcal{M}_{7,22} = 0, \qquad \mathcal{M}_{8,j} = -\mathcal{M}_{7,j+8},$$

$$\mathcal{M}_{8,j+4} = -\mathcal{M}_{7,j+12}, \qquad \mathcal{M}_{8,j+8} = \mathcal{M}_{7,j}, \qquad \mathcal{M}_{8,j+12} = \mathcal{M}_{7,j+4}, \quad \mathcal{M}_{8,17} = 0,$$

$$\mathcal{M}_{8,18} = \mathcal{M}_{8,19} = 0, \quad \mathcal{M}_{8,20} = h_1^m(\mu_1 \bar{r}_0), \quad \mathcal{M}_{8,21} = h_2^m(\mu_2 \bar{r}_0), \quad \mathcal{M}_{8,22} = h_3^m(\mu_3 \bar{r}_0),$$

$$\mathcal{M}_{9,j} = -i\delta_j^{p2} F_{p,j}^{m,1}\left(k_j \bar{r}_0\right), \quad \mathcal{M}_{9,j+4} = -i\delta_j^{p2} F_{p,j}^{m,2}\left(k_j \bar{r}_0\right), \quad \mathcal{M}_{9,j+8} = i\delta_j^{p1} G_{p,j}^{m,1}\left(k_j \bar{r}_0\right),$$

$$\mathcal{M}_{9,j+12} = i\delta_j^{p1} G_{p,j}^{m,2}\left(k_j \bar{r}_0\right), \quad \mathcal{M}_{9,17} = \mathcal{M}_{9,18} = \mathcal{M}_{9,19} = 0, \quad \mathcal{M}_{9,20} = g_1^m(\mu_1 \bar{r}_0),$$

$$\mathcal{M}_{9,21} = g_2^m(\mu_2 \bar{r}_0), \qquad \mathcal{M}_{9,22} = g_3^m(\mu_3 \bar{r}_0),$$

$$\mathcal{M}_{10,j} = \mathcal{M}_{9,j+8}, \qquad \mathcal{M}_{10,j+4} = \mathcal{M}_{9,j+12}, \qquad \mathcal{M}_{10,j+8} = -\mathcal{M}_{9,j},$$

$$\mathcal{M}_{10,j+12} = -\mathcal{M}_{9,j+4}, \quad \mathcal{M}_{10,17} = l_1^m(\mu_1 \bar{r}_0), \quad \mathcal{M}_{10,18} = l_2^m(\mu_2 \bar{r}_0),$$

$$\mathcal{M}_{10,19} = l_3^m(\mu_3 \bar{r}_0), \quad \mathcal{M}_{10,20} = \mathcal{M}_{10,21} = \mathcal{M}_{10,22} = 0,$$

$$\mathcal{M}_{11,j} = K_{p,j}^{m,1}\left(k_j\bar{r}_0\right), \quad \mathcal{M}_{11,j+4} = K_{p,j}^{m,2}\left(k_j\bar{r}_0\right), \quad \mathcal{M}_{11,j+8} = \mathcal{M}_{11,j+12} = 0,$$

$$\mathcal{M}_{11,17} = n_1^m(\mu_1\bar{r}_0), \quad \mathcal{M}_{11,18} = n_2^m(\mu_2\bar{r}_0), \quad \mathcal{M}_{11,19} = n_3^m(\mu_3\bar{r}_0),$$

$$\mathcal{M}_{11,20} = \mathcal{M}_{11,21} = \mathcal{M}_{11,22} = 0,$$

$$\mathcal{M}_{12,j} = 0, \quad \mathcal{M}_{12,j+4} = 0, \quad \mathcal{M}_{12,j+8} = \mathcal{M}_{11,j}, \quad \mathcal{M}_{12,j+12} = \mathcal{M}_{11,j+4},$$

$$\mathcal{M}_{12,17} = \mathcal{M}_{12,18} = \mathcal{M}_{12,19} = 0, \quad \mathcal{M}_{12,20} = n_1^m(\mu_1\bar{r}_0), \quad \mathcal{M}_{12,21} = n_2^m(\mu_2\bar{r}_0),$$

$$\mathcal{M}_{12,22} = n_3^m(\mu_3\bar{r}_0),$$

$$\mathcal{M}_{13,j} = W_{p,j}^{m,1}\left(k_j\bar{r}_0\right), \quad \mathcal{M}_{13,j+4} = W_{p,j}^{m,2}\left(k_j\bar{r}_0\right), \quad \mathcal{M}_{13,j+8} = \mathcal{M}_{13,j+12} = 0,$$

$$\mathcal{M}_{13,17} = \mathcal{M}_{13,18} = \mathcal{M}_{13,19} = \mathcal{M}_{13,20} = \mathcal{M}_{13,21} = \mathcal{M}_{13,22} = 0,$$

$$\mathcal{M}_{14,j} = \mathcal{M}_{14,j+4} = 0, \quad \mathcal{M}_{14,j+8} = \mathcal{M}_{13,j}, \quad \mathcal{M}_{14,j+12} = \mathcal{M}_{13,j+4},$$

$$\mathcal{M}_{14,17} = \mathcal{M}_{14,18} = \mathcal{M}_{14,19} = \mathcal{M}_{14,20} = \mathcal{M}_{14,21} = \mathcal{M}_{14,22} = 0,$$

$$\mathcal{M}_{15,j} = P_{p,j}^{m,1}\left(k_j\bar{r}_1\right), \quad \mathcal{M}_{15,j+4} = P_{p,j}^{m,2}\left(k_j\bar{r}_1\right), \quad \mathcal{M}_{15,j+8} = Q_{p,j}^{m,1}\left(k_j\bar{r}_1\right), \quad \mathcal{M}_{15,j+12} = Q_{p,j}^{m,2}\left(k_j\bar{r}_1\right),$$

$$\mathcal{M}_{15,17} = \mathcal{M}_{15,18} = \mathcal{M}_{15,19} = \mathcal{M}_{15,20} = \mathcal{M}_{15,21} = \mathcal{M}_{15,22} = 0,$$

$$\mathcal{M}_{16,j} = -\mathcal{M}_{15,j+8}, \quad \mathcal{M}_{16,j+4} = -\mathcal{M}_{15,j+12}, \quad \mathcal{M}_{16,j+8} = \mathcal{M}_{15,j}, \quad \mathcal{M}_{16,j+12} = \mathcal{M}_{15,j+4},$$

$$\mathcal{M}_{16,17} = \mathcal{M}_{16,18} = \mathcal{M}_{16,19} = \mathcal{M}_{16,20} = \mathcal{M}_{16,21} = \mathcal{M}_{16,22} = 0,$$

$$\mathcal{M}_{17,j} = R_{p,j}^{m,1}\left(k_j\bar{r}_1\right), \quad \mathcal{M}_{17,j+4} = R_{p,j}^{m,2}\left(k_j\bar{r}_1\right), \quad \mathcal{M}_{17,j+8} = S_{p,j}^{m,1}\left(k_j\bar{r}_1\right), \quad \mathcal{M}_{17,j+12} = S_{p,j}^{m,2}\left(k_j\bar{r}_1\right),$$

$$\mathcal{M}_{17,17} = \mathcal{M}_{17,18} = \mathcal{M}_{17,19} = \mathcal{M}_{17,20} = \mathcal{M}_{17,21} = \mathcal{M}_{17,22} = 0,$$

$$\mathcal{M}_{18,j} = -\mathcal{M}_{17,j+8}, \quad \mathcal{M}_{18,j+4} = -\mathcal{M}_{17,j+12}, \quad \mathcal{M}_{18,j+8} = \mathcal{M}_{17,j}, \quad \mathcal{M}_{18,j+12} = \mathcal{M}_{17,j+4},$$

$$\mathcal{M}_{18,17} = \mathcal{M}_{18,18} = \mathcal{M}_{18,19} = \mathcal{M}_{18,20} = \mathcal{M}_{18,21} = \mathcal{M}_{18,22} = 0,$$

$$\mathcal{M}_{19,j} = T_{p,j}^{m,1}\left(k_j\bar{r}_1\right), \quad \mathcal{M}_{19,j+4} = T_{p,j}^{m,2}\left(k_j\bar{r}_1\right), \quad \mathcal{M}_{19,j+8} = U_{p,j}^{m,1}\left(k_j\bar{r}_1\right), \quad \mathcal{M}_{19,j+12} = U_{p,j}^{m,2}\left(k_j\bar{r}_1\right),$$

$$\mathcal{M}_{19,17} = \mathcal{M}_{19,18} = \mathcal{M}_{19,19} = \mathcal{M}_{19,20} = \mathcal{M}_{19,21} = \mathcal{M}_{19,22} = 0,$$

$$\mathcal{M}_{20,j} = -\mathcal{M}_{19,j+8}, \quad \mathcal{M}_{20,j+4} = -\mathcal{M}_{19,j+12}, \quad \mathcal{M}_{20,j+8} = \mathcal{M}_{19,j}, \quad \mathcal{M}_{20,j+12} = \mathcal{M}_{19,j+4},$$

$$\mathcal{M}_{20,17} = \mathcal{M}_{20,18} = \mathcal{M}_{20,19} = \mathcal{M}_{20,20} = \mathcal{M}_{20,21} = \mathcal{M}_{20,22} = 0,$$

$$\mathcal{M}_{21,j} = W_{p,j}^{m,1}\left(k_j\bar{r}_1\right), \quad \mathcal{M}_{21,j+4} = W_{p,j}^{m,2}\left(k_j\bar{r}_1\right), \quad \mathcal{M}_{21,j+8} = \mathcal{M}_{21,j+12} = 0,$$

$$\mathcal{M}_{21,17} = \mathcal{M}_{21,18} = \mathcal{M}_{21,19} = \mathcal{M}_{21,20} = \mathcal{M}_{21,21} = \mathcal{M}_{21,22} = 0,$$

$$\mathcal{M}_{22,j} = \mathcal{M}_{22,j+4} = 0, \quad \mathcal{M}_{22,j+8} = \mathcal{M}_{21,j}, \quad \mathcal{M}_{22,j+12} = \mathcal{M}_{21,j+4},$$

$$\mathcal{M}_{22,17} = \mathcal{M}_{22,18} = \mathcal{M}_{22,19} = \mathcal{M}_{22,20} = \mathcal{M}_{22,21} = \mathcal{M}_{22,22} = 0,$$

for j=1,2,3,4, where

$$G_{p,j}^{m,l}\left(k_jx\right) = \Omega\frac{m}{x}\zeta^{m,l}\left(k_jx\right), \quad F_{p,j}^{m,l}\left(k_jx\right) = \Omega\frac{\partial\zeta^{m,l}\left(k_jx\right)}{\partial x}, \quad K_{p,j}^{m,l}\left(k_jx\right) = \Omega\delta_j^{p3}\zeta^{m,l}\left(k_jx\right),$$

$$f_l^m(\mu_lx) = \sigma_l^{i1}\frac{\partial I^m(\mu_lx)}{\partial x}, \quad l = 1,2, \qquad f_3^m(\mu_3x) = \frac{m}{x}I^m(\mu_3x),$$

$$h_l^m(\mu_lx) = f_l^m(\mu_lx), \quad l = 1,2, \qquad h_3^m(\mu_3x) = -f_3^m(\mu_3x),$$

$$g_l^m(\mu_lx) = \sigma_l^{i1}\frac{m}{x}I^m(\mu_lx), \quad l = 1,2, \qquad g_3^m(\mu_3x) = -\frac{\partial I^m(\mu_3x)}{\partial x},$$

$$l_l^m(\mu_lx) = g_l^m(\mu_lx), \quad l = 1,2, \qquad l_3^m(\mu_3x) = -g_3^m(\mu_3x),$$

$$n_l^m(\mu_lx) = -\sigma_l^{i2}I^m(\mu_lx), \quad l = 1,2, \qquad n_3^m(\mu_3x) = 0.$$

D.I. FOTIADIS
Dept. of Computer Science, University of Ioannina, GR 451 10 Ioannina, Greece

G. FOUTSITZI and C.V. MASSALAS
Dept. of Mathematics, University of Ioannina, GR 451 10 Ioannina, Greece

I E LAGARIS, A LIKAS, D I FOTIADIS AND C V MASSALAS

A hardware implementable non-linear method for the solution of ordinary, partial and integrodifferential equations

1 Introduction

Basis set methods have been successfully used in the past for solving differential (ordinary or partial) and integrodifferential equations. The solution is usually written as a linear combination of the basis functions and the coefficients are determined so that the resulting approximation satisfies both the equation and the boundary conditions. It is quite common to use basis sets containing orthonormal functions and it is highly desirable if their functional form is such that the required integrals, usually involved, can be calculated analytically. However all these desired convenient properties pose very tough constraints in the construction of the approximate solution. For example, linearity may impose on the size of the basis-set. Analytic integrability may not be compatible with orthonormality and then one is forced to follow an orthogonalization process such as Gram-Schmidt that complicates the whole scheme. There is always a temptation to use a basis set such that each element satisfies the boundary conditions by construction, so that the problem becomes easier. Such a choice however may not possess some of the other properties and so on so forth. For large problems two are the key factors that one is mainly concerned to reduce. The number of the parameters, and the time needed for evaluating a trial approximation. To achieve the first goal one may sacrifice the linearity. When this is done, usually the orthonormality is also sacrificed. To achieve the second goal one may pick functions that satisfy the boundary conditions by construction and be such that their evaluation may be performed with extremely high speed by employing specialised hardware. If also the analytical integrability of the resulting integrands can be retained, then one may be able to solve very large problems. However this last possibility is heavily problem dependent since the problem itself may contain functions that cannot be integrated analytically. We will concentrate in the construction of non-linear hardware implementable methods that explicitly satisfy the BC's and we will use optimization techniques to determine a set of parameters such that the model becomes an approximate solution of the differential equation. Artificial neural-networks (both sigmoid multilayer perceptrons and radial basis function networks) [11] have the property of the hardware implementability, are known to be universal approximators and they can be the core elements for such a model. We show that the boundary conditions may be satisfied by construction in a rather simple manner.

2 Description of the method

The proposed approach will be illustrated in terms of the following general type of problem:

$$G(\vec{x}, \Psi(\vec{x}), \nabla\Psi(\vec{x}), \nabla^2\Psi(\vec{x})) = 0, \vec{x} \in D \tag{1}$$

subject to certain boundary conditions (BCs) (for instance Dirichlet and/or Neumann), where $\vec{x} = (x_1, \ldots, x_n) \in R^n$ and $D \subset R^n$ denotes the solution domain and $\Psi(\vec{x})$ is the solution to be computed.

To obtain a solution to the above equation the collocation method is adopted which assumes a discretization $\{\vec{x_i}\}$, of the domain D and its boundary ∂D. The problem is then transformed into the following system of equations:

$$G(\vec{x_i}, \Psi(\vec{x_i}), \nabla\Psi(\vec{x_i}), \nabla^2\Psi(\vec{x_i})) = 0, \quad \forall \vec{x_i} \in D \tag{2}$$

subject to the constraints imposed by the B.Cs.

If by $\Psi_t(\vec{x}, \vec{p})$ we denote a trial solution with adjustable parameters $\vec{p}$ the problem is transformed to:

$$min_{\vec{p}} \sum_{\vec{x_i}} G(\vec{x_i}, \Psi_t(\vec{x_i}, \vec{p}), \nabla\Psi(\vec{x_i}, \vec{p}), \nabla^2\Psi(\vec{x_i}, \vec{p}))^2 \tag{3}$$

subject to the constraints imposed by the B.Cs.

In the proposed approach the trial solution Ψ_t employs a feedforward neural network and the parameters $\vec{p}$ correspond to the weights and biases of the neural architecture. Fig. 1 displays a typical architecture with one hidden layer. We choose a form for the trial function $\Psi_t(\vec{x})$ such that by construction satisfies the BCs. This is achieved by writing it as a sum of two terms:

$$\Psi_t(\vec{x}) = A(\vec{x}) + F(\vec{x}, N(\vec{x}, \vec{p})) \tag{4}$$

where $N(\vec{x}, \vec{p})$ is a single-output feedforward neural network with parameters $\vec{p}$ and n input units fed with the input vector $\vec{x}$.

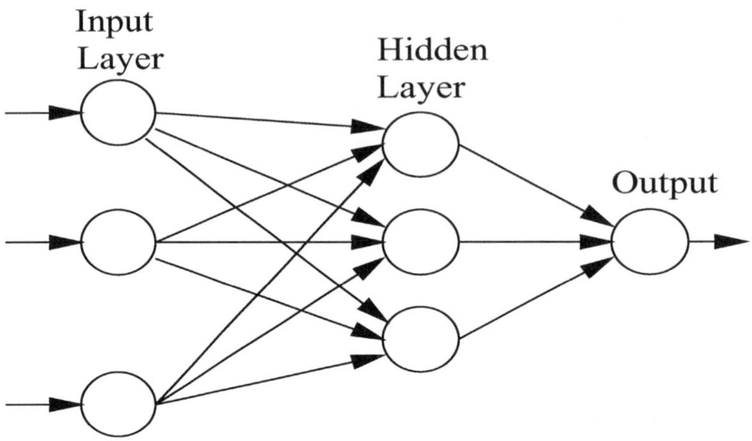

Figure 1: Feedforward neural network with one hidden layer.

The term $A(\vec{x})$ contains no adjustable parameters and satisfies the BCs. The second term F is constructed so as not to contribute to the BCs, since $\Psi_t(\vec{x})$ must also satisfy them. This term employs a neural network whose weights and biases are to be adjusted in order to deal with the minimization problem. Note at this point that the problem has been reduced from the original constrained optimization problem to an unconstrained one (which is much easier to handle) due to the choice of the form of the trial solution that satisfies by construction the B.Cs.

In the next section we present a systematic way to construct the trial solution, i.e. the functional forms of both A and F. We treat several common cases that one frequently encounters in various scientific fields. As indicated by our experiments, the approach based on the above formulation is very effective and provides in reasonable computing time accurate solutions with impressive generalization (interpolation) properties.

The efficient minimization of equation (3) can be considered as a procedure of training the neural network where the error corresponding to each input vector $\vec{x_i}$ is the value $G(\vec{x_i})$ which has to become zero. Computation of this error value involves not only the network output (as is the case in conventional training) but also the derivatives of the output with respect to any of its inputs. Therefore, in computing the gradient of the error with respect to the network weights, we need to compute not only the gradient of the network but also the gradient of the network derivatives with respect to its inputs.

Consider a multilayer perceptron (MLP) with n input units, one hidden layer with H sigmoid units and a linear output unit. For a given input vector $\vec{x} = (x_1, \ldots, x_n)$ the output of the network is $N = \sum_{i=1}^{H} v_i \sigma(z_i)$ where $z_i = \sum_{j=1}^{N} w_{ij} x_j + u_i$, w_{ij} denotes the weight from the input unit i to the hidden unit j, v_i denotes the weight from the hidden unit i to the output, u_i denotes the bias of hidden unit i and $\sigma(z)$ is the sigmoid transfer function.

It is straightforward to show that:

$$\frac{\partial^k N}{\partial x_j^k} = \sum_{i=1}^{H} v_i w_{ij}^k \sigma_i^{(k)} \tag{5}$$

where $\sigma_i = \sigma(z_i)$ and $\sigma^{(k)}$ denotes the kth order derivative of the sigmoid. Moreover it is readily verifiable that:

$$\frac{\partial^{\lambda_1}}{\partial x_1^{\lambda_1}} \frac{\partial^{\lambda_2}}{\partial x_2^{\lambda_2}} \cdots \frac{\partial^{\lambda_n}}{\partial x_2^{\lambda_n}} N = \sum_{i=1}^{n} v_i P_i \sigma_i^{(\Lambda)} \tag{6}$$

where

$$P_i = \prod_{k=1}^{n} w_{ki}^{\lambda_k} \tag{7}$$

and $\Lambda = \sum_{i=1}^{n} \lambda_i$.

Equation (6) indicates that the derivative of the network with respect to any of its inputs is equivalent to a feedforward neural network $N_g(\vec{x})$ with one hidden layer, having the same values for the weights w_{ij} and thresholds u_i and with each weight v_i being replaced with $v_i P_i$. Moreover the transfer function of each hidden unit is replaced with the Λth order derivative of the sigmoid.

Therefore the gradient of N_g with respect to the parameters of the original network can be easily obtained as:

$$\frac{\partial N_g}{\partial v_i} = P_i \sigma_i^{(\Lambda)} \tag{8}$$

$$\frac{\partial N_g}{\partial u_i} = v_i P_i \sigma_i^{(\Lambda+1)} \tag{9}$$

$$\frac{\partial N_g}{\partial w_{ij}} = x_j v_i P_i \sigma_i^{(\Lambda+1)} + v_i \left(\prod_{k=1, k \neq j} w_{ki}^{\lambda_k} \right) \sigma_i^{(\Lambda)} \tag{10}$$

In the case of a radial basis function (RBF) network we have: $N(\vec{x}) = \sum_{j=1}^{M} v_j \phi_j(\vec{x})$, with $\phi_j(\vec{x}) = e^{-\frac{1}{2\sigma_j^2}(|\vec{x} - \vec{\mu_j}|^2}$ for $j = 1, 2, \ldots, M$ and corresponding formulas for its derivatives may be easily derived as well. Once the derivative of the error with respect to the network parameters has been defined it is then straightforward to employ almost any minimization technique. For example it is possible to use either the traditional backpropagation algorithm (or any of its variants), the conjugate gradient method, quasi Newton methods, or other techniques proposed in the literature. In our experiments we have employed a new (unpublished) beta-version of the Merlin [9] software package, that contains a host of efficient minimization algorithms and which has demonstrated excellent performance.

Another class of problems that can be treated is given by:

$$H\Psi(\vec{x}) = \epsilon\Psi(\vec{x}), \quad \vec{x} \in D$$

with the BC: $\Psi(\vec{x}) = 0, \vec{x} \in \partial D$. This is an eigenvalue problem for the operator H, with Dirichlet BCs. The trial solution takes the form: $\Psi_t(\vec{x}) = B(\vec{x}, \vec{\lambda})N(\vec{x}, \vec{p})$, where $B(\vec{x}, \vec{\lambda})$ vanishes on the boundary for a range of values of $\vec{\lambda}$ and hence the Boundary conditions are satisfied by construction. Again by discretizing, the problem is transformed to minimizing the following error quantity:

$$\frac{\sum_i [H\Psi_t(\vec{x_i}) - \epsilon\Psi_t(\vec{x_i})]^2}{\sum_i [\Psi_t(\vec{x_i})]^2}$$

The denominator is necessary in order to avoid the trivial solution $\Psi_t(\vec{x}) = 0$. The eigenvalue ϵ, is determined by requiring that the above error is a minimum with respect to variations in ϵ. Namely:

$$\epsilon = \frac{\sum_i \Psi_t(\vec{x_i}) H\Psi_t(\vec{x_i})}{\sum_i \Psi_t(\vec{x_i})^2}$$

We apply this approach in problems taken from the field of Quantum Mechanics in which case H is a linear (Integro)-differential Hermitian operator.

3 Illustration of the method

3.1 Solution of single ODEs and Systems of coupled ODEs

To illustrate the method, we consider the *first order ODE*:

$$\frac{d\Psi(x)}{dx} = f(x, \Psi)$$

with $x \in [0,1]$ and with the initial condition (IC) $\Psi(0) = A$.
A trial solution is written as:

$$\Psi_t(x) = A + xN(x, \vec{p}) \tag{11}$$

where $N(x, \vec{p})$ is the output of an ANN with one input unit for x and weights $\vec{p}$. Note that $\Psi_t(x)$ satisfies the IC by construction. The error quantity to be minimized is given by:

$$E[\vec{p}] = \sum_i \{\frac{d\Psi_t(x_i)}{dx} - f(x_i, \Psi_t(x_i))\}^2 \tag{12}$$

where the x_i's are points in $[0,1]$.

The same procedure can be applied to the *second order ODE*:

$$\frac{d^2\Psi(x)}{dx^2} = f(x, \Psi, \frac{d\Psi}{dx})$$

For the *initial value* problem: $\Psi(0) = A$ and $\frac{d}{dx}\Psi(0) = A'$, the trial solution can be cast as:

$$\Psi_t(x) = A + A'x + x^2 N(x, \vec{p}) \tag{13}$$

For the *two point Dirichlet* BCs: $\Psi(0) = A$ and $\Psi(1) = B$, the trial solution is written as:

$$\Psi_t(x) = A(1-x) + Bx + x(1-x)N(x, \vec{p}) \tag{14}$$

In the above two cases of second order ODEs the error function to be minimized is given by:

$$E[\vec{p}] = \sum_i \{\frac{d^2\Psi_t(x_i)}{dx^2} - f(x_i, \Psi_t(x_i), \frac{d\Psi_t(x_i)}{dx})\}^2 \tag{15}$$

For *systems of K first order ODEs*

$$\frac{d\Psi_i}{dx} = f_i(x, \Psi_1, \Psi_2, \ldots \Psi_K)$$

with $\Psi_i(0) = A_i$, $(i = 1, \ldots, K)$ we consider one neural network for each trial solution Ψ_{t_i} $(i = 1, \ldots, K)$ which is written as:

$$\Psi_{t_i}(x) = A_i + xN_i(x, \vec{p}) \tag{16}$$

and we minimize the following error quantity:

$$E[\vec{p}] = \sum_{k=1}^{K} \sum_i \{\frac{d\Psi_{t_k}(x_i)}{dx} - f_k(x_i, \Psi_{t_1}, \Psi_{t_2}, \ldots, \Psi_{t_K})\}^2 \tag{17}$$

3.2 Solution of single PDEs

We treat here two–dimensional problems only. However it is possible to extend the method to more dimensions. For example consider the *Poisson equation*:

$$\frac{\partial^2}{\partial x^2}\Psi(x,y) + \frac{\partial^2}{\partial y^2}\Psi(x,y) = f(x,y)$$

$x \in [0,1], y \in [0,1]$ with *Dirichlet* BC: $\Psi(0,y) = f_0(y)$, $\Psi(1,y) = f_1(y)$ and $\Psi(x,0) = g_0(x)$, $\Psi(x,1) = g_1(x)$. The trial solution is written as:

$$\Psi_t(x,y) = A(x,y) + x(1-x)y(1-y)N(x,y,\vec{p}) \tag{18}$$

where $A(x,y)$ is chosen so as to satisfy the BC, namely:

$$\begin{aligned} A(x,y) &= (1-x)f_0(y) + xf_1(y) + (1-y)\{g_0(x) - [(1-x)g_0(0) + xg_0(1)]\} \\ &+ y\{g_1(x) - [(1-x)g_1(0) + xg_1(1)]\} \end{aligned} \tag{19}$$

For *mixed boundary conditions* of the form: $\Psi(0,y) = f_0(y)$, $\Psi(1,y) = f_1(y)$, $\Psi(x,0) = g_0(x)$ and $\frac{\partial}{\partial y}\Psi(x,1) = g_1(x)$ (i.e. *Dirichlet* on part of the boundary and *Neumann* elsewhere), the trial solution is written as:

$$\Psi_t(x,y) = B(x,y) + x(1-x)y[N(x,y,\vec{p}) - N(x,1,\vec{p}) - \frac{\partial}{\partial y}N(x,1,\vec{p})] \tag{20}$$

and $B(x,y)$ is again chosen so as to satisfy the BCs:

$$B(x,y) = (1-x)f_0(y)+xf_1(y)+g_0(x)-[(1-x)g_0(0)+xg_0(1)]+y\{g_1(x)-[(1-x)g_1(0)+xg_1(1)]\} \tag{21}$$

Note that the second term of the trial solution does not affect the boundary conditions since it vanishes at the part of the boundary where Dirichlet BCs are imposed and its gradient component normal to the boundary vanishes at the part of the boundary where Neumann BCs are imposed.

4 Examples

In this section we report on the solution of a number of model problems. For some of the test problems the exact analytic solution $\Psi_a(\vec{x})$ was known in advance. Therefore we test the accuracy of the obtained solutions by computing the deviation $\Delta\Psi(\vec{x}) = \Psi_t(\vec{x}) - \Psi_a(\vec{x})$. In the first two problems (concerning ODEs) we have

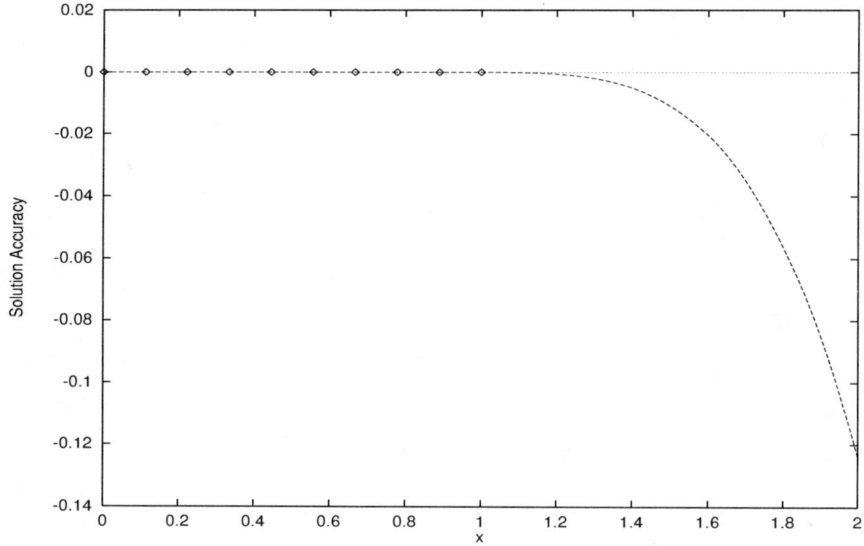

Figure 2: Problem 1: Accuracy of the computed solution.

conducted expriments using both MLPs and RBFs having the same number of parameters (i.e. same number of units in the hidden layer). Both types of networks were equally effective in solving the corresponding equations and we think that more detailed experimentation is needed in order to assess the relative advantages of each architecture in the particular task. The accuracy results presented in the following subsections are those obtained with MLPs.

4.1 ODEs and systems of ODEs

4.1.1 Problem 1

$$\frac{d}{dx}\Psi + (x + \frac{1+3x^2}{1+x+x^3})\Psi = x^3 + 2x + x^2\frac{1+3x^2}{1+x+x^3}$$

with $\Psi(0) = 1$ and $x \in [0,1]$. The exact solution is: $\Psi(x) = \frac{e^{-x^2/2}}{1+x+x^3} + x^2$. The trial neural form of the solution is taken to be: $\Psi_t(x) = 1 + xN(x, \vec{p})$. The network was trained using a grid of 10 equidistant points in [0,1]. Fig. 2 provides results concerning the interpolation as well as the extrapolation cababilities of the solution $\Psi_t(x)$. It is clear that the solution is of high accuracy, although training was performed using a small number of points.

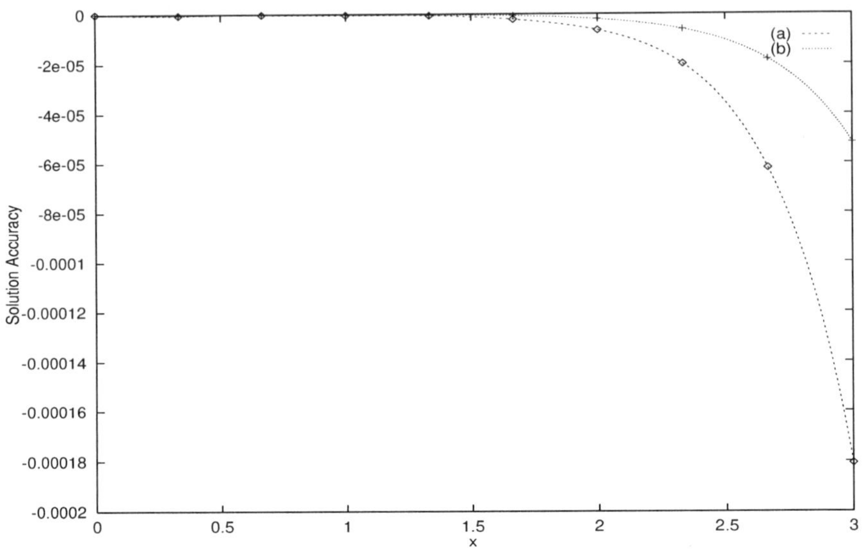

Figure 3: Problem 2: Accuracy of the computed solutions.

4.1.2 Problem 2

Consider the system of two coupled first order ODEs:

$$\frac{d}{dx}\Psi_1 = cos(x) + \Psi_1^2 + \Psi_2 - (1 + x^2 + sin^2(x))$$

$$\frac{d}{dx}\Psi_2 = 2x - (1 + x^2)sin(x) + \Psi_1\Psi_2$$

with $x \in [0,1]$ and $\Psi_1(0) = 0$ and $\Psi_2(0) = 1$. The exact solutions are: $\Psi_1(x) = sin(x)$ and $\Psi_2(x) = 1 + x^2$.

The trial neural solutions employ two neural networks N_1 and N_2: $\Psi_{t_1}(x) = xN_1(x, \vec{p}_1)$ and $\Psi_{t_2}(x) = 1 + xN_2(x, \vec{p}_2)$ The accuracy of the computed solutions is displayed in Fig. 3.

4.2 PDEs

We consider boundary value problems with Dirichlet and Neumann BCs. All subsequent problems were defined on the domain $[0,1] \times [0,1]$ and in order to perform training we consider a mesh of 100 points obtained by dividing the domain $[0,1]$ of each variable into 10 equidistant points. In analogy with the previous cases the neural architecture was considered to be a MLP with two inputs (accepting the coordinates x and y of each point).

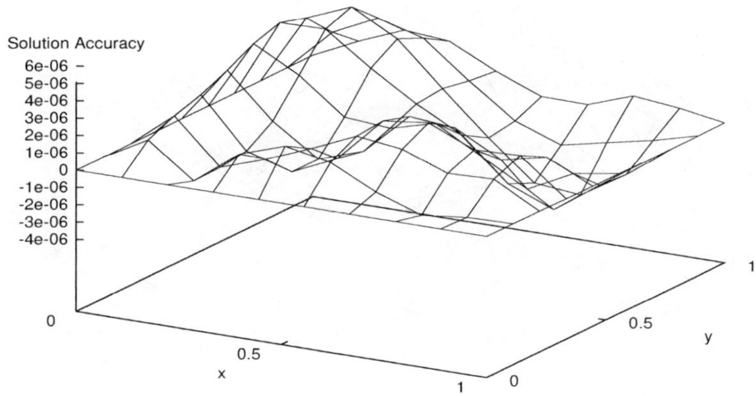

Figure 4: Problem 3: Accuracy of the computed solution at the training points.

4.2.1 Problem 3

$$\nabla^2 \Psi(x, y) = (2 - \pi^2 y^2) sin(\pi x)$$

with $x, y \in [0, 1]$ and with mixed BCs: $\Psi(0, y) = 0$, $\Psi(1, y) = 0$ and $\Psi(x, 0) = 0$, $\frac{\partial}{\partial y}\Psi(x, 1) = 2sin(\pi x)$. The exact solution is : $\Psi(x, y) = y^2 sin(\pi x)$. The trial neural form is specified by:

$$\Psi_t(x, y) = B(x, y) + x(1 - x)y[N(x, y, \vec{p}) - N(x, 1, \vec{p}) - \frac{\partial}{\partial y} N(x, 1, \vec{p})]$$

where $B(x, y)$ is obtained by direct substitution in equation (21).

The accuracy of the neural solution is depicted in Figures 4 and 5 for training and interpolation points respectively.

4.2.2 Problem 4

This is an example of a *non-linear* PDE.

$$\nabla^2 \Psi(x, y) + \Psi(x, y) \frac{\partial}{\partial y} \Psi(x, y) = sin(\pi x)(2 - \pi^2 y^2 + 2y^3 sin(\pi x))$$

with the same mixed BCs as in the previous problem. The exaxt solution is again $\Psi(x, y) = y^2 sin(\pi x)$ and the parametrization of the trial neural form is the same as

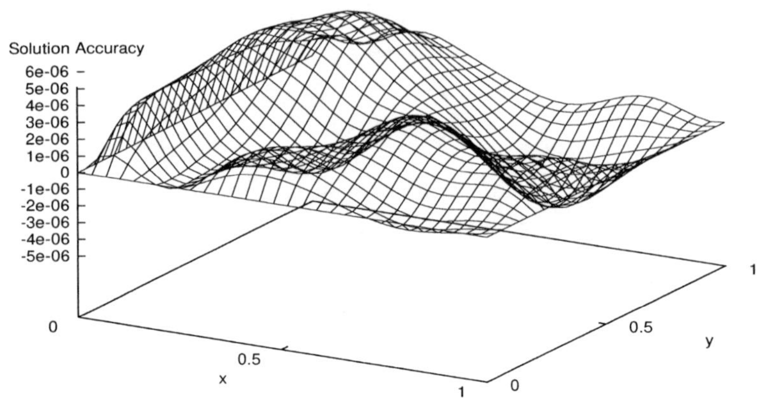

Figure 5: Problem 3: Accuracy of the computed solution at the interpolation points.

in problem 3. No plots of the accuracy are presented since they are almost the same with those of problem 3.

4.3 Eigenvalue problems

4.3.1 Non-Local Schrödinger equation for the $n + \alpha$ system

We consider here the non-local Schrödinger equation :

$$-\frac{\hbar^2}{2\mu}\frac{d^2\phi}{dr^2}(r) + V(r)\phi(r) + \int_0^\infty K_0(r,r')\phi(r')dr' = \epsilon\phi(r)$$

with $V(r) = -V_0 e^{-\beta r^2}$, where $V_0 = 41.28386$, $\beta = 0.2751965$ and $K_0(r,r') = -Ae^{-\gamma(r^2+r'^2)}(e^{2krr'} - e^{-2krr'})$ with $A = -62.03772$, $\gamma = -0.8025$, $k = 0.46$. This describes the $n + \alpha$ system and is derived in the framework of the Resonating Group Method [4], μ is the system's reduced mass given by: $\frac{1}{\mu} = \frac{1}{m_n} + \frac{1}{2m_n+2m_p}$.

We parametrized the trial wavefunction as:

$$\phi_t(r) = re^{-\beta r}N(r,\vec{u},\vec{w},\vec{v}), \quad \beta > 0$$

and minimized the following error quantity:

$$\frac{\sum_i \left\{-\frac{\hbar^2}{2\mu}\frac{d^2}{dr^2}\phi_t(r_i) + V(r_i)\phi_t(r_i) + \int_0^\infty K_0(r_i,r')\phi_t(r')dr' - \epsilon\phi_t(r_i)\right\}^2}{\sum_i \phi_t^2(r_i)}$$

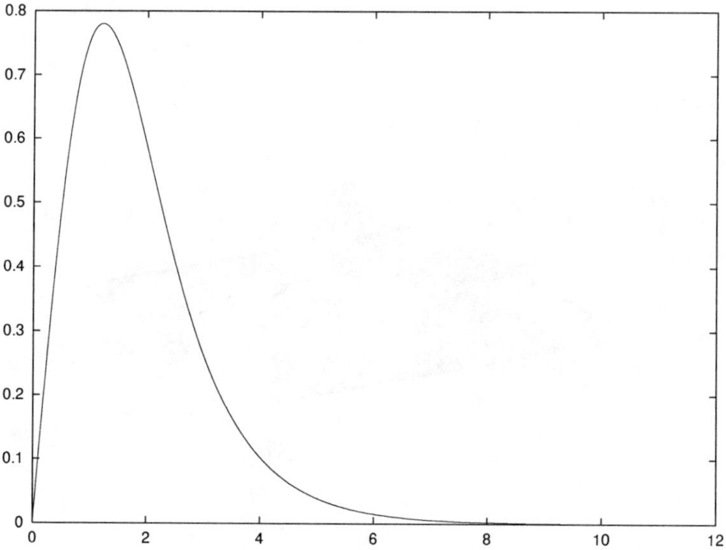

Figure 6: Ground state of the non-local Schrödinger equation for the $n + \alpha$ system ($\epsilon = -24.07644$).

We have considered 100 equidistant points in $[0, 12]$ and the computed ground state is depicted in Fig. 6, while the corresponding eigenvalue was found equal to -24.07644, in agreement with previous calculations [5].

4.4 Two dimensional Schrödinger equation

We consider here the well studied [5] example of the Henon-Heiles potential.
The Hamiltonian is written as:

$$H = -\frac{1}{2}(\frac{\partial^2}{\partial x^2} + \frac{\partial^2}{\partial y^2}) + V(x, y)$$

with $V(x, y) = \frac{1}{2}(x^2 + y^2) + \frac{1}{4\sqrt{5}}(xy^2 - \frac{1}{3}x^3)$.
We parametrize the trial solution as:

$$\phi_t(x, y) = e^{-\lambda(x^2+y^2)} N(x, y, \vec{u}, \vec{w}^{(x)}, \vec{w}^{(y)}, \vec{v}), \quad \lambda > 0$$

where N is a feedforward neural network with one hidden layer and two input nodes (accepting the x and y values).

$$N(x, y, \vec{u}, \vec{w}^{(x)}, \vec{w}^{(y)}, \vec{v}) = \sum_{j=1}^{m} v_j \sigma(x w_j^{(x)} + y w_j^{(y)} + u_j)$$

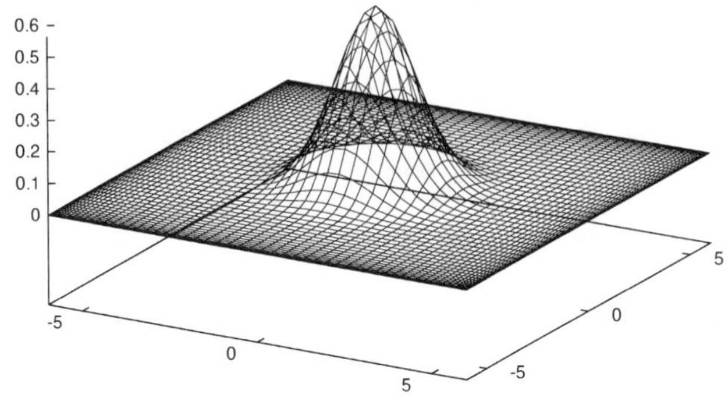

Figure 7: Ground state of the Henon-Heiles problem ($\epsilon = 0.99866$).

We have considered a grid of 20×20 points in $[-6, 6] \times [-6, 6]$. The quantity minimized is:

$$\sum_{i,j} [H\phi_t(x_i, y_j) - \epsilon\phi_t(x_i, y_j)]^2 / \sum_{ij} \phi_t^2(x_i, y_j) \tag{22}$$

For this problem we calculate not only the ground state but a few more levels. The way we followed is the extraction from the trial wavefunction of the already computed levels If for example by $\phi_0(x, y)$ we denote the normalized ground state, the trial wavefunction to be used for the computation of another level would be:

$$\phi_t(x, y) = \tilde{\phi}_t(x, y) - \phi_0(x, y) \int_{-\infty}^{\infty} \int_{-\infty}^{\infty} \phi_0(x', y')\tilde{\phi}_t(x', y')dx'dy'$$

where $\tilde{\phi}_t(x, y)$ is parametrized in the same way as before.

Note that $\phi_t(x, y)$ is orthogonal to $\phi_0(x, y)$ by construction. Following this procedure we calculated the first four levels for the Henon-Heiles Hamiltonian. Our results are reported in Figs. 7-10.

4.5 Three Coupled Anharmonic Oscillators

As a three-dimensional example we consider the potential for the three coupled sextic anharmonic oscillators [3]:

$$V(x, y, z) = V(x) + V(y) + V(z) + xy + xz + yz$$

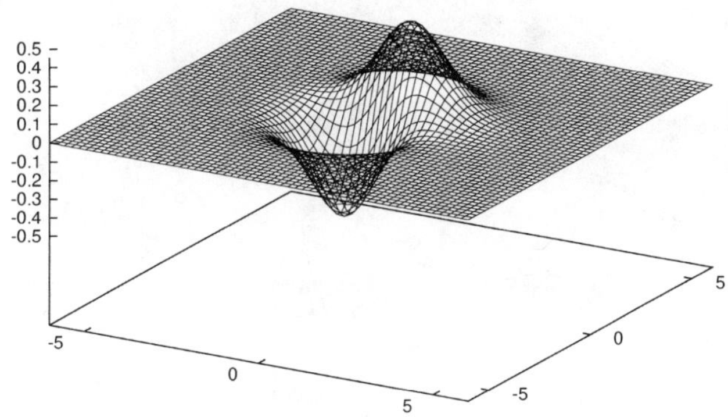

Figure 8: First excited state of the Henon-Heiles problem ($\epsilon = 1.990107$).

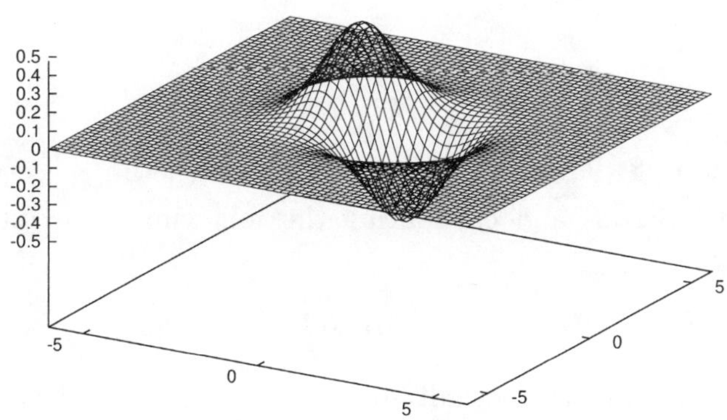

Figure 9: Second excited state (degenerate) of the Henon-Heiles problem ($\epsilon = 1.990107$).

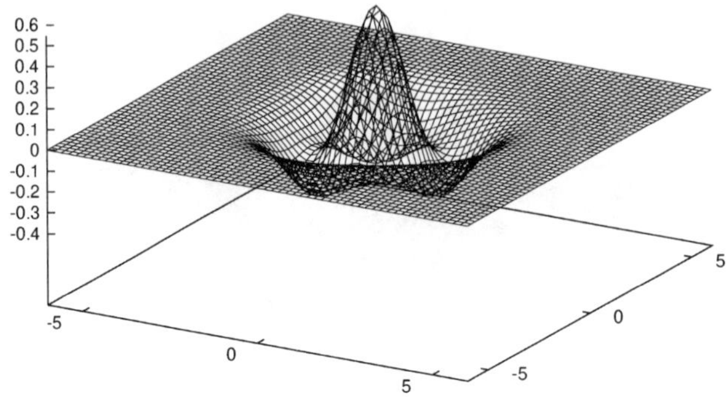

Figure 10: Third excited state of the Henon-Heiles problem ($\epsilon = 2.957225$).

where

$$V(x) = \frac{1}{2}x^2 + 2x^4 + \frac{1}{2}x^6$$

The trial solution $\phi_t(x, y, z)$ is parametrized as:

$$\phi_t(x, y, z) = e^{-\lambda(x^2+y^2+z^2)} N(x, y, z, \vec{u}, \vec{w}^{(x)}, \vec{w}^{(y)}, \vec{w}^{(z)}, \vec{v}), \quad \lambda > 0$$

where N is a feedforward neural network with one hidden layer and three input nodes (accepting the values of x, y and z):

$$N(x, y, z, \vec{u}, \vec{w}^{(x)}, \vec{w}^{(y)}, \vec{w}^{(z)}, \vec{v}) = \sum_{j=1}^{m} v_j \sigma(x w_j^{(x)} + y w_j^{(y)} + z w_j^{(z)} + u_j)$$

We have considered a 28×28×28 grid in the $[-4, 4] \times [-4, 4] \times [-4, 4]$ domain both for computing the integrals and calculating the following error quantity that was minimized:

$$\sum_{i,j,k} [H\phi_t(x_i, y_j, z_k) - \epsilon\phi_t(x_i, y_j, z_k)]^2 / \sum_{ijk} \phi_t^2(x_i, y_j, z_k)$$

where the energy is calculated as usually by:

$$\epsilon = \frac{\sum_{ijk} \phi_t(x_i, y_j, z_k) H\phi_t(x_i, y_j, z_k)}{\sum_{ijk} \phi_t^2(x_i, y_j, z_k)}$$

The ground state was computed and the corresponding eigenvalue was found equal to 2.9783, in agreement with the highly accurate result obtained by Kaluza [3].

5 Conclusions

We presented a method for solving a broad class of problems such as ordinary and partial differential equations and eigevalue problems for differential and integrodifferential operators. The method offers a compact solution, that is expressed in a closed analytic form, which in addition is differentiable, and posseses superior interpolation properties. The method can be implemented on special chips called "neuroprocessors" or "DSP" (Digital Signal Processing) processors that are designed to perform this kind of calculations very efficiently, namely a few orders of magnitude faster than conventional cpus. This last property alone makes the method a very attractive one, since it will render the solution of large complex problems possible.

References

[1] Evangelakis, G.A. and Rizos, J.P. and Lagaris, I.E. and Demetropoulos, I.N., *MERLIN - A portable system for multidimensional minimization*, Computer Physics Communications, vol. 46, pp 401-415, 1987.

[2] Fletcher R., *Practical methods of Optimization*, second edition, John Wiley 1987.

[3] Kaluza M. Comput. Phys. Commun. 79 (1994) 425.

[4] Kaneko T., LeMere M., Tang Y. C., Phys. Rev. C, vol 44, p. 1588, 1991.

[5] Lagaris I. E., Papageorgiou D. G., Braun M. and Sofianos S. A., J. Comput. Phys. 126 (1996) 229.

[6] Lee, H. and Kang, I., *Neural algorithms for solving differential equations*, Journal of Computational Physics, 91, 110, 1990.

[7] Meade Jr, A.J. and Fernadez, A.A., *The numerical solution of Linear Ordinary Differential Equations by Feedforward Neural networks*, Math. Comput. Modelling, vol. 19, no. 12, pp. 1-25, 1994.

[8] Meade Jr, A.J. and Fernadez, A.A., *Solution of Nonlinear Ordinary Differential Equations by Feedforward Neural networks*, Math. Comput. Modelling, vol. 20, no. 9, pp. 19-44, 1994.

[9] Papageorgiou D. G., Chassapis C. S. and Lagaris I. E., *MERLIN–2.0 Enhanced and Programmable Version*, Computer Physics Communications, vol. 52, pp. 241-247, 1988.

[10] Wang, L. and Mendel J.M., *Structured trainable networks for matrix algebra*, IEEE Int. Joint Conference on Neural Networks, vol. 2, pp. 125-129, 1990.

[11] Williamson, R.C. and Helmke, U., *Existence and Uniqueness Results for Neural Network Approximations*, IEEE Trans. on Neural Networks, vol. 6, no. 1, pp. 2-13, 1995.

[12] Yentis, R. and Zaghoul, M.E., *VLSI Implementation of Locally Connected Neural Network for Solving Partial Differential Equations*, IEEE Trans on Circuits and Systems-I, vol. 43, no. 8, pp. 687-690, 1996.

I. E. Lagaris, A. Likas and D. I. Fotiadis
Dept. of Computer Science
Univ. of Ioannina, 45110, Greece

C.V. Massalas
Dept. of Mathematics
Univ. of Ioannina, 45110, Greece

T MILOH AND P A TYVAND

Initial free convection about an isothermal sphere in a porous medium

1 Introduction

Transient free convection around a three-dimensional body with a sudden change in temperature was first studied by Elliott (1970). However, there are some basic inconsistencies in his asymptotic solution. In the two-dimensional case, Elliott's work was followed up by Gupta & Pop (1977). Recently, a small-time consistent matched-asymptotic theory for the circular cylinder problem has been published by Sano & Kuribayashi (1992). Foster (1965) studied this type of transient convection in the context of Rayleigh-Benard instability.

So far, it seems that no work have been published to remedy the inconsistencies of the early asymptotic analysis of Elliott (1970) for the three-dimensional case. Here a first step is made in this direction by considering the flow in a porous medium governed by the Darcy-Boussinesq equation, which is mathematically simpler than the common Newtonian fluid model. We consider a spherical shape embedded in a porous medium. Thereby we carry out a three-dimensional analogy to the recent work by Pop, Ingham & Cheng (1993, subsequently referred to as PIC). The present problem has recently been investigated numerically by Yan ,Pop & Ingham (1997). They gave only the first-order inner solution in the asymptotic expansion without presenting the outer and higher order solutions. In this respect the present solution may be considered as a valuable addition to the YPI paper.

Transient free convection around a sphere is a basic problem with many applications, e.g. in metallurgy, glassworks, food industry, etc. The case where the sphere is embedded in a porous medium is of special interest in insulation techniques, storage of nuclear waste, and geothermal engineering. The present work will give information on the early stages of development of both the temperature and the flow field, but not on the nonlinear instabilities that eventually may lead to a plume formation. Sano & Okihara (1994) have recently studied the present problem for small Rayleigh numbers, except that they apply a constant heat flux instead of a constant temperature condition on the surface of the sphere.

2 Mathematical Problem

Let us consider an impermeable sphere of radius (a) surrounded by an infinite porous medium with permeability K. A Cartesian coordinate system (x, y, z) is introduced, with an origin at the sphere centre and where the z-axis is directed upwards in the gravity field. We also introduce a spherical coordinate system (r, θ, ϕ), where ϕ is the azimuthal angle around the z-axis. We consider only cases of axisymmetric flows, so there is no dependence on ϕ. The upward direction is defined as $\theta = 0$ and the downward direction as $\theta = \pi$.

At negative times there is a constant temperature T_∞ everywhere. At time $t = 0$ the temperature at the sphere wall is suddenly changed to the constant value T_w and is later maintained at this value. This induces free transient convection in the porous medium.

In the following we will apply a notation and dimensionless quantities similar to PIC. Our length scale unit is the sphere radius (a). The dimensionless Darcy-Boussinesq equations for the Stokes stream function ψ and the temperature field T may be written in the form:

$$d^2\psi = \frac{\partial T}{\partial \theta} \cos\,\theta + \frac{\partial T}{\partial r}\, r\, \sin\theta \tag{2.1}$$

$$\frac{\partial T}{\partial t} + \frac{\epsilon}{r^2\,\sin\,\theta}\,\frac{\partial(T,\psi)}{\partial(r,\theta)} = \alpha^*\,\epsilon^2\,\nabla^2\,T \tag{2.2}$$

The Stokes stream function is defined by the velocity components in the r and θ directions:

$$u_r = \frac{1}{r^2\,\sin\,\theta}\,\frac{\partial\psi}{\partial\theta}\ ;\ u_\theta = -\frac{1}{r\,\sin\,\theta}\,\frac{\partial\psi}{\partial r} \tag{2.3}$$

Two second-order differential operators have been introduced above, the Laplacian and a related operator:

$$\nabla^2 = \frac{1}{r^2}\,\frac{\partial}{\partial r}\,(r^2\,\frac{\partial}{\partial r}) + \frac{1}{r^2\,\sin\,\theta}\,\frac{\partial}{\partial\theta}\,(\sin\,\theta\,\frac{\partial}{\partial\theta}) \tag{2.4}$$

$$d^2 = \frac{1}{\sin\,\theta}\,\frac{\partial^2}{\partial r^2} + \frac{1}{r^2}\,\frac{\partial}{\partial\theta}\,(\frac{1}{\sin\,\theta}\,\frac{\partial}{\partial\theta}) \tag{2.5}$$

The initial boundary conditions for the resulting free convection problem are:

$$\psi = 0,\ \ T = 0 \text{ everywhere in the porous medium for } t < 0, \tag{2.6}$$

$$\psi = 0, \quad T = 1 \text{ for } r = 1 \text{ when } t \geq 0, \tag{2.7}$$

$$\psi \text{ is finite}, \quad T = 0 \text{ as } r \to \infty \text{ for } t \geq 0. \tag{2.8}$$

As in PIC we introduce a characteristic time scale t_0. The unit of dimensionless time is thus $(\sigma\ t_0)$, where σ denotes the ratio of heat capacity in the saturated medium versus that of the fluid. So far, all our definitions and notations are kept as close as possible to the 2D version of PIC. But in eq. (2.1) there is a sign discrepency, which appears as if the direction of their buoyancy term (PIC eq. 1) is unintentionally taken downwards. This error seems to arise because their unusual definition of the polar coordinate system.

Two important dimensionless parameters ϵ and α^* have been introduced in the heat equation (2.2): α^* is a dimensionless diffusivity which is of order one by definition (in accordance with PIC's choice of time unit).

The formal expansion parameter ϵ may then be defined by:

$$\epsilon = (\alpha^*\ Ra)^{-1} \tag{2.9}$$

where it is assumed that $\epsilon << 1$. Here Ra represents the Rayleigh number for free convection in the porous medium (PIC, Katto & Masuoka 1967). It must be of order larger than one for the present theory to have any relevance. A similar problem where the Rayleigh number is small has been solved by Sano & Okihara (1994).

Let us next introduce the diffusion time, which is the characteristic (dimensional) time for a temperature change to diffuse out one characteristic length in the absence of convection, i.e,

$$t_{\text{diff}} = a^2/\alpha \tag{2.10}$$

where α is the thermal diffusivity of the saturated porous medium. The time span for the present theory to be valid (see PIC) can be expressed as (t' denoting dimensional time):

$$0 < t' << t_{\text{diff}}/Ra \tag{2.11}$$

An estimate of the relative magnitude of the two consecutive terms in the asymptotic expansion is then;

$$Ra\ t'/t_{\text{diff}} << 1 \tag{2.12}$$

3 Matched Asymptotic Expansions and First-Order Solution

The outer (superscript "o") asymptotic expansions are formally expressed as:

$$T^{(o)} = \sum_{n=1}^{\infty} \epsilon^{n-1} T_n^{(o)}(r, \theta, t) \ ; \ \psi^{(o)} = \sum_{n=1}^{\infty} \epsilon^n \psi_n^{(o)}(r, \theta, t) \tag{3.1}$$

The outer temperature field is identicaly zero, implying that the outer stream function satisfies the following equation:

$$d^2 \psi_n^{(o)} = 0 \quad (n = 1, 2, ..) \tag{3.2}$$

We postulate the general solution:

$$\psi_n^{(o)}(r, \theta, t) = \sum_{k=1}^{\infty} A_n^{(k)}(t)\ r^{-k} \int_{-1}^{\mu} P_k(\lambda)\ d\lambda \tag{3.3}$$

where $P_n(\mu)$ is the n-th degree Legendre polynomial with $\mu = \cos\theta$. Imposing the axisymmetry condition (for any n) yields;

$$\psi_n^{(o)}(r, 0, t) = \psi_n^{(o)}(r, \pi, t) = 0 \tag{3.4}$$

In order to define the inner expansion (superscript "i") we introduce the stretched normal coordinate R:

$$R = (r - 1)/\epsilon \tag{3.5}$$

The inner asymptotic expansion is then defined by a power series in ϵ:

$$T^{(i)} = \sum_{n=1}^{\infty} \epsilon^{n-1} T_n^{(i)}(R, \theta, t) \ ; \ \psi^{(i)} = \sum_{n=1}^{\infty} \epsilon^n \psi_n^{(i)}(R, \theta, t) \tag{3.6}$$

To the leading- order we get no effects of convection and boundary curvature from the heat equation (2.2):

$$\frac{\partial T_1^{(i)}}{\partial t} - \alpha^* \frac{\partial^2 T_1^{(i)}}{\partial R^2} = 0 \tag{3.7}$$

The solution can be written as a complementary error function:

$$T_1^{(i)} = \operatorname{erfc}(\eta) \tag{3.8}$$

where the boundary-layer similarity coordinate η is defined by;

$$\eta = \frac{(r-1)\sqrt{\sigma}}{2\sqrt{(t'/t_{\text{diff}})}} = (R/2)\ (\alpha^* t)^{-1/2}. \tag{3.9}$$

Here σ is the ratio of heat capacity in the saturated porous medium to that of the fluid (see PIC). The leading-order inner stream function is governed by the equation:

$$\frac{\partial^2 \psi_1^{(i)}}{\partial R^2} = \sin^2\theta \, \frac{\partial T_1^{(i)}}{\partial R} \tag{3.10}$$

The solution for the inner stream function satisfying (2.7) is:

$$\psi_1^{(i)} = 2(\alpha^* t)^{1/2}(\eta \text{ erfc } (\eta) - \pi^{-1/2}e^{-\eta^2} + \pi^{-1/2}) \sin^2\theta \tag{3.11}$$

The matching condition is:

$$\lim_{R\to\infty} \psi_n^{(i)} = \lim_{(r-1)\to 0} \psi_n^{(o)} \tag{3.12}$$

Matching with $n = 1$ produces the amplitude of the leading-order outer stream function (where δ_{jk} is the Kronecker delta):

$$A_1^{(k)} = -4(\alpha^* t/\pi)^{1/2}\delta_{1k} \tag{3.13}$$

This prescribes a pure dipole flow;

$$\psi_1^{(o)} = 2(\alpha^* t/\pi)^{1/2} r^{-1} \sin^2\theta \tag{3.14}$$

Note the square in the angular dependence, as compared with the first power in the corresponding two-dimensional problem (PIC). The discrepency in the buoyancy term (see PIC eq.1) causes an opposite sign in PIC compared with our eqs (3.11) and (3.14). Note that our sign in eq. (3.14) implies downward outer flow along $\theta = 0$, which is necessary in order to compensate for the upward inner flow near the wall.

4 Second-Order Solution

The second-order solution takes into account the leading-order effects of the boundary-layer curvature and the convective heat transfer. From the heat equation (2.2) we get for the inner temperature field:

$$\begin{aligned} \frac{\partial T_2^{(i)}}{\partial t} \quad &- \quad \alpha^* \frac{\partial^2 T_2^{(i)}}{\partial R^2} = 2\,\alpha^* \frac{\partial T_1^{(i)}}{\partial R} - \frac{1}{\sin\theta}\,\frac{\partial T_1^{(i)}}{\partial R}\,\frac{\partial \psi^{(i)}}{\partial \theta} \\ &+ \quad \frac{1}{\sin\theta}\,\frac{\partial T_1^{(i)}}{\partial \theta}\,\frac{\partial \psi^{(i)}}{\partial R} = -2\,\pi^{-1/2}\,(\alpha^*/t)^{1/2}\,e^{-\eta^2} \\ &+ \quad 4(\pi^{-1/2}\,\eta\,e^{-\eta^2}\,\text{erfc}(\eta) - \pi^{-1}\,e^{-2\eta^2} + \pi^{-1}\,e^{-\eta^2})\cos\,\theta \end{aligned} \tag{4.1}$$

The second-order inner temperature field satisfying condition (2.7) is given then by:

$$T_2^{(i)} = -2\ (\alpha^* t)^{1/2}\ \eta\ \mathrm{erfc}(\eta) + 2\ t f_0(\eta)\ \cos\ \theta \tag{4.2}$$

The first term here is due to the leading curvature effect in the thermal boundary layer. It slows down the outward thermal diffusion, as the minus sign shows and is twice as strong in 3D as in 2D (PIC eq.24a). The second term is due to the leading-order convective heat transfer. It is represented by the function $f_0(\eta)$, formally identical to that introduced by PIC (but note that the similarity coordinate η has a different meaning in three dimensions);

$$\begin{aligned} f_0(\eta) &= (1/2+\eta^2)\ \mathrm{erfc}^2(\eta) - \pi^{-1/2}\ \eta\ e^{-\eta^2}\mathrm{erfc}(\eta) + 4/(3\pi)e^{-\eta^2} \\ &- [1/2+4/(3\pi)][(1+2\eta^2)\mathrm{erfc}(\eta) - 2\ \pi^{-1/2}\eta\ e^{-\eta^2}] \end{aligned} \tag{4.3}$$

This leading-order convective effect on the temperature field tends to reduce the distance between the isotherms along the lower hemisphere, and increase their mutual distance along the upper hemisphere according to eq.(4.2). The leading convective effect of the heated fluid along the sphere will cause an inward flow along the lower hemisphere and an outward flow along the upper hemisphere. The sign discrepancy found in eq. (2.1) vs. PIC (eq.1) also persists when our eq. (4.2) is compared with PIC eq. (24a).

Let us keep the spatial coordinate fixed and pick out the asymptotically leading terms as $t \to 0$ (with R fixed, so that $\eta \to \infty$):

$$f_0(\eta) \to 4/(3\pi)\ e^{-\eta^2}, \qquad \text{as } \eta \to \infty \tag{4.4}$$

This is an example of the type of asymptotic approximation that will be performed in the third-order analysis. It is not easy to see immediately from eq.(4.3) that this term is indeed the asymptotically dominating one. This becomes clear from eq.(4.1), where it represents the last term in the r.h.s. inhomogeneity. The point of making this approximation (4.4) of the exact solution (4.3), is to indicate how good our approximate third-order solution will be. This is important because the exact third-order solution is too complex to be practicable. In Table 1 we investigate the relative value of the asymptotic expression (4.4) compared with the exact formula (4.3). An obvious disadvantage with the asymptotic approximation (4.4) is that it cannot satisfy the boundary condition at the cylinder, since it is a far-field approximation for fixed time.

The exact second-order inner stream function is governed by:

$$\frac{\partial^2 \psi_2^{(i)}}{\partial R^2} = \sin^2\theta \left(\frac{\partial T_2^{(i)}}{\partial R} + R\,\frac{\partial T_1^{(i)}}{\partial R} \right) + \sin\theta\ \cos\theta\ \frac{\partial T_1^{(i)}}{\partial \theta} \tag{4.5}$$

Since, by virtue of (3.8), the last term in (4.5) vanishes, the solution can be written as:

$$\begin{aligned} \psi_2^{(i)} &= (\alpha^* t)[2\pi^{-1/2}\ \eta\ e^{-\eta^2} - \mathrm{erfc}(\eta) - 2\ \eta^2 \mathrm{erfc}(\eta) + 1]\sin^2\theta \\ &+\ 4(\alpha^* t)^{1/2}\ t\ f_1(\eta)\sin^2\theta\ \cos\theta \end{aligned} \tag{4.6}$$

Our function $f_1(\eta)$ is essentially different from the corresponding function in the 2D case (PIC eq.25b), although it contains some similar terms.

$$\begin{aligned} f_1(\eta) &= [(8/9)\ \pi^{-3/2} + (1/3)\pi^{-1/2}](1+\eta^2)\ e^{-\eta^2} + (1/3)\pi^{-1}\ \eta\ e^{-2\eta^2} \\ &-\ (2/3)\ \pi^{-1/2}\ \ \mathrm{erfc}(\eta) - (2/3)\ (2/\pi)^{1/2}\ [1 - \mathrm{erfc}(\sqrt{2}\eta)] \\ &-\ (7/6)\pi^{-1/2}\ e^{-\eta^2}\ \mathrm{erfc}(\eta) - [1/2 + (4/3)\pi^{-1}]\ \eta\ \mathrm{erfc}(\eta) \\ &-\ (2/3)\pi^{-1/2}\ \eta^2\ e^{-\eta^2}\ \mathrm{erfc}(\eta) - [1/3 + (8/9)\pi^{-1}]\ \eta^3\ \mathrm{erfc}(\eta) \\ &+\ (1/2)\ \eta\mathrm{erfc}^2(\eta) + (1/3)\ \eta^3\mathrm{erfc}^2(\eta) - (8/9)\pi^{-3/2} + (3/2)\pi^{-1/2} \end{aligned} \tag{4.7}$$

The matching condition (3.12) with $n = 2$ is next applied to the second-order outer stream function given by eq.(3.3). This determines the relevant amplitudes in (3.3):

$$A_2^{(1)}(t) = -2\alpha^* t\ \ ;\ \ A_2^{(2)}(t) = -8(\alpha^* t)^{1/2} t\ f_1(\infty) \tag{4.8}$$

Thus, the second-order outer stream function is obtained:

$$\begin{aligned} \psi_2^{(o)} &= \alpha^* t r^{-1}\ \sin^2\theta \\ &-\ 4\pi^{-1/2}(-3/2 + 2\sqrt{2}/3 + 8\pi^{-1}/9)(\alpha^* t)^{1/2}\ t r^{-2}\ \sin^2\theta\ \cos\theta \end{aligned} \tag{4.9}$$

The first term is a symmetrical dipole flow around the top of the sphere which adds to the amplitude of the first-order flow. The second quadrupole term makes the outer flow field asymmetric with a concentration of streamlines near the top of the sphere.

5 Approximate Third-Order Solution

The third-order inner temperature field is determined by:

$$\frac{\partial T_3^{(i)}}{\partial t} - \alpha^* \frac{\partial^2 T_3^{(i)}}{\partial R^2} = 2\,\alpha^* \left(\frac{\partial T_2^{(i)}}{\partial R} - R\,\frac{\partial T_1^{(i)}}{\partial R} \right) \tag{5.1}$$

$$+\frac{1}{\sin\theta}\left(\frac{\partial T_2^{(i)}}{\partial \theta}\,\frac{\partial \psi_1^{(i)}}{\partial R} - \frac{\partial T_2^{(i)}}{\partial R}\,\frac{\partial \psi_1^{(i)}}{\partial \theta} - \frac{\partial T_1^{(i)}}{\partial R}\,\frac{\partial \psi_2^{(i)}}{\partial \theta} \right) + \frac{2\,R}{\sin\theta}\,\frac{\partial T_1^{(i)}}{\partial R}\,\frac{\partial \psi_1^{(i)}}{\partial \theta}$$

We have omitted here terms containing the angular derivative of the first-order temperature (identically zero). In the Appendix the full r.h.s. of this equation is given. We will not attempt to solve it generally, but we want to include all different types of θ dependence. For each different sine or cosine term in θ, we include only the asymptotically leading contributions as $t \to 0$ and $\theta \to \infty$. This gives the simplified third-order heat equation:

$$\begin{aligned}
\frac{\partial T_3^{(i)}}{\partial T} \quad - \quad & \alpha^* \frac{\partial^2 T_3^{(i)}}{\partial R^2} = 8\,\pi^{-1/2}\alpha^*\;\eta\;\exp(-\eta^2) - 2\,\alpha^*\;\mathrm{erfc}(\eta) \\
- \quad & (88/3)\;\pi^{-1}\;\sqrt{(\alpha^* t)}\;\eta\;\exp(-\eta^2)\;\cos\theta \\
+ \quad & (32/3)\pi^{-3/2}\;t\;\eta\;\exp(-\eta^2)\;\cos^2\theta \\
+ \quad & [(32/9)\pi^{-2} - 6\;\pi^{-1} + (2^{7/2}/3)\pi^{-1}]\;t\;\exp(-\eta^2)\sin^2\theta
\end{aligned} \tag{5.2}$$

The asymptotically dominating contribution is the first one, which is independent of θ, and where no approximation has been made. In the terms with angular dependence we have taken only the asymptotically leading term. These angular dependent terms are ranked according to their asymptotic order of magnitude. We have here neglected several terms larger than terms retained but with different angular dependence.

$$\begin{aligned}
T_3^{(i)} \;=\; & 4\;\alpha^*\;t\eta^2\;\exp(-\eta^2) - (176/15)\;\pi^{-1}\sqrt{\alpha^*}\;t^{3/2}\;\eta\;\exp(-\eta^2)\cos\theta \\
+ \; & (32/9)\;\pi^{-3/2}\;t^2\eta\;\exp(-\eta^2)\;\cos^2\theta \\
+ \; & [(64/45)\;\pi^{-2} - (12/5)\pi^{-2} + (2^{9/2}/15)\;\pi^{-1}]\;t^2\;\;exp(-\eta^2)\;\sin^2\theta
\end{aligned} \tag{5.3}$$

This formula gives the full third-order temperature field independent of the angle, plus the leading far-field contributions for each different angular dependence. Alternatively, we may consider these terms as the asymptotically dominating ones at a fixed radial position, as time goes to zero. Since we have included the full inhomogeneity independent of θ, we are here able to satisfy the boundary condition (2.7) at the sphere, only as far as the terms independent of the angle are concerned. The angle dependent terms cannot be expected to satisfy any boundary condition except for the far-field condition. Thus, only an approximate evaluation of the third -order Nusselt number is possible by using this simplified theory.

The third-order inner stream function is governed by the equation:

$$\begin{aligned}\frac{\partial^2 \psi_3^{(i)}}{\partial R^2} &= \sin^2\,\theta\left(\frac{\partial T_3^{(i)}}{\partial R} + R\,\frac{\partial T_2^{(i)}}{\partial R}\right) + \sin\theta\,\cos\theta\,\frac{\partial T_2^{(i)}}{\partial \theta}\\ &- \sin\theta\frac{\partial}{\partial\theta}\left(\frac{1}{\sin\theta}\,\frac{\partial \psi_i^{(i)}}{\partial\theta}\right)\end{aligned} \tag{5.4}$$

Since our knowledge of the third-order temperature field is not sufficient, we will refrain here from solving this equation. PIC claim in their eqs. (27a-b) to have found the full third-order inner solution. This can not be correct, since the angular dependence is over simplified compared with our solution (5.2) and some double angle trigonometric functions should be present. There is also a more complicated radial dependence than what they have accounted for which arise from the error- function term in their eq. (25b).

6 Composite Second-Order Solution and Nusselt Number

Due to space limitations, the complete derivation of the third-order solution will be presented elsewhere. Here we choose to present the second-order composite solutions for both the temperature and flow fields, which are given by the sum of the inner and outer solutions minus their common part (e.g., Van Dyke 1964). Thus, combining (3.11), (3.14), (4.6) and (4.9) render the following small-time second-order composite solution for the stream-function;

$$\begin{aligned}\psi_2^{(c)} &= 2\left(\frac{t^*}{Ra}\right)^{1/2}\left[\eta\,\mathrm{erfc}(\eta) - \pi^{-1/2}e^{-\eta^2} + \pi^{-1/2}r^{-1}\right]\sin^2\theta\\ &+ \left(\frac{t^*}{Ra}\right)\left[2\pi^{-1/2}\eta e^{-\eta^2} - (1+2\eta^2)\,\mathrm{erfc}(\eta) + r^{-1}\right]\sin^2\theta\\ &+ 4t^*\left(\frac{t^*}{Ra}\right)^{1/2}\left[f_1(\eta) + \pi^{-1/2}\left(1-r^{-2}\right)\left(-3/2 + 2\sqrt{2}/3 + 8\pi^{-1}/9\right)\right]\end{aligned} \tag{6.1}$$

$\sin^2\theta\cos\theta + \cdots$

where $t^* = \epsilon t$ and $f_1(\eta)$ is defined by (4.7).

Similarly, the second-order composite small-time expansion for the temperature field is given by,

$$T_2^{(c)} = \text{erfc}(\eta) - 2\left(\frac{t^*}{Ra}\right)^{1/2} \eta\, \text{erfc}(\eta) + 2t^* f_0(\eta)\cos\theta + \cdots \qquad (6.2)$$

To this order of approximation, the local Nusselt number is

$$Nu = -\left.\frac{\partial T_2^{(c)}}{\partial r}\right|_{r=1} = \left(\frac{Ra}{\pi t^*}\right)^{1/2} + 1 + \left(\frac{Rat^*}{\pi}\right)^{1/2}\left(1 - \frac{16}{3\pi}\right)\cos\theta + \cdots \qquad (6.3)$$

Next, we present an expansion for the local Nusselt number, which is asymptotically correct to third- order by combining (3.8), (4.2) and (5.3).

$$\begin{aligned} Nu &= \left(\frac{Ra}{\pi t^*}\right)^{1/2} + 1 + \left(\frac{Rat^*}{\pi}\right)^{1/2}\left(1 - \frac{16}{3\pi}\right)\cos\theta + \\ &+ \frac{88}{15\pi}t^*\cos\theta - \frac{16}{9Ra}\left(\frac{Rat^*}{\pi}\right)^{3/2}\cos^2\theta + \end{aligned} \qquad (6.4)$$

Finally, the average Nusselt number $\overline{Nu}$ over the isothermal surface of the sphere is computed as:

$$\overline{Nu} = \frac{1}{2}\int_0^{\pi} Nu(\theta)\sin\theta d\theta = \left(\frac{Ra}{\pi t^*}\right)^{1/2} + 1 - \frac{16}{27Ra}\left(\frac{Rat^*}{\pi}\right)^{3/2} + \cdots \qquad (6.5)$$

Comparing (6.5) with the corresponding average Nusselt number for a 2D cylinder (i.e. PIC eq. 37) show that the two expressions are identical only to the leading order term proportional to $(t^*)^{-1/2}$, that is the first term in the the r.h.s of (6.5). However, they differ in the higher order terms in the dimensionless temperature t^*.. Due to space limitation the detailed solutions,correct to third-order, for both the temperature field and the stream function for various Rayleigh numbers, will be presented in the full version of this paper to appear elsewhere together with discussions of some pertinent physical aspects of the resulting convective flows.

Appendix: The Exact Third-Order Heat Equation

In the main text we work only with the asymptotically leading contributions to the r.h.s. of eq.(5.1) as $t \to 0$, i.e. as $\eta \to \infty$. The exact version of the third-order heat equation (5.1) is:

$$
\begin{aligned}
\frac{\partial T_3^{(i)}}{\partial t} \quad - \quad \alpha^* \frac{\partial^2 T_3^{(i)}}{\partial R^2} = \\
= \quad \sum_{m=0}^{3}\sum_{n=0}^{3}\sum_{p=0}^{3}\sum_{q=0}^{3} \eta^m \ \exp(-n\ \eta^2)\ \mathrm{erfc}^p(\eta) t^{q/2} (A_{mnpq}\ \cos^q\theta + B_{mnpq}\ \sin^q\theta) \\
+ \quad 2^{9/2}/(3\pi)\ \exp(-\eta^2)\ \mathrm{erfc}(\eta\sqrt{2}) t\ \cos^2\theta \\
- \quad 2^{7/2}/(3\pi)\ \exp(-\eta^2)\ \mathrm{erfc}(\eta\sqrt{2}) t\ \sin^2\theta
\end{aligned}
$$

The nonzero elements of the matrices A_{mnpq} and B_{mnpq} are:

$$
\begin{aligned}
A_{1100} &= 8\ \pi^{-1/2}\alpha^* \\
A_{0101} &= (32/3)\pi^{-3/2}\sqrt{\alpha^*} + 6\pi^{-1}\sqrt{\alpha*} \\
A_{1201} &= 32\ \pi^{-1}\sqrt{\alpha*} \\
A_{1101} &= -\ (88/3)\pi^{-1}\sqrt{\alpha*} \\
A_{0202} &= (256/9)\ \pi^{-2} + (32/3)\ \pi^{-1} \\
A_{2202} &= (64/9)\pi^{-2} + (8/3)\ \pi^{-1} \\
A_{1302} &= A_{1102} = (32/3)\ \pi^{-3/2} \\
A_{1202} &= -\ (56/3)\ \pi^{-3/2} \\
A_{0102} &= 4\ \pi^{-1} - (2^{9/2}/3)\ \pi^{-1} - (256/9)\ \pi^{-2} \\
A_{0010} &= -\ 2\ \alpha^* \\
A_{1011} &= -\ 4\ \sqrt{\alpha^*} - (32/3)\ \pi^{-1}\ \sqrt{\alpha^*} \\
A_{0011} &= 4\ \pi^{-1}\ \sqrt{\alpha^*} \\
A_{0111} &= -\ 12\ \pi^{-1/2}\ \sqrt{\alpha^*} \\
A_{2111} &= -\ 32\ \pi^{-1/2}\ \sqrt{\alpha^*} \\
A_{1012} &= (64/3)\ \pi^{-3/2} + 8\ \pi^{-1/2} \\
A_{1112} &= -\ (160/3)\ \pi^{-3/2} - 20\pi^{-1/2} \\
A_{3112} &= -\ (64/9)\ \pi^{3/2} - (8/3)\ \pi^{-1/2} \\
A_{0212} &= A_{2212} = -(64/3)\ \pi^{-1}
\end{aligned}
$$

$$
\begin{aligned}
A_{0112} &= (20/3)\ \pi^{-1} \\
A_{2112} &= (56/3)\ \pi^{-1} \\
A_{1021} &= 8\sqrt{\alpha *} \\
A_{2022} &= 8 + (64/3)\ \pi^{-1} \\
A_{1022} &= -\ 8\ \pi^{-1/2} \\
A_{1122} &= 24\ \pi^{-1/2} \\
A_{3122} &= (32/3)\ \pi^{1/2} \\
A_{2032} &= -\ 8 \\
B_{0102} &= (32/9)\ \pi^{-2} - 6\ \pi^{-1} + (2^{7/2}/3)\ \pi^{-1} \\
B_{2202} &= -\ (32/9)\ \pi^{-2} - (4/3)\ \pi^{-1} \\
B_{1302} &= -\ (4/3)\pi^{-3/2} \\
B_{0202} &= -\ (32/9)\ \pi^{-2} - (4/3)\ \pi^{-1} \\
B_{3112} &= (32/9)\pi^{-3/2} + (4/3)\ \pi^{-1/2} \\
B_{0212} &= (14/3)\ \pi^{-1} \\
B_{2212} &= (8/3)\ \pi^{-1} \\
B_{0022} &= 1 + (8/3)\ \pi^{-1} \\
B_{2022} &= 2 + (16/3)\ \pi^{-1} \\
B_{3122} &= -\ (4/3)\ \pi^{-1/2} \\
B_{0032} &= -\ 1 \\
B_{2032} &= -\ 2
\end{aligned}
$$

We have here omitted terms containing the angular derivative of the first-order temperature (identically zero).

References

Elliott, L. (1970) Free convection on a two-dimensional or axisymmetric body. Q. J. Mech. Appl. Math. 23, 153-162.

Foster, T.D. (1965) Stability of a homogeneous fluid cooled uniformly from above. Phys. Fluids 8, 1249-1257.

Gupta, A.S. & I. Pop (1977) Effects of curvature on unsteady free convection past a circular cylinder. Phys. Fluids 20, 162-163.

Katto, Y. & T. Masuoka (1967) Criterion for the onset of convective flow in a fluid in a porous medium. Int. J. Heat Mass Transfer 10, 297-309.

Pop, I., D.B. Ingham & P. Cheng (1993) Transient free convection about a horizontal circular cylinder in a porous medium. Fluid Dyn. Res. 12, 295-305 (Referred to in the text as "PIC").

Sano, T. & K. Kuribayashi (1992) Transient natural convection around a horizontal circular cylinder. Fluid Dyn. Res. 10, 25-37.

Sano, T. & R. Okihara (1994) Natural convection around a sphere immersed in a porous medium at small Rayleigh numbers. Fluid Dyn. Res. 13, 39-44.

Van Dyke, M. (1964) Perturbation Methods in Fluid Mechanics, Academic Press.

Yan, B.,I. Pop & D.B. Ingham (1997) A numerical study of unsteady free convection from a sphere in a porous medium. Int. J. Heat & Mass Transf. 40, 893-903.

TABLE 1: The ratio between the asymptotic formula (4.4) for the function $f_0(\eta)$ and its exact formula (4.3), for increasing values of η.

η	2	3	5	10
ratio	1.0999	1.0358	1.0089	1.0019

T. MILOH
[1] Department of Fluid Mechanics and Heat Transfer
Tel Aviv University, Ramat Aviv, Tel Aviv 69978, Israel

P. A. TYVAND
Department of Agricultural Engineering,
Agricultural University of Norway, Box 5065 1432 AS, Norway

ZDISŁAW NANIEWICZ[1] AND PANAGIOTIS D PANAGIOTOPOULOS[2]

Optimality conditions for some nonconvex problem related to liquid crystals with variable degree of orientation

1 Introduction

In the static theory of nematic liquid crystals formulated by Ericksen in [5] the problem is to find the stable equilibrium configuration of a nematic liquid crystal characterized by the orientation vector $\mathbf{n} : \Omega \to \mathbf{R}^3$ with $|\mathbf{n}| = 1$ and the degree of orientation $s : \Omega \to [-\frac{1}{2}, 1]$. In the simplest case of the, "so called", one-constant approximation, the energy functional can be written as

$$(\mathbf{n}, s) \longrightarrow \int\limits_{\Omega} \left[(k|\nabla s|^2 + s^2|\nabla \mathbf{n}|^2)\right] d\Omega + \int\limits_{\Omega} \sigma(s) d\Omega,$$

where $\nabla = (\partial_j)$ is the gradient operator, $k0$ is a positive constant and $\sigma : [-\frac{1}{2}, 1] \to \mathbf{R} \cup \{+\infty\}$ is De Gennes potential. The mathematical study of the corresponding optimization problem concerning the existence of solutions, the necessary conditions and the regularity of minimizers started in [1] – [4], [6]. See also [8] for the application of hemivariational inequalities in the theory of liquid crystals.

In this paper we consider more general functional involving lower order term depending on $\mathbf{n}$ and s, namely

$$(\mathbf{n}, s) \longrightarrow \int\limits_{\Omega} \left[k|\nabla s|^2 + s^2|\nabla \mathbf{n}|^2\right] d\Omega + \int\limits_{\Omega} \widetilde{h}(\mathbf{n}, s) d\Omega,$$

where $\widetilde{h} : \mathbf{R}^3 \times \mathbf{R} \to \mathbf{R}$ is a locally Lipschitz function in $\mathbf{R}^3 \times \mathbf{R} \setminus \{(\mathbf{0}, 0)\}$. Since the term

$$(\mathbf{n}, s) \longrightarrow \int\limits_{\Omega} \left[k|\nabla s|^2 + s^2|\nabla \mathbf{n}|^2\right] d\Omega$$

is not coercive in $H^1(\Omega; \mathbf{R}^3) \times H^1(\Omega)$, it will be more convenient to introduce a new variable $\mathbf{u} = s\mathbf{n}$ which due to the constraint $|\mathbf{u}| = |s|$ yields

$$|\nabla \mathbf{u}|^2 = |\nabla s|^2 + s^2|\nabla \mathbf{n}|^2.$$

---

[1]Warsaw University, Institute of Applied Mathematics and Mechanics, Banacha 2, 02-097, Warsaw, Poland, E-mail: znaniew@mimuw.edu.pl

[2]Aristotle University of Thessaloniki, Department of Civil Engineering, GR-54006 Thessaloniki, Greece and Faculty of Mathematics and Physics, RWTH, Aachen, Germany, E-mail: pdpana@heron.civil.auth.gr

Therefore we are led to study the constrained optimization problem for a nonconvex functional of the form

$$\mathcal{F}(\mathbf{u}, s) = \int_{\Omega} \left[|\nabla \mathbf{u}|^2 + (k-1)|\nabla s|^2\right] d\Omega + \int_{\Omega} h(\mathbf{u}, s) d\Omega, \tag{1.1}$$

with the nonconvex constraint

$$|\mathbf{u}| = |s| \quad \text{a.e. in } \Omega, \tag{1.2}$$

and the convex one

$$s \in [-\frac{1}{2}, 1] \quad \text{a.e. in } \Omega. \tag{1.3}$$

The lower order term expressed by $h(\cdot,\cdot)$ will be assumed to satisfy some additional regularity properties specified later.

The aim of this paper is to show that (1.1) possesses at least one minimizer and to derive the necessary optimality conditions which due to the nonsmoothness of h are expressed by variational-hemivariational inequalities. Our approach relies on the full characterization and the calculation of the contingent cone $T_{\mathcal{A}}(\cdot,\cdot)$ to the admissible set $\mathcal{A} = \{(\mathbf{x}, r) \in \mathbf{R}^3 \times \mathbf{R} \colon |\mathbf{x}| = |r|\}$ which is the main difference compared to the method presented in [3].

2 Existence of minimizers

Throughout the paper the inner product and the tensor product of any two vectors $\mathbf{u}$ and $\mathbf{v}$ will be denoted by $\mathbf{u}\cdot\mathbf{v}$ and $\mathbf{u}\otimes\mathbf{v}$, respectively. For any tensors $\mathbf{N}$ and $\mathbf{M}$ we shall use the symbols: $\mathbf{N}^T$ – the transpose of $\mathbf{N}$, $\operatorname{tr}\mathbf{N}$ – the trace of $\mathbf{N}$, $\mathbf{N}\cdot\mathbf{M} := \operatorname{tr}\mathbf{N}^T\mathbf{M}$ – the inner product of $\mathbf{N}$ and $\mathbf{M}$ and $|\mathbf{M}| := \sqrt{\mathbf{M}\cdot\mathbf{M}}$ – the magnitude of $\mathbf{M}$.

In order to specify the space for minimizers of $\mathcal{F}$ we define

$$\mathcal{A} = \{(\mathbf{u}, s) \in H^1(\Omega; \mathbf{R}^3) \times H^1(\Omega) \colon |\mathbf{u}| = |s| \text{ a.e. in } \Omega\} \tag{2.1}$$

and

$$\mathcal{B} = \left\{(\mathbf{u}, s) \in H^1(\Omega; \mathbf{R}^3) \times H^1(\Omega) \colon s \in [-\frac{1}{2}, 1]\right\}. \tag{2.2}$$

The intersection $\mathcal{A} \cap \mathcal{B}$ will be referred to as the admissible set. Thus our problem now reads

$$\inf_{(\mathbf{v},r)\in\mathcal{A}\cap\mathcal{B}} \{\mathcal{F}(\mathbf{v}, r)\}. \tag{2.3}$$

The following properties of $\mathcal{A}$ and $\mathcal{B}$ can be pointed out.

Proposition 2.1 *For* $(\mathbf{u}, s) \in \mathcal{A}$ *define* $\mathbf{n} : \Omega \to \mathbf{S}^2 \cup \{\mathbf{0}\}$, $\mathbf{S}^2 = \{y \in \mathbf{R}^3 : |y| = 1\}$ *being a unit sphere in* $\mathbf{R}^3$, *by*

$$\mathbf{n} = \begin{cases} \frac{\mathbf{u}}{s} & \text{in } \Omega \setminus \mathcal{S}(\mathbf{u}) \\ \mathbf{0} & \text{in } \mathcal{S}(\mathbf{u}), \end{cases} \tag{2.4}$$

where $\mathcal{S}(\mathbf{u}) = \{x \in \Omega : \mathbf{u}(x) = \mathbf{0}\}$. *Then*

(i) $\mathcal{A}$ *is weakly closed in* $H^1(\Omega; \mathbf{R}^3) \times H^1(\Omega)$;

(ii) *For each* $(\mathbf{u}, s) \in \mathcal{A}$ *there exists* $\mathbf{N} \in L^2(\Omega; \mathbf{R}^{3\times 3})$ *such that*

$$\begin{aligned} &\nabla \mathbf{u} = \mathbf{n} \otimes \nabla s + \mathbf{N} \\ &\mathbf{N}^T \mathbf{n} = \mathbf{0} \\ &|\nabla \mathbf{u}|^2 = |\nabla s|^2 + |\mathbf{N}|^2. \end{aligned} \tag{2.5}$$

Proof. (i) To show that $\mathcal{A}$ is weakly closed choose a sequence $(\mathbf{u}_n, s_n) \in \mathcal{A}$ weakly converging to $(\mathbf{u}, s)$ in $H^1(\Omega; \mathbf{R}^3) \times H^1(\Omega)$. By the compact imbedding $H^1(\Omega) \hookrightarrow L^2(\Omega)$ the strong convergence of $(\mathbf{u}_n, s_n)$ to $(\mathbf{u}, s)$ in $L^2(\Omega; \mathbf{R}^3) \times L^2(\Omega)$ follows and without loss of generality we may suppose that $(\mathbf{u}_n, s_n) \to (\mathbf{u}, s)$ a.e. in Ω (passing to a subsequence if necessary). Since $(\mathbf{u}_n, s_n) \in \mathcal{A}$, so $|\mathbf{u}_n| = |s_n|$. The almost everywhere convergence clearly forces the equality $|\mathbf{u}| = |s|$ a.e. in Ω which means that $(\mathbf{u}, s) \in \mathcal{A}$, as desired.

(ii) We decompose the proof of this fact into two steps.
Step 1: For $\mathbf{u} \in H^1(\Omega; \mathbf{R}^3)$ define $\mathbf{m} : \Omega \to \mathbf{S}^2 \cup \{\mathbf{0}\}$ by

$$\mathbf{m} = \begin{cases} \frac{\mathbf{u}}{|\mathbf{u}|} & \text{in } \Omega \setminus \mathcal{S}(\mathbf{u}) \\ \mathbf{0} & \text{in } \mathcal{S}(\mathbf{u}). \end{cases} \tag{2.6}$$

It will be shown now that there exists $\mathbf{N} \in L^2(\Omega; \mathbf{R}^{3\times 3})$ such that

$$\begin{aligned} &\text{(a)} \quad \nabla \mathbf{u} = \mathbf{m} \otimes \nabla(|\mathbf{u}|) + \mathbf{N} \\ &\text{(b)} \quad \mathbf{N}^T \mathbf{m} = \mathbf{0} \\ &\text{(c)} \quad |\nabla \mathbf{u}|^2 = |\nabla(|\mathbf{u}|)|^2 + |\mathbf{N}|^2. \end{aligned} \tag{2.7}$$

We check that a tensor field $\mathbf{N}$ given by the formula

$$\mathbf{N} := \nabla \mathbf{u} - \mathbf{m} \otimes \nabla(|\mathbf{u}|) \tag{2.8}$$

satisfies all the requirements (a) – (c). Easy calculations show that $\mathbf{N} \in L^2(\Omega; \mathbf{R}^{3\times 3})$. Choose a smooth sequence $\mathbf{u}_n \in C^\infty(\overline{\Omega}; \mathbf{R}^3)$ with $\mathbf{u}_n \to \mathbf{u}$ in $H^1(\Omega; \mathbf{R}^3)$ and set $v_n := \sqrt{\sum_i |u_i|^2 + \frac{1}{n}}$. Next define a sequence $(\mathbf{m}_n)$ by means of the relation

$$\mathbf{m}_n := \frac{\mathbf{u}_n}{v_n}. \tag{2.9}$$

The sequence $(\mathbf{m}_n)$ is bounded in $L^\infty(\Omega; \mathbf{R}^3)$, more precisely,

$$|\mathbf{m}_n| \leq 1. \tag{2.10}$$

By the strong convergence in H^1-spaces of $(\mathbf{u}_n)$ and (v_n) we can suppose (by passing to a subsequence if necessary) that $\mathbf{u}_n \to \mathbf{u}$ and $v_n \to |\mathbf{u}|$ a.e. in Ω whereas $\mathbf{m}_n \to \mathbf{m}$ a.e. in $\Omega \setminus \mathcal{S}(\mathbf{u})$. Using the smoothness of $(\mathbf{u}_n)$ one obtains

$$\nabla \mathbf{u}_n = \mathbf{m}_n \otimes \nabla v_n + v_n \nabla \mathbf{m}_n. \tag{2.11}$$

Since $\nabla v_n \to \nabla(|\mathbf{u}|)$ a.e. in Ω, $\mathbf{m}_n \to \mathbf{m}$ a.e. in $\Omega \setminus \mathcal{S}(\mathbf{u})$, $\nabla(|\mathbf{u}|) = \mathbf{0}$ a.e. in $\mathcal{S}(\mathbf{u})$ and (2.10) holds, we obtain immediately that

$$\mathbf{m}_n \otimes \nabla v_n \to \mathbf{m} \otimes \nabla(|\mathbf{u}|) \quad \text{a.e. in} \quad \Omega \tag{2.12}$$

which by the dominated convergence yields

$$\mathbf{m}_n \otimes \nabla v_n \to \mathbf{m} \otimes \nabla(|\mathbf{u}|) \;\text{ in }\; L^2(\Omega; \mathbf{R}^{3\times 3}). \tag{2.13}$$

Thus, by (2.8) and (2.12), from (2.11) it follows

$$v_n \nabla \mathbf{m}_n \to \mathbf{N} \;\text{ in }\; L^2(\Omega; \mathbf{R}^{3\times 3}) \tag{2.14}$$

and (c) holds. Analogously we obtain

$$v_n \nabla \mathbf{m}_n^T \mathbf{m}_n \to \mathbf{N}^T \mathbf{m} \;\text{ in }\; L^2(\Omega; \mathbf{R}^3). \tag{2.15}$$

In order to show that $\mathbf{N}^T \mathbf{m} = \mathbf{0}$ we establish first that

$$v_n \nabla \mathbf{m}_n^T \mathbf{m}_n \to \mathbf{0} \;\text{ in }\; L^2(\Omega; \mathbf{R}^3).$$

For this purpose the density of $C_0^\infty(\Omega; \mathbf{R}^3)$ in $L^2(\Omega; \mathbf{R}^3)$ will be used. Let $\varphi \in C_0^\infty(\Omega; \mathbf{R}^3)$ be arbitrarily given. Consider the following sequence

$$\int_\Omega v_n \nabla \mathbf{m}_n^T \mathbf{m}_n \cdot \varphi d\Omega.$$

By recalling that $\frac{1}{2}\nabla(|\mathbf{m}_n|^2) = \nabla \mathbf{m}_n^T \mathbf{m}_n$ we have

$$\begin{aligned} \int_\Omega v_n \nabla \mathbf{m}_n^T \mathbf{m}_n \cdot \varphi d\Omega &= \tfrac{1}{2} \int_\Omega v_n \textstyle\sum_j^3 \partial_j(|\mathbf{m}_n|^2) \varphi_j d\Omega \\ &= -\tfrac{1}{2} \int_\Omega |\mathbf{m}_n|^2 \textstyle\sum_j^3 \partial_j(v_n \varphi_j) d\Omega \\ &\to -\tfrac{1}{2} \int_\Omega \textstyle\sum_j^3 \partial_j(|\mathbf{u}|\varphi_j) d\Omega = 0 \quad \text{as } n \to \infty. \end{aligned}$$

The convergence above is guaranteed by: (2.13), the convergence of $|\mathbf{m}_n| \to 1$ a.e. in $\Omega \setminus \mathcal{S}(\mathbf{u})$ and $\nabla(v_n\varphi) \to \nabla(|\mathbf{u}|\varphi)$ in $L^2(\Omega; \mathbf{R}^3)$, and by the fact that $\nabla(|\mathbf{u}|\varphi) = \mathbf{0}$ a.e. in $\mathcal{S}(\mathbf{u})$. These properties combined with (2.15) yield

$$\int_\Omega \mathbf{N}^T \mathbf{n} \cdot \varphi = 0 \quad \forall \;\; \varphi \in C_0^\infty(\Omega; \mathbf{R}^3).$$

Thus (b) has been established. To prove (c) we use (a) and (b) from which one has

$$|\nabla \mathbf{u}|^2 = \operatorname{tr}(\nabla \mathbf{u}^T \nabla \mathbf{u}) = \operatorname{tr}\Big[\big(\nabla(|\mathbf{u}|) \otimes \mathbf{m} + \mathbf{N}^T\big)\big(\mathbf{m} \otimes \nabla(|\mathbf{u}|) + \mathbf{N})\Big] =$$
$$\operatorname{tr}[(\nabla(|\mathbf{u}|) \otimes \mathbf{m})(\mathbf{m} \otimes \nabla(|\mathbf{u}|)) + (\nabla(|\mathbf{u}|) \otimes \mathbf{m})\mathbf{N} + \mathbf{N}^T \mathbf{m} \nabla(|\mathbf{u}|) + \mathbf{N}^T\mathbf{N}] =$$
$$|\nabla(|\mathbf{u}|)|^2 + |\mathbf{N}|^2.$$

Accordingly, (a) – (c) of (2.7) have been established.
Step 2: Now the assertion will be deduced from (2.7). Indeed, since $\mathbf{n} = \operatorname{sgn}(s)\mathbf{m}$ and $s = \operatorname{sgn}(s)|\mathbf{u}|$, we have $\nabla s = \operatorname{sgn}(s)\nabla(|\mathbf{u}|)$. Whence $\nabla \mathbf{u} = \mathbf{m} \otimes \nabla(|\mathbf{u}|) + \mathbf{N} = \operatorname{sgn}(s)\mathbf{m} \otimes \nabla(\operatorname{sgn}(s)|\mathbf{u}|) + \mathbf{N} = \mathbf{n} \otimes \nabla s + \mathbf{N}$. Further, $\mathbf{0} = \mathbf{N}^T\mathbf{m} = \operatorname{sgn}(s)\mathbf{N}^T\mathbf{n}$ which implies $\mathbf{N}^T\mathbf{n} = \mathbf{0}$. We also have $|\nabla(|\mathbf{u}|)| = |\nabla s|$ thanks to $s = \operatorname{sgn}(s)|\mathbf{u}|$. This completes the proof of (ii). □

The next result is related to the coercivity and weakly lower semicontinuity of $\mathcal{F}$ in $H^1(\Omega; \mathbf{R}^3) \times H^1(\Omega)$.

Proposition 2.2 *Assume that $h : \mathbf{R}^3 \times \mathbf{R} \to \mathbf{R} \cup \{+\infty\}$ is a bounded below and lower semicontinuous function and that $k0$ is a positive constant. Then the functional*

$$\mathcal{F}(\mathbf{u}, s) = \int_\Omega \Big[|\nabla \mathbf{u}|^2 + (k-1)|\nabla s|^2\Big] d\Omega + \int_\Omega h(\mathbf{u}, s) d\Omega \tag{2.16}$$

is weakly lower semicontinuous and coercive on $\mathcal{A} \cap \mathcal{B}$ with respect to $H^1(\Omega; \mathbf{R}^3) \times H^1(\Omega)$-norm.

Proof. To prove the coerciveness of $\mathcal{F}$ it is enough to consider the case $0 < k \leq 1$ because for $k1$ this property is obvious. For our purpose we shall need the estimates

$$\int_\Omega |\mathbf{u}|^2 d\Omega \leq \operatorname{meas} \Omega$$

and

$$\int_\Omega |s|^2 d\Omega \leq \operatorname{meas} \Omega$$

which follow directly from the fact that $|\mathbf{u}| = |s|$ and $s(x) \in [-\frac{1}{2}, 1]$ a.e. in Ω. Moreover, from (2.5) it follows

$$|\nabla \mathbf{u}|^2 \geq |\nabla s|^2.$$

Thus we get

$$\mathcal{F}(\mathbf{u}, s) \;=\; \int_\Omega \Big[|\nabla \mathbf{u}|^2 + (k-1)|\nabla s|^2\Big] d\Omega + \int_\Omega h(\mathbf{u}, s) d\Omega$$

$$
\begin{aligned}
&\geq \int_\Omega \Big[(1-k)(|\nabla \mathbf{u}|^2 - |\nabla s|^2\Big] d\Omega + \int_\Omega k|\nabla \mathbf{u}|^2 d\Omega - C \\
&\geq \int_\Omega \frac{k}{2}|\nabla \mathbf{u}|^2 d\Omega + \int_\Omega \frac{k}{2}|\nabla s|^2 d\Omega - C \\
&\geq \int_\Omega \Big\{\frac{k}{2}|\nabla \mathbf{u}|^2 + |\mathbf{u}|^2\Big\} d\Omega + \int_\Omega \Big\{\frac{k}{2}|\nabla s|^2 + |s|^2\Big\} d\Omega - 2\,\mathrm{meas}\,\Omega - C \\
&\geq c||\mathbf{u}||^2_{H^1(\Omega;\mathbf{R}^3)} + c||s||^2_{H^1(\Omega)} - 2\,\mathrm{meas}\,\Omega - C,
\end{aligned}
$$

where $c = \min(1, \frac{k}{2})$, C is a constant. Thus the coerciveness of $\mathcal{F}$ on $\mathcal{A} \cap \mathcal{B}$ follows.

Now it will be proved that $\mathcal{F}$ is weakly lower semicontinuous on $\mathcal{A}$. To this end let us suppose that a sequence $(\mathbf{u}_n, s_n) \in \mathcal{A}$ weakly converges in $H^1(\Omega;\mathbf{R}^3) \times H^1(\Omega)$ to $(\mathbf{u}, s)$. From Proposition 2.1 we have

$$
\begin{aligned}
\nabla \mathbf{u}_n &= \mathbf{n}_n \otimes \nabla s_n + \mathbf{N}_n \\
|\nabla \mathbf{u}_n|^2 &= |\nabla s_n|^2 + |\mathbf{N}_n|^2
\end{aligned}
\tag{2.17}
$$

where $\mathbf{n}_n = \frac{\mathbf{u}_n}{s_n}$ a.e. in $\Omega \backslash \mathcal{S}(\mathbf{u}_n)$ and $\mathbf{n}_n = \mathbf{0}$ a.e. in $\mathcal{S}(\mathbf{u}_n)$. By the compact imbedding $H^1(\Omega) \hookrightarrow L^2(\Omega)$ the strong convergence of $(\mathbf{u}_n, s_n)$ to $(\mathbf{u}, s)$ in $L^2(\Omega;\mathbf{R}^3) \times L^2(\Omega)$ follows and without loss of generality we may suppose that $(\mathbf{u}_n, s_n) \to (\mathbf{u}, s)$ a.e. in Ω (passing to a subsequence if necessary). This amounts to $\mathbf{n}_n \to \mathbf{n}$ a.e. in $\Omega \setminus \mathcal{S}(\mathbf{u})$. Because of the boundedness of $\{\mathbf{n}_n\}$ in $L^\infty(\Omega;\mathbf{R}^3)$ we have

$$
\mathbf{n}_n \to \mathbf{n} \quad \text{strongly in} \quad L^2(\Omega \setminus \mathcal{S}(\mathbf{u}); \mathbf{R}^3).
$$

Hence by virtue of the weak convergence $\nabla s_n \rightharpoonup \nabla s$ in $L^2(\Omega;\mathbf{R}^3)$ we conclude the weak convergence of $\mathbf{n}_n \otimes \nabla s_n$ to $\mathbf{n} \otimes \nabla s$ in $L^2(\Omega \setminus \mathcal{S}(\mathbf{u}); \mathbf{R}^{3\times3})$. This in turn yields

$$
\mathbf{N}_n \to \mathbf{N} = \nabla \mathbf{u} - \mathbf{n} \otimes \nabla s \quad \text{weakly in} \quad L^2(\Omega \setminus \mathcal{S}(\mathbf{u}); \mathbf{R}^{3\times3}). \tag{2.18}
$$

From (2.5) we have

$$
\begin{aligned}
\mathcal{F}(\mathbf{u}_n, s_n) &= \int_\Omega \Big[|\nabla \mathbf{u}_n|^2 + (k-1)|\nabla s_n|^2\Big] d\Omega + \int_\Omega h(\mathbf{u}_n, s_n) d\Omega \\
= \int_{\Omega\backslash\mathcal{S}(\mathbf{u})} \Big[k|\nabla s_n|^2 + |\mathbf{N}_n|^2)\Big] d\Omega &+ \int_{\mathcal{S}(\mathbf{u})} \Big[k|\nabla s_n|^2 + |\mathbf{N}_n|^2\Big] d\Omega + \int_\Omega h(\mathbf{u}_n, s_n) d\Omega \\
&\geq \int_{\Omega\backslash\mathcal{S}(\mathbf{u})} \Big[k|\nabla s_n|^2 + |\mathbf{N}_n|^2)\Big] d\Omega + \int_\Omega h(\mathbf{u}_n, s_n) d\Omega.
\end{aligned}
$$

By convexity with respect to variables ∇s_n and $\mathbf{N}_n$ and (2.18) we obtain

$$
\liminf_{n\to\infty} F(\mathbf{u}_n, s_n) \geq \liminf_{n\to\infty} \Big\{ \int_{\Omega\backslash\mathcal{S}(\mathbf{u})} \Big[k|\nabla s_n|^2 + |\mathbf{N}_n|^2\Big] d\Omega + \int_\Omega h(\mathbf{u}_n, s_n) d\Omega \Big\}
$$

$$\geq \int\limits_{\Omega\setminus\mathcal{S}(\mathbf{u})} \left[k|\nabla s|^2 + |\mathbf{N}|^2)\right] d\Omega + \int\limits_{\Omega} h(\mathbf{u}, s) d\Omega$$

$$= \int\limits_{\Omega\setminus\mathcal{S}(\mathbf{u})} \left[|\nabla \mathbf{u}|^2 + (k-1)|\nabla s|^2)\right] d\Omega + \int\limits_{\Omega} h(\mathbf{u}, s) d\Omega = \mathcal{F}(\mathbf{u}, s),$$

because $\nabla \mathbf{u} = \mathbf{0}$ and $\nabla s = \mathbf{0}$ a.e. in $\mathcal{S}(\mathbf{u})$. The proof is complete. □

Proposition 2.2 implies the existence result.

Theorem 2.3 *Assume that $h : \mathbf{R}^3 \times \mathbf{R} \to \mathbf{R} \cup \{+\infty\}$ is a bounded below and lower semicontinuous function and that $k0$ is a positive constant. Then the functional $\mathcal{F}$ defined by (1.1) possesses at least one minimizer on $\mathcal{A} \cap \mathcal{B}$, i.e. there exists at least one $(\mathbf{u}, s) \in \mathcal{A} \cap \mathcal{B}$ such that*

$$\mathcal{F}(\mathbf{u}, s) = \inf_{(\mathbf{v}, r) \in \mathcal{A} \cap \mathcal{B}} \mathcal{F}(\mathbf{v}, r). \tag{2.19}$$

3 Geometrical properties of admissible set

Throughout this Section we assume that $(\mathbf{u}, s) \in \mathcal{A}$. Let us denote by $T_{\mathcal{A}}(\mathbf{u}, s)$ the contingent cone to $\mathcal{A}$ at $(\mathbf{u}, s) \in \mathcal{A}$, i.e.

$$\begin{aligned} T_{\mathcal{A}}(\mathbf{u}, s) = \{(\mathbf{v}, r) \in H^1(\Omega; \mathbf{R}^3) \times H^1(\Omega) : \exists \lambda_k \to 0_+, \exists (\mathbf{v}_k, r_k) \to (\mathbf{v}, r) \\ \text{in } H^1(\Omega; \mathbf{R}^3) \times H^1(\Omega) \text{ such that } (\mathbf{u}, s) + \lambda_k(\mathbf{v}_k, r_k) \in \mathcal{A}\}. \end{aligned} \tag{3.1}$$

Proposition 3.1. *The necessary condition for $(\mathbf{v}, r)$ to belong to the contingent cone $T_{\mathcal{A}}(\mathbf{u}, s)$ is that*

$$\mathbf{v} \cdot \mathbf{u} - rs = 0 \quad a.e. \ in \ \Omega. \tag{3.2}$$

If meas $\mathcal{S}(\mathbf{u})0$ *then the additional requirement is necessary, namely*

$$|\mathbf{v}| = |r| \quad a.e. \ in \ \mathcal{S}(\mathbf{u}). \tag{3.3}$$

Proof. The assertion follows from the following calculations:

$$\begin{aligned} |\mathbf{u} + \lambda_k \mathbf{v}_k|^2 &= |s + \lambda_k r_k|^2 \\ |\mathbf{u}|^2 + 2\lambda_k \mathbf{u} \cdot \mathbf{v}_k + \lambda_k^2 |\mathbf{v}_k|^2 &= |s|^2 + 2\lambda_k s r_k + \lambda_k^2 |r_k|^2 \\ 2\mathbf{u} \cdot \mathbf{v}_k + \lambda_k |\mathbf{v}_k|^2 &= 2 s r_k + \lambda_k |r_k|^2, \end{aligned} \tag{3.4}$$

where λ_k and $(\mathbf{v}_k, r_k)$ are taken from (3.1). Thus by letting $k \to \infty$ we arrive easily at (3.2). To get (3.3) it is enough to observe that (3.4) implies

$$|\mathbf{v}_k|^2 = |r_k|^2 \quad \text{a.e. in } \mathcal{S}(\mathbf{u}).$$

Consequently, (3.3) follows. This completes the proof of Proposition 3.1. □

Now we proceed to examine the structure of $T_{\mathcal{A}}(\mathbf{u}, s)$. First result in this direction reads.

Proposition 3.2. *If* $(\mathbf{v}, r) \in T_{\mathcal{A}}(\mathbf{u}, s)$ *then*

$$\begin{aligned} &(\mathbf{v} - \mathbf{u}, r - s) \in T_{\mathcal{A}}(\mathbf{u}, s) \\ &(\mathbf{v} + \mathbf{u}, r + s) \in T_{\mathcal{A}}(\mathbf{u}, s). \end{aligned} \tag{3.5}$$

Proof. Let $(\mathbf{v}, r) \in T_{\mathcal{A}}(\mathbf{u}, s)$. Then $\exists \lambda_k \to 0_+$, $\exists (\mathbf{v}_k, r_k) \to (\mathbf{v}, r)$ strongly in $H^1(\Omega; \mathbf{R}^3) \times H^1(\Omega)$, such that

$$|\mathbf{u} + \lambda_k \mathbf{v}_k|^2 = |s + \lambda_k r_k|^2.$$

Let μ_k be related to λ_k by $\lambda_k = \frac{\mu_k}{1-\mu_k}$. Substituting $\frac{\mu_k}{1-\mu_k}$ into λ_k yields

$$|\mathbf{u} + \frac{\mu_k}{1 - \mu_k} \mathbf{v}_k|^2 = |s + \frac{\mu_k}{1 - \mu_k} r_k|^2.$$

This implies in turn that

$$|\mathbf{u} + \mu_k(\mathbf{v}_k - \mathbf{u})|^2 = |s + \mu_k(r_k - s)|^2,$$

which is equivalent to the statement that $(\mathbf{v} - \mathbf{u}, r - s) \in T_{\mathcal{A}}(\mathbf{u}, s)$. To obtain the remaining assertion we proceed analogously, choosing $\frac{\mu_k}{1+\mu_k}$ instead of λ_k. □

Proposition 3.3. *If* $\varphi \in H^1(\Omega)$ *is such that* $(\varphi\mathbf{u}, \varphi s) \in H^1(\Omega; \mathbf{R}^3) \times H^1(\Omega)$ *then*

$$(\varphi\mathbf{u}, \varphi s) \in T_{\mathcal{A}}(\mathbf{u}, s). \tag{3.6}$$

Proof. It is easily seen that if $(\mathbf{v}, r) \in H^1(\Omega; \mathbf{R}^3) \times H^1(\Omega)$, $\mathbf{u} \cdot \mathbf{v} - sr = 0$ and $|\mathbf{v}| = |r|$ a.e. in Ω then for any positive λ it holds

$$(\mathbf{u} + \lambda \mathbf{v}, s + \lambda r) \in \mathcal{A}.$$

Since for $\varphi \in H^1(\Omega)$ with $(\varphi\mathbf{u}, \varphi s) \in H^1(\Omega; \mathbf{R}^3) \times H^1(\Omega)$, we check immediately that $\mathbf{u} \cdot (\varphi\mathbf{u}) - s(\varphi s) = 0$ and $|\varphi\mathbf{u}| = |\varphi s|$ a.e. in Ω, so, in particular,

$$(\mathbf{u} + \lambda\varphi\mathbf{u}, s + \lambda\varphi s) \in \mathcal{A}. \tag{3.7}$$

Accordingly, the assertion follows. □

Proposition 3.4 *If* $\mathbf{w} \in H^1(\Omega; \mathbf{R}^3)$ *is such that* $\mathbf{w} \cdot \mathbf{u} = 0$ *a.e. in* Ω *and* $s\mathbf{w} \in H^1(\Omega; \mathbf{R}^3)$ *then*

$$(s\mathbf{w}, 0) \in T_{\mathcal{A}}(\mathbf{u}, s). \tag{3.8}$$

Proof. We divide the proof into two steps.
Step 1. Let $\mathbf{w} \in H^1(\Omega; \mathbf{R}^3) \cap L^\infty(\Omega; \mathbf{R}^3)$ be such that $\mathbf{w} \cdot \mathbf{u} = 0$ a.e. in Ω and let $\lambda_k \to 0_+$. To show the assertion we shall construct $(\mathbf{y}_k) \subset H^1(\Omega; \mathbf{R}^3)$ with $\mathbf{y}_k \to \mathbf{0}$ in $H^1(\Omega; \mathbf{R}^3)$ as $k \to \infty$, such that

$$\Big(\mathbf{u} + \lambda_k(s\mathbf{w} + \mathbf{y}_k), s\Big) \in \mathcal{A}. \tag{3.9}$$

For this purpose it is necessary to solve the equation

$$|\mathbf{u} + \lambda_k(s\mathbf{w} + \mathbf{y}_k)|^2 = |s|^2 \quad \text{a.e. in } \Omega \tag{3.10}$$

with $\mathbf{y}_k$ as the main unknown. Whence we get

$$|\mathbf{u}|^2 + 2\lambda_k s\mathbf{u} \cdot \mathbf{w} + 2\lambda_k \mathbf{u} \cdot \mathbf{y}_k + \lambda_k^2 |s\mathbf{w} + \mathbf{y}_k|^2 = |s|^2.$$

Since $\mathbf{u} \cdot \mathbf{w} = 0$ and $|\mathbf{u}| = |s|$, one obtains

$$2\mathbf{u} \cdot \mathbf{y}_k + \lambda_k |s\mathbf{w} + \mathbf{y}_k|^2 = 0$$

after reducing similar terms. By the introduction of a new variable t related to $\mathbf{y}_k$ by

$$\mathbf{y}_k = \lambda_k t\mathbf{u}$$

we arrive at

$$\begin{gathered} 2\lambda_k t|\mathbf{u}|^2 + \lambda_k |s\mathbf{w} + \lambda_k t\mathbf{u}|^2 = 0, \\ 2t|\mathbf{u}|^2 + s^2|\mathbf{w}|^2 + 2\lambda_k ts\mathbf{u} \cdot \mathbf{w} + \lambda_k^2 t^2 |\mathbf{u}|^2 = 0, \\ 2t|\mathbf{u}|^2 + s^2|\mathbf{w}|^2 + \lambda_k^2 t^2 |\mathbf{u}|^2 = 0, \end{gathered}$$

which after eliminating $s^2 = |\mathbf{u}|^2$ yields

$$\lambda_k^2 t^2 + 2t + |\mathbf{w}|^2 = 0.$$

To solve the last equation let us calculate its determinant

$$\Delta = 4(1 - \lambda_k^2 |\mathbf{w}|^2).$$

Hence, the following solution for t follows

$$t = \frac{\sqrt{1 - \lambda_k^2 |\mathbf{w}|^2} - 1}{\lambda_k^2} = \frac{-|\mathbf{w}|^2}{\sqrt{1 - \lambda_k^2 |\mathbf{w}|^2} + 1}.$$

Now taking into account the relation between $\mathbf{y}_k$ and t we arrive at the solution $\mathbf{y}_k$ of the equation (3.10) which takes the form

$$\mathbf{y}_k = -\lambda_k \mathbf{u} \frac{|\mathbf{w}|^2}{\sqrt{1-\lambda_k^2|\mathbf{w}|^2}+1}. \tag{3.11}$$

It is clear that $\mathbf{y}_k \in H^1(\Omega;\mathbf{R}^3)$ for sufficiently small λ_k and that $\mathbf{y}_k \to \mathbf{0}$ in $H^1(\Omega;\mathbf{R}^3)$. Accordingly, for $\mathbf{w} \in H^1(\Omega;\mathbf{R}^3) \cap L^\infty(\Omega;\mathbf{R}^3)$ with $\mathbf{w}\cdot\mathbf{u} = 0$ a.e. in Ω we have established that $(s\mathbf{w}, 0) \in T_{\mathcal{A}}(\mathbf{u}, s)$.

Step 2. Now we consider the case in which $\mathbf{w} \in H^1(\Omega;\mathbf{R}^3)$ is such that $\mathbf{w}\cdot\mathbf{u} = 0$ a.e. in Ω and $s\mathbf{w} \in H^1(\Omega;\mathbf{R}^3)$. The method relies on constructing a sequence $\varepsilon_n \in H^1(\Omega;[0,1])$ with the properties that $(1-\varepsilon_n)\mathbf{w} \in H^1(\Omega;\mathbf{R}^3) \cap L^\infty(\Omega;\mathbf{R}^3)$ and $(1-\varepsilon_n)\mathbf{w}s \to \mathbf{w}s$ in $H^1(\Omega;\mathbf{R}^3)$ as $n \to \infty$. Such a truncation procedure for vector-valued Sobolev spaces has been introduced in [7].

For this purpose let us recall that if $v \in H^1(\Omega)$ then $|v| \in H^1(\Omega)$. Moreover, if $\theta : R \to R$ is a Lipschitz function, differentiable except a finite number of points $\{t_1,\ldots,t_l\}$, then $\theta(v) \in H^1(\Omega)$ and

$$\nabla(\theta(v)) = \begin{cases} \theta'(v)\nabla(v) & \text{a.e. in } \{x \in \Omega :\ v(x) \notin \{t_1,\ldots,t_l\}\} \\ 0 & \text{a.e. in } \{x \in \Omega :\ v(x) \in \{t_1,\ldots,t_l\}\}. \end{cases} \tag{3.12}$$

Let $\theta_n(t) = \frac{(|t|-n)^+}{|t|}$, $t \in R$, ([3]). It can be easily seen that $\theta_n(\cdot)$ is a Lipschitz function, differentiable except $t_n = n$. Indeed, $\theta_n'(t) = \frac{n}{t^2}$ for tn and $\theta_n'(t) = 0$ for $t < n$. Thus if we set

$$\varepsilon_n = \theta_n(|w_1|+|w_2|+|w_3|) = \frac{(|w_1|+|w_2+|w_3|-n)^+}{|w_1|+|w_2|+|w_3|},$$

then $\{\varepsilon_n\} \subset H^1(\Omega)$, $0 \le \varepsilon_n \le 1$ and $\{(1-\varepsilon_n)\mathbf{w}\} = \{(1-\varepsilon_n)(w_1,w_2,w_3)\} \subset L^\infty(\Omega;\mathbf{R}^3)$.

Due to (3.12) in order to show that $\{(1-\varepsilon_n)\mathbf{w}\} \subset H^1(\Omega;\mathbf{R}^3)$ and $(1-\varepsilon_n)\mathbf{w}s \to \mathbf{w}s$ in $H^1(\Omega;\mathbf{R}^3)$ it is enough to invoke the estimates

$$|\nabla(\varepsilon_n)| \le \frac{|\nabla(|w_1|+|w_2|+|w_3|)|}{|w_1|+|w_2|+|w_3|} \quad \text{in} \quad A_n,$$

where $A_n = \{x \in \Omega : |w_1(x)|+|w_2(x)|+|w_3(x)|n\}$ and

$$\begin{aligned} |\nabla(\varepsilon_n w_i s)| &\le |\nabla(\varepsilon_n)||w_i s| + |\nabla(w_i s)| \\ &\le \frac{|\nabla(|w_1|+|w_2|+|w_3|)|}{|w_1|+|w_2|+|w_3|}|w_i| + |\nabla(w_i s)| \\ &\le (|\nabla(w_1)|+|\nabla(w_2)|+|\nabla(w_3)| \\ &\quad +|\nabla(w_1 s)|+|\nabla(w_2 s)|+|\nabla(w_3 s)|) \ \text{ in } A_n, \ \ i=1,2,3. \end{aligned}$$

[3] $a^+ = \max(a,0)$

Hence we obtain $\{\varepsilon_n w_i s\} \in H^1(\Omega)$, $i = 1,2,3$, which implies $\{(1-\varepsilon_n)\mathbf{w}s\} \subset H^1(\Omega;\mathbf{R}^3)$. The convergence $(1-\varepsilon_n)\mathbf{w}s \to \mathbf{w}s$ in $H^1(\Omega;\mathbf{R}^3)$ as $n \to +\infty$ is an immediately consequence of the estimates

$$||\varepsilon_n \mathbf{w}s||^2_{H^1(\Omega;\mathbf{R}^3)} = \int_{A_n} \Big(\sum_j^3 (|\varepsilon_n w_j s|^2 + |\nabla(\varepsilon_n w_j s)|^2)\Big) d\Omega$$

$$\leq \int_{A_n} 2^2 3^2 \Big(\sum_j^3 (|w_j|^2 + |\nabla(w_j)|^2 + |\nabla(w_j s)|^2)\Big) d\Omega \longrightarrow 0 \text{ as } n \to \infty.$$

Accordingly the convergence $(1-\varepsilon_n)\mathbf{w}s \to \mathbf{w}s$ in $H^1(\Omega;\mathbf{R}^3)$ as $n \to \infty$ has been established.

To complete the proof it is enough to observe that $(1-\varepsilon_n)\mathbf{w}\cdot\mathbf{u} = 0$ a.e. in Ω and we are allowed to benefit from Step 1 writing $((1-\varepsilon_n)s\mathbf{w}, 0) \in T_{\mathcal{A}}(\mathbf{u}, s)$. Since $T_{\mathcal{A}}(\mathbf{u}, s)$ is closed, we get (3.8), as desired. □

Let us introduce linear, closed subspaces

$$H^1_{(s)}(\Omega;\mathbf{R}^3) := \overline{sH^1(\Omega;\mathbf{R}^3) \cap H^1(\Omega;\mathbf{R}^3)}^{H^1(\Omega;\mathbf{R}^3)} \tag{3.13}$$

and

$$H^1_{(s)}(\Omega) := \overline{sH^1(\Omega) \cap H^1(\Omega)}^{H^1(\Omega)}. \tag{3.14}$$

Further we introduce a linear continuous operator $\Phi : H^1(\Omega;\mathbf{R}^3) \to L^2(\Omega)$ defined by

$$\Phi(\mathbf{v}) := \mathbf{v}\cdot\mathbf{u} \quad \mathbf{v} \in H^1(\Omega;\mathbf{R}^3) \tag{3.15}$$

and denote by $\operatorname{Ker}\Phi$ its kernel, i.e. $\operatorname{Ker}\Phi = \{\mathbf{v} \in H^1(\Omega;\mathbf{R}^3) : \mathbf{v}\cdot\mathbf{u} = 0\}$. For the restriction of Φ to $H^1_{(s)}(\Omega;\mathbf{R}^3)$ the symbol $\Phi_{(s)}$ will be used, i.e. $\Phi_{(s)} = \Phi|_{H^1_{(s)}(\Omega;\mathbf{R}^3)}$. Moreover, we use the symbol $(\operatorname{Ker}\Phi_{(s)})^\perp$ to denote the annihilator of $\operatorname{Ker}\Phi_{(s)}$, i.e.

$$(\operatorname{Ker}\Phi_{(s)})^\perp := \Big\{f \in H^1_{(s)}(\Omega;\mathbf{R}^3)^\star : \langle f, \mathbf{v}\rangle_{H^1(\Omega;\mathbf{R}^3)} = 0 \quad \forall\, \mathbf{v} \in \operatorname{Ker}\Phi_{(s)}\Big\}. \tag{3.16}$$

Recall that the classical annihilator theorem states that

$$(\operatorname{Ker}\Phi_{(s)})^\perp = \overline{\operatorname{Im}\Phi^\star_{(s)}}^{H^1(\Omega;\mathbf{R}^3)^\star}, \tag{3.17}$$

where $\operatorname{Im}\Phi^\star_{(s)}$ stands for the image of the transpose operator $\Phi^\star_{(s)} : L^2(\Omega) \to H^1_{(s)}(\Omega;\mathbf{R}^3)^\star$ defined for any $\lambda \in L^2(\Omega)$ by

$$\langle \Phi^\star_{(s)}(\lambda), \mathbf{v}\rangle_{H^1(\Omega;\mathbf{R}^3)} = \int_\Omega \lambda\mathbf{u}\cdot\mathbf{v}\, d\Omega \quad \forall\, \mathbf{v} \in H^1_{(s)}(\Omega;\mathbf{R}^3). \tag{3.18}$$

The first problem is whether for any $\varphi \in H^1(\Omega)$ with $\varphi\mathbf{u} \in H^1(\Omega;\mathbf{R}^3)$ it follows that $\varphi\mathbf{u} \in H^1_{(s)}(\Omega;\mathbf{R}^3)$. With this respect the following result can be stated:

Proposition 3.5 *Let $\varphi \in H^1(\Omega)$ be such that $\varphi\mathbf{u} \in H^1(\Omega;\mathbf{R}^3)$ then $\varphi\mathbf{u} \in H^1_{(s)}(\Omega;\mathbf{R}^3)$.*

Proof. We divide the proof into two steps.
Step 1: First it will be shown that if $\varphi \in H^1(\Omega) \cap L^\infty(\Omega)$ then

$$k\beta_k(s)\varphi\mathbf{u} \to \varphi\mathbf{u} \quad \text{strongly in } H^1(\Omega;\mathbf{R}^3) \text{ as } k \to \infty, \tag{3.19}$$

where β_k, $k = 1, 2, \ldots$, is defined by

$$\beta_k(t) = \begin{cases} \frac{1}{k} & \text{if } t \geq \frac{1}{k} \\ t & \text{if } |t| < \frac{1}{k} \\ -\frac{1}{k} & \text{if } t \leq -\frac{1}{k}. \end{cases}$$

Indeed, we get the estimates

$$\int_\Omega \Big(|k\beta_k(s)\varphi\mathbf{u} - \varphi\mathbf{u}|^2 + \sum_1^3 |\partial_i(k\beta_k(s)\varphi\mathbf{u}) - \partial_i(\varphi\mathbf{u})|^2\Big) d\Omega$$

$$\leq \int_{A_k} \Big(|ks\varphi\mathbf{u} - \varphi\mathbf{u}|^2 + \sum_1^3 |k\partial_i(s)\varphi\mathbf{u} + ks\partial_i(\varphi\mathbf{u}) - \partial_i(\varphi\mathbf{u})|^2\Big) d\Omega$$

$$\leq \int_{A_k} \Big(4|\varphi\mathbf{u}|^2 + 2\sum_1^3 (|\partial_i(s)|^2|\varphi|^2 + 2|\partial_i(\varphi\mathbf{u})|^2)\Big) d\Omega,$$

where $A_k = \{x \in \Omega : |s(x)| < \frac{1}{k}\}$ is a decreasing family of sets, i.e. $A_{k+1} \subset A_k$, $k = 1, 2, \ldots$, with

$$\bigcap_{k\geq 1} A_k = \mathcal{S}(\mathbf{u}).$$

Since $|\varphi\mathbf{u}| = |\varphi s| = 0$ in $\mathcal{S}(\mathbf{u})$,

$$\int_{A_k} \Big(4|\varphi\mathbf{u}|^2 + 2\sum_1^3 (|\partial_i(s)|^2|\varphi|^2 + 2|\partial_i(\varphi\mathbf{u})|^2)\Big) d\Omega$$

$$\longrightarrow \int_{\mathcal{S}(\mathbf{u})} \Big(4|\varphi\mathbf{u}|^2 + 2\sum_1^3 (|\partial_i(s)|^2|\varphi|^2 + 2|\partial_i(\varphi\mathbf{u})|^2)\Big) d\Omega = 0 \quad \text{as } k \to \infty,$$

and the convergence (3.19) follows.

Now to show the assertion it suffices to notice that the approximate sequence $\{k\beta_k(s)\varphi\mathbf{u}\}$ can be written equivalently as

$$k\beta_k(s)\varphi\mathbf{u} = s\mathbf{y}_k,$$

with

$$\mathbf{y}_k = \begin{cases} \varphi\mathbf{n} & \text{if } |s| \geq \frac{1}{k} \\ k\varphi\mathbf{u} & \text{if } |s| < \frac{1}{k}, \end{cases}$$

where $\mathbf{u} = s\mathbf{n}$. Since, as it can be easily checked, $\mathbf{y}_k \in H^1(\Omega; \mathbf{R}^3)$, we have $s\mathbf{y}_k \to \varphi\mathbf{u}$ as $k \to \infty$ and the assertion results.

Step 2. To show the result in the general case, i.e. when $\varphi \in H^1(\Omega)$ and $\varphi\mathbf{u} \in H^1(\Omega; \mathbf{R}^3)$, it is enough to apply the truncation procedure of the form $(1 - \varepsilon_n)\varphi \to \varphi$ in $H^1(\Omega)$ with

$$\varepsilon_n := \theta_n(\varphi),$$

where $\theta_n(t) := \frac{(|t|-n)^+}{|t|}$, $t \in R$. Since $(1 - \varepsilon_n)\varphi\mathbf{u} \in H^1_{(s)}(\Omega; \mathbf{R}^3)$, following the lines of the proof of Proposition 3.4 we conclude that $\varphi\mathbf{u} \in H^1_{(s)}(\Omega; \mathbf{R}^3)$. Therefore the details will be omitted here. □

Lemma 3.6. *Let* $\Lambda \in \left(\operatorname{Ker} \Phi_{(s)}\right)^\perp$ *be such that*

$$\langle \Lambda, \varphi\mathbf{u}\rangle_{H^1(\Omega;\mathbf{R}^3)} = 0 \quad \forall\, \varphi \in H^1(\Omega; \mathbf{R}^3) \text{ with } \varphi\mathbf{u} \in H^1(\Omega; \mathbf{R}^3). \tag{3.20}$$

Then $\Lambda = 0$.

Proof. From (3.17) and (3.18) it follows that there exists a sequence (λ_n) in $L^2(\Omega)$ with the property that

$$\Phi^\star_{(s)}(\lambda_n) \to \Lambda \text{ in } H^1_{(s)}(\Omega; \mathbf{R}^3)^\star. \tag{3.21}$$

Thus

$$\sup\Big\{\Big|\int_\Omega \lambda_n \mathbf{u} \cdot \mathbf{v} d\Omega - \langle \Lambda, \mathbf{v}\rangle_{H^1(\Omega;\mathbf{R}^3)}\Big| : \mathbf{v} \in H^1_{(s)}(\Omega; \mathbf{R}^3) \text{ and } ||\mathbf{v}||_{H^1(\Omega;\mathbf{R}^3)} \le 1\Big\} \longrightarrow 0 \quad \text{as } n \to \infty.$$

Inserting into the foregoing relation $\mathbf{v} = \varphi\mathbf{u} \in H^1_{(s)}(\Omega; \mathbf{R}^3)$ yields

$$\sup\Big\{\Big|\int_\Omega \lambda_n \varphi s^2 d\Omega\Big| : \varphi \in H^1(\Omega) \text{ and } ||\varphi\mathbf{u}||_{H^1(\Omega;\mathbf{R}^3)} \le 1\Big\} \longrightarrow 0 \quad \text{as } n \to \infty. \tag{3.22}$$

Now suppose the assertion of the lemma is false. By the Riesz representation theorem there exists $\mathbf{w} \in H^1_{(s)}(\Omega; \mathbf{R}^3)$ with $\mathbf{w} \neq \mathbf{0}$ such that $\Lambda = \mathcal{R}(\mathbf{w})$, where $\mathcal{R} : H^1_{(s)}(\Omega; \mathbf{R}^3) \to H^1_{(s)}(\Omega; \mathbf{R}^3)^\star$ stands for the Riesz isomorphism between the Hilbert space $H^1_{(s)}(\Omega; \mathbf{R}^3)$ and its dual $H^1_{(s)}(\Omega; \mathbf{R}^3)^\star$, defined by

$$\langle \mathcal{R}(\mathbf{w}), \mathbf{v}\rangle_{H^1(\Omega;\mathbf{R}^3)} = \int_\Omega \Big(\nabla\mathbf{w} \cdot \nabla\mathbf{v} + \mathbf{w} \cdot \mathbf{v}\Big) d\Omega \quad \forall \mathbf{v} \in H^1_{(s)}(\Omega; \mathbf{R}^3).$$

Moreover, by (3.21) one can find a sequence $\{\mathbf{w}_n\} \subset H^1_{(s)}(\Omega; \mathbf{R}^3)$ converging strongly in $H^1(\Omega; \mathbf{R}^3)$ to $\mathbf{w}$, with the property that $\mathcal{R}(\mathbf{w}_n) = \Phi^\star_{(s)}(\lambda_n)$. This implies that $\mathcal{R}(\mathbf{w}_n) \to \mathcal{R}(\mathbf{w}) = \Lambda$ in $H^1_{(s)}(\Omega; \mathbf{R}^3)^\star$ as $n \to \infty$, and therefore

$$\int_\Omega \lambda_n \mathbf{u} \cdot \mathbf{v} d\Omega = \int_\Omega \Big(\nabla\mathbf{w}_n \cdot \nabla\mathbf{v} + \mathbf{w}_n \cdot \mathbf{v}\Big) d\Omega \longrightarrow \int_\Omega \Big(\nabla\mathbf{w} \cdot \nabla\mathbf{v} + \mathbf{w} \cdot \mathbf{v}\Big) d\Omega$$

as $n \to \infty$, for any $\mathbf{v} \in H^1_{(s)}(\Omega; \mathbf{R}^3)$. Since $\mathbf{w} \in H^1_{(s)}(\Omega; \mathbf{R}^3)$, there exists a sequence $s\mathbf{w}_k \to \mathbf{w}$ in $H^1(\Omega; \mathbf{R}^3)$ such that $\{\mathbf{w}_k\} \subset H^1(\Omega; \mathbf{R}^3)$. By applying the truncation procedure analogous to that of Proposition 3.4, without loss of generality we can assume that $\{\mathbf{w}_k\} \subset H^1(\Omega; \mathbf{R}^3) \cap L^\infty(\Omega; \mathbf{R}^3)$. Thus there exists $\overline{\mathbf{w}} \in H^1(\Omega; \mathbf{R}^3) \cap L^\infty(\Omega; \mathbf{R}^3)$ with the property that

$$\int_\Omega \lambda_n \mathbf{u} \cdot s\overline{\mathbf{w}} d\Omega \geq \frac{1}{2} ||\mathbf{w}||^2_{H^1(\Omega;\mathbf{R}^3)} 0 \tag{3.23}$$

for sufficiently large n. Now, as in Proposition 3.5, $\mathbf{u}$ can be approximated by a sequence $\{s\mathbf{y}_k\}$ with

$$\mathbf{y}_k = \begin{cases} \mathbf{n} & \text{if } \ |s| \geq \frac{1}{k} \\ k\mathbf{u} & \text{if } \ |s| < \frac{1}{k}. \end{cases}$$

This allows us to extract a sequence k_n for which

$$\int_\Omega \lambda_n s\mathbf{y}_{k_n} \cdot s\overline{\mathbf{w}} d\Omega \geq \frac{1}{4} ||\mathbf{w}||^2_{H^1(\Omega;\mathbf{R}^3)} 0. \tag{3.24}$$

Now we show that $\{(\mathbf{y}_k \cdot \overline{\mathbf{w}})\mathbf{u}\}$ is bounded in $H^1(\Omega; \mathbf{R}^3)$. First of all we have to be sure that $\{(\mathbf{y}_k \cdot \overline{\mathbf{w}})\mathbf{u}\} \subset H^1(\Omega; \mathbf{R}^3)$. But this is a consequence of the fact that $\mathbf{y}_k$, $\overline{\mathbf{w}}$ and $\mathbf{u}$ are elements of $H^1(\Omega; \mathbf{R}^3) \cap L^\infty(\Omega; \mathbf{R}^3)$. The desired boundedness can be obtained from the estimates

$$\int_\Omega \Big(|(\mathbf{y}_k \cdot \overline{\mathbf{w}})\mathbf{u}|^2 + \sum_{i=1}^3 |\partial_i((\mathbf{y}_k \cdot \overline{\mathbf{w}})\mathbf{u})|^2 \Big) d\Omega$$

$$\leq \int_\Omega |\overline{\mathbf{w}}|^2 d\Omega + \sum_{i=1}^3 \Big\{ \int_{\Omega\setminus A_k} |\partial_i((\mathbf{n} \cdot \overline{\mathbf{w}})\mathbf{u})|^2 d\Omega + \int_{A_k} |\partial_i((k\mathbf{u} \cdot \overline{\mathbf{w}})\mathbf{u})|^2 d\Omega \Big\}$$

$$\leq \int_\Omega |\overline{\mathbf{w}}|^2 d\Omega + \sum_{i=1}^3 \Big\{ \int_{\Omega\setminus A_k} |(s\partial_i(\mathbf{n}) \cdot \overline{\mathbf{w}})\mathbf{n} + (\mathbf{n} \cdot \partial_i(\overline{\mathbf{w}}))\mathbf{u} + (\mathbf{n} \cdot \overline{\mathbf{w}})\partial_i(\mathbf{u})|^2 d\Omega \Big\}$$

$$+ \sum_{i=1}^3 \Big\{ \int_{A_k} |(\partial_i(\mathbf{u}) \cdot \overline{\mathbf{w}})k\mathbf{u} + (k\mathbf{u} \cdot \partial_i(\overline{\mathbf{w}}))\mathbf{u} + (k\mathbf{u} \cdot \overline{\mathbf{w}})\partial_i(\mathbf{u})|^2 d\Omega \Big\}$$

$$\leq \int_\Omega |\overline{\mathbf{w}}|^2 d\Omega + \sum_{i=1}^3 \Big\{ \int_{\Omega\setminus A_k} 3\Big(|s\partial_i(\mathbf{n})|^2 |\overline{\mathbf{w}}|^2 + 3|\partial_i(\overline{\mathbf{w}})|^2 + 3|\overline{\mathbf{w}}|^2 |\partial_i(\mathbf{u})|^2 \Big) d\Omega \Big\}$$

$$+ \sum_{i=1}^3 \Big\{ \int_{A_k} 3\Big(|\partial_i(\mathbf{u})|^2 |\overline{\mathbf{w}}|^2 + |\partial_i(\overline{\mathbf{w}})|^2 + |\overline{\mathbf{w}}|^2 |\partial_i(\mathbf{u})|^2 \Big) d\Omega \Big\}$$

$$\leq \int_\Omega |\overline{\mathbf{w}}|^2 d\Omega + \sum_{i=1}^3 \Big\{ \int_\Omega 3\Big(|\mathbf{N}_i|^2 |\overline{\mathbf{w}}|^2 + 3|\partial_i(\overline{\mathbf{w}})|^2 + 3|\overline{\mathbf{w}}|^2 |\partial_i(\mathbf{u})|^2 \Big) d\Omega \Big\}$$

$$+\sum_{i=1}^{3}\Big\{\int_{\Omega} 3\big(|\partial_i(\mathbf{u})|^2|\overline{\mathbf{w}}|^2 + |\partial_i(\overline{\mathbf{w}})|^2 + |\overline{\mathbf{w}}|^2|\partial_i(\mathbf{u})|^2\big)d\Omega\Big\},$$

where $A_k = \{x \in \Omega : |s(x)| < \frac{1}{k}\}$ and $\mathbf{N} = (\mathbf{N}_i)$ is defined by (2.5). Accordingly we are allowed to insert $\varphi = (\mathbf{y}_k \cdot \overline{\mathbf{w}})$ into (3.22) which in view of (3.24) leads to a contradiction. The proof of Lemma 3.6 is complete. □

An immediate consequence of Lemma 3.6 is the conclusion that $\big(s\,\mathrm{Ker}\,\Phi + \mathbf{u}H^1(\Omega)\big) \cap H^1(\Omega;\mathbf{R}^3)$ is dense in $H^1_{(s)}(\Omega;\mathbf{R}^3)$. Taking into account Propositions 3.3 and 3.4 we obtain:

Proposition 3.7. *Let* $(\mathbf{u}, s) \in \mathcal{A}$. *Then*

$$\overline{T_{\mathcal{A}}(\mathbf{u},s) + T_{\mathcal{A}}(\mathbf{u},s)}^{H^1(\Omega;\mathbf{R}^3)} \supset H^1_{(s)}(\Omega;\mathbf{R}^3) \times H^1_{(s)}(\Omega).$$

4 Necessary optimality conditions

Let us denote by $I_{[-\frac{1}{2},1]} : \mathbf{R} \to \mathbf{R} \cup \{+\infty\}$ the indicator function of the interval $[-\frac{1}{2}, 1]$, i.e. $I_{[-\frac{1}{2},1]}(r) = 0$ if $r \in [-\frac{1}{2}, 1]$, or $I_{[-\frac{1}{2},1]}(r) = +\infty$, otherwise.

From now on, the function $h : \mathbf{R}^3 \times \mathbf{R} \to \mathbf{R}$ is assumed to satisfy the conditions:

(H1) h is locally Lipschitz in $\mathcal{O} \setminus \{(\mathbf{0}, 0)\}$, $\mathcal{O}$ being a neighbourhood of $\mathcal{D} := \{(\mathbf{x}, a) \in \mathbf{R}^3 \times \mathbf{R} : |\mathbf{x}| = |a|,\ a \in [-\frac{1}{2}, 1]\}$,

(H2) h is lower semicontinuous at $(\mathbf{0}, 0)$,

(H3) If $(\mathbf{x}, a) \in \mathcal{D} \setminus \{(\mathbf{0}, 0)\}$ then

$$|h^0(\mathbf{x}, a; a\mathbf{y}, ab)| \le K(|\mathbf{y}| + |b|) \quad \forall\,(\mathbf{y}, b) \in \mathbf{R}^3 \times \mathbf{R}, \qquad K = \text{constant}.$$

Here we used the symbol $h^0(\mathbf{x}, a; \mathbf{y}, b)$ to denote the directional Clarke differential of h at $(\mathbf{x}, a) \ne (\mathbf{0}, 0)$ in the direction $(\mathbf{y}, b)$, i.e.

$$h^0(\mathbf{x}, a; \mathbf{y}, b) = \limsup_{\substack{(\mathbf{h},r)\to(\mathbf{0},0)\\ t\searrow 0}} \frac{h(\mathbf{x} + \mathbf{h} + t\mathbf{y}, a + r + tb) - h(\mathbf{x} + \mathbf{h}, a + r)}{t}.$$

By the convention we set $h^0(\mathbf{0}, 0; \mathbf{y}, b) = 0$. The indicator function of $\mathrm{Ker}\,\Phi$ will be denoted by $\mathcal{I}_{\mathrm{Ker}\,\Phi} : H^1(\Omega;\mathbf{R}^3) \to \mathbf{R} \cup \{+\infty\}$, i.e. $\mathcal{I}_{\mathrm{Ker}\,\Phi}(\mathbf{w}) = 0$ if $\mathbf{w} \in \mathrm{Ker}\,\Phi$, or $\mathcal{I}_{\mathrm{Ker}\,\Phi}(\mathbf{w}) = +\infty$, otherwise. Now we can state the necessary optimality conditions.

Theorem 4.1 *Assume that* $h : \mathbf{R}^3 \times \mathbf{R} \to \mathbf{R}$ *fulfills the hypotheses (H1) – (H3). Then the functional*

$$\mathcal{F}(\mathbf{v}, r) = \int\limits_{\Omega} \Big[|\nabla \mathbf{v}|^2 + (k-1)|\nabla r|^2\Big] d\Omega + \int\limits_{\Omega} h(\mathbf{v}, r) d\Omega, \tag{4.1}$$

possesses at least one minimizer on $\mathcal{A} \cap \mathcal{B}$, *i.e. there exists* $(\mathbf{u}, s) \in \mathcal{A} \cap \mathcal{B}$ *such that*

$$\mathcal{F}(\mathbf{u}, s) = \inf_{(\mathbf{v},r) \in \mathcal{A} \cap \mathcal{B}} \mathcal{F}(\mathbf{v}, r). \tag{4.2}$$

Moreover, the following variational–hemivariational inequalities hold:

$$2\int\limits_{\Omega} \Big[\nabla \mathbf{u} \cdot \Big(\nabla(\varphi \mathbf{u}) - \nabla \mathbf{u}\Big) + (k-1)\nabla s \cdot \Big(\nabla(\varphi s) - \nabla s\Big)\Big] d\Omega + \int\limits_{\Omega} h^0\Big(\mathbf{u}, s; \varphi \mathbf{u} - \mathbf{u}, \varphi s - s\Big) d\Omega + \int\limits_{\Omega} I_{[-\frac{1}{2},1]}(\varphi s) d\Omega \geq 0 \quad \forall\, \varphi \in H^1(\Omega) \tag{4.3}$$

and

$$2\int\limits_{\Omega} \nabla \mathbf{u} \cdot \nabla(s\mathbf{w}) d\Omega + \int\limits_{\Omega} h^0\Big(\mathbf{u}, s; s\mathbf{w}, 0\Big) d\Omega + \mathcal{I}_{\operatorname{Ker}\Phi}(s\mathbf{w}) \geq 0 \quad \forall\, \mathbf{w} \in H^1(\Omega; \mathbf{R}^3) \tag{4.4}$$

where only such φ*'s and* $\mathbf{w}$*'s are allowed that* $s\varphi \in H^1(\Omega)$ *and* $s\mathbf{w} \in H^1(\Omega; \mathbf{R}^3)$.

Proof. The existence of a minimizer for $\mathcal{F}$ follows directly from Theorem 2.3.

Now we start with showing (4.3). Choose $(\varphi \mathbf{u}, \varphi s)$ as in Proposition 3.3. From (3.7) with $\lambda = \frac{\lambda_k}{1-\lambda_k}$, $\lambda_k \to 0_+$ as $k \to \infty$, we get

$$\Big(\mathbf{u} + \lambda_k(\varphi \mathbf{u} - \mathbf{u}), s + \lambda_k(\varphi s - s)\Big) \in \mathcal{A}.$$

Thus, by optimality,

$$\mathcal{F}\Big(\mathbf{u} + \lambda_k(\varphi \mathbf{u} - \mathbf{u}), s + \lambda_k(\varphi s - s)\Big) - \mathcal{F}(\mathbf{u}, s) \geq 0$$

holds for each $\varphi \in H^1(\Omega)$ with $(\varphi \mathbf{u}, \varphi s) \in H^1(\Omega; \mathbf{R}^3) \times H^1(\Omega)$. Dividing this inequality by λ_k and letting $\lambda_k \to 0$, with the help of Fatou's lemma and convexity of $I_{[-\frac{1}{2},1]}(\cdot)$, we are easy led to (4.3).

To get (4.4) we let $\mathbf{w} \in \operatorname{Ker}\Phi \cap L^\infty(\Omega; \mathbf{R}^3)$. From (3.9) it follows

$$\Big(\mathbf{u} + \lambda_k(s\mathbf{w} + \mathbf{y}_k), s\Big) \in \mathcal{A}$$

for $\mathbf{y}_k$ given by (3.11). Thus, again, by optimality

$$\mathcal{F}\Big(\mathbf{u}+\lambda_k(s\mathbf{w}+\mathbf{y}_k),s\Big)-\mathcal{F}(\mathbf{u},s)\geq 0.$$

This can be written as

$$\frac{\mathcal{F}\Big(\mathbf{u}+\lambda_k^2\overline{\mathbf{y}}_k+\lambda_k s\mathbf{w},s\Big)-\mathcal{F}(\mathbf{u}+\lambda_k^2\overline{\mathbf{y}}_k,s)}{\lambda_k}+\frac{\mathcal{F}(\mathbf{u}+\lambda_k^2\overline{\mathbf{y}}_k,s)-\mathcal{F}(\mathbf{u},s)}{\lambda_k}\geq 0,$$

where

$$\overline{\mathbf{y}}_k:=-\mathbf{u}\frac{|\mathbf{w}|^2}{\sqrt{1-\lambda_k^2|\mathbf{w}|^2}+1},$$

with the later term tending to 0 as $k\to\infty$. Applying Fatou's lemma with respect to the integral involving h we easily arrive at (4.4). To get the validity of (4.4) for $\mathbf{w}\in\operatorname{Ker}\Phi$ with $s\mathbf{w}\in H^1(\Omega;\mathbf{R}^3)$ we use the truncation procedure as described in Proposition 3.4. The details will be omitted here. The proof is complete. □

Remark 4.2. Taking into account that $|\mathbf{u}|^2=|s|^2$ we get

$$\nabla\mathbf{u}\cdot\nabla(\varphi\mathbf{u})=\varphi|\nabla\mathbf{u}|^2+\nabla\mathbf{u}\cdot(\mathbf{u}\otimes\nabla\varphi)=\varphi|\nabla\mathbf{u}|^2+s\nabla s\cdot\nabla\varphi,$$

from which the equivalent formulation of (3.16) results

$$2\int_\Omega[|\nabla\mathbf{u}|^2(\varphi-1)+s\nabla s\cdot(\nabla\varphi-\nabla 1)+(k-1)\nabla s\cdot(\nabla(\varphi s)-\nabla s)]d\Omega$$
$$+\int_\Omega h^0\Big(\mathbf{u},s;\varphi\mathbf{u}-\mathbf{u},\varphi s-s\Big)d\Omega+\int_\Omega I_{[-\frac{1}{2},1]}(\varphi s)d\Omega\geq 0. \qquad (4.3')$$

which has to hold for any $\varphi\in H^1(\Omega)$ with $\varphi\mathbf{u}\in H^1(\Omega,\mathbf{R}^3)$.

Remark 4.3. As far as the theory of liquid crystals is concerned the case in which $\widetilde{h}(\cdot,\cdot)$ can be expressed as $\widetilde{h}(\mathbf{n},s)=g(\mathbf{n})+\sigma(s)$ for some locally Lipschitz $g(\cdot)$ and $\sigma(\cdot)$ is of importance. Note that if $g(\mathbf{0})\leq g(\mathbf{x})$ for any $\mathbf{x}\in\mathbf{R}^3$ with $|\mathbf{x}|=1$, then the function

$$h(\mathbf{u},s):=g(\frac{\mathbf{u}}{s})+\sigma(s)$$

fulfills (H1) – (H3) and under the convention $g^0(\mathbf{0};\cdot)=0$ the necessary conditions (4.3) and (4.4) take the form:

$$2\int_\Omega\Big[\nabla(s\mathbf{n})\cdot\Big(\nabla(\varphi s\mathbf{n})-\nabla(s\mathbf{n})\Big)+(k-1)\nabla s\cdot\Big(\nabla(\varphi s)-\nabla s\Big)\Big]d\Omega$$
$$+\int_\Omega\sigma^0\Big(s,\varphi s-s\Big)d\Omega+\int_\Omega I_{[-\frac{1}{2},1]}(\varphi s)d\Omega\geq 0$$
$$\forall\,\varphi\in H^1(\Omega) \qquad (4.5)$$

and

$$2\int_{\Omega} \nabla(s\mathbf{n}) \cdot \nabla(s\mathbf{w}) d\Omega + \int_{\Omega} g^0\Big(\mathbf{n}; \mathbf{w}\Big) d\Omega + \mathcal{I}_{\mathrm{Ker}\,\Phi}(s\mathbf{w}) \geq 0 \qquad \forall\, \mathbf{w} \in H^1(\Omega; \mathbf{R}^3) \tag{4.6}$$

where only such φ's and $\mathbf{w}$'s are allowed that $s\varphi \in H^1(\Omega)$ and $s\mathbf{w} \in H^1(\Omega; \mathbf{R}^3)$.

References

[1] F.H Lin, Nonlinear theory of defects in nematic liquid crystals; Phase transition and flow phenomena, Comm. Pure Appl. Math. XLII (1989) 789-814.

[2] L. Ambrosio, E.G. Virga, A boundary-value problem for nematic liquid crystals with variable degree of orientation, Arch. Rational Mech. Anal. 114 (1991) 335-347.

[3] L. Ambrosio, Existence of minimal energy configurations of nematic liquid crystals with variable degree of orientation, Manuscripta Math. 68 (1990) 215-228.

[4] F.H.Lin, On nematic liquid crystals with variable degree of orientation, Comm. Pure Appl. Math. XLIV (1991) 453-468.

[5] J.L. Ericksen, Liquid crystals with variable degree of orientation, Arch. Rational Mech. Anal. 113 (1991) 97-120.

[6] E.G. Virga, *Variational Theories for Liquid Crystals*, Chapman & Hall London 1994.

[7] Z. Naniewicz, Hemivariational Inequalities as Necessary Conditions for Optimality for a Class of Nonsmooth Nonconvex Functionals, Nonlinear World, 4 (1997) 117-133.

[8] D. Motreanu and P.D. Panagiotopoulos, Double eigenvalue problems for hemivariational inequalities, Arch. Rational Mech. Anal. (in press).

A C PAYATAKES, G N CONSTANTINIDES AND M S VALAVANIDES

Hierarchical theoretical models: an informal introduction

1. Introduction

Many physical and physicochemical processes of great practical interest do *not* yield to "one-shot" theoretical analysis and modeling. This intransigence may be due to one or both of the following causes. First, the vast magnitude of the computational effort that "one-shot" modeling/simulation would require (e.g., use of molecular dynamics to predict the macroscopic flow properties of polymer melts or concentrated solutions based on the molecular structure of the macromolecules). Second, the fact that the process itself may be affected strongly by factors residing at several different length and/or occuring over widely different time scales. In such cases a rational and cost effective (not to say "the only feasible") approach is Hierarchical Theoretical Modeling (HTM).

In recent years HTM has become the object of *systematic* development. A particularly important case is the hierarchical procedure that has been developed to model flows of fluids that have internal microstructure, such as polymer solutions and polymer melts, an excellent presentation of which is given in [2]. Another very important case is the hierarchies of theoretical and simulation methods that are used to span a wide spectrum of time-and-length scales in order to achieve quantitative prediction of the macroscopic mechanical properties of polymeric and other materials from their molecular structure [9].

In the present work we will present a substantially different HTM approach, which has been designed to deal with the flow of two immiscible Newtonian fluids through a permeable porous medium. The reason that we have selected this particular example is twofold. First, the physical process itself is of considerable practical interest, since it is encountered in oil recovery from reservoir rocks, in soil and acquifer contamination by liquid organic wastes, in certain processes of soil reconstitution, etc. Second, the example demonstrates the crucial fact that the HTM method to be used in each case depends on the particular features of the physicochemical process being modeled. Consequently, one has to custom-design the HTM for the particular physical system or process.

The main stages of the proposed hierarchical model are: (a) pore-scale models, (b) a microflow network-ype simulator, and (c) the ganglion population balance equations (integrodifferential). The basic physics is incorporated into the pore-scale models which contain the system parameters. The fundamental information on the cooperative flow behavior of the two fluid phases at the scale of a few thousand pores is codified into a few system factors, which are calculated as functions of the system parameters using the simulator. Finally, the system factors are utilized by the population balance equations which are integrated to predict the macroscopic behavior of the process.

2. The Physical Process: Steady-State Two-Phase Flow in Porous Media (SS2φFPM)

Consider the simultaneous flow of two immiscible fluids through a porous medium. One of the fluids is wetting ("water") and the other non-wetting ("oil"). To fix ideas, we will choose as porous medium a typical reservoir rock, say, Berea sandstone. This rock is the result of diagnetic consolidation of sand grains that had accumulated through sedimentation, and is permeated by a three-dimensional network of pores. One class of the pores – those at the nodes of the network – are relatively large (average volumetric diameter ca 100 μm) and are called *chambers*. The pores that connect the chambers are narrower (*ca* 0.5 to 20 μm) and are called *throats*. The coordination number (number of throats connected to one chamber) is in the range from 1 to *ca* 7, with an average value *ca* 5. The porosity (pore volume fraction) is *ca* 20%.

The flow mode that we will consider is simultaneous forced flow of both oil and water at preselected flowrates. [Other possible modes could be: (a) imbibition, that is, displacement of oil by water, and (c) drainage, that is, displacement of water by oil, but those will not be considered here to avoid complicating the issue.]

3. Hierarchical Theoretical Modeling of SS2φFPM

3.1 Outline

A diagram of the HTM approach that we have developed for SS2φFPM is given in Fig. 1. Note that it would be foolhardy to attempt theoretical modeling and simulation of such a complex process without guidance from experimental observations and measurements. This is the reason that the "boxes" representing experimental work are *integral* parts of the overall scheme.

To begin with, a prerequisite to the entire process (not even indicated in Fig. 1) is the thorough characterization of the porous medium, that is the determination of (a) the topological characteristics of the pore skeleton, (b) the geometrical characteristics of the chambers and throats, and (c) any correlational characteristics of the pore network. The procedure for such characterization is described in [7], [10], [11], [12], and [14]. This information is used, in turn, to develop a theoretical pore network model that is a reasonable idealization of the actual porous medium [3].

Experiments are required on at least two different length scales, Fig. 1. Pore scale experiments span the region from, say, one pore to *ca* 10^5 pores and are used to determine the pore scale flow mechanism and the concomitant cooperative effects. The observations and quantitative measurements from these experiments provide the basis for the development of analytical (or semi-analytical) *pore-scale models* of the diverse individual flow phenomena (see below). The set of physical and geometrical parameters that appear in the pore-scale models comprise the set of *system parameters*.

The cooperative effects of the pore-scale flow phenomenon are modeled with a pore network simulator (Ganglion Dynamics Simulator) which predicts the collective behavior of a large number of fluid elements, so that statistically valid results can be obtained. The collective behavior is codified in terms of the *system factors*.

The system factors, in turn, appear as parameters in a system of two integrodifferential population balance equations which, when integrated, predict the macroscopic behavior of the process as a function of all the system parameters. Core-scale experiments are required to validate the macroscopic theoretical predictions.

A more detailed discussion follows.

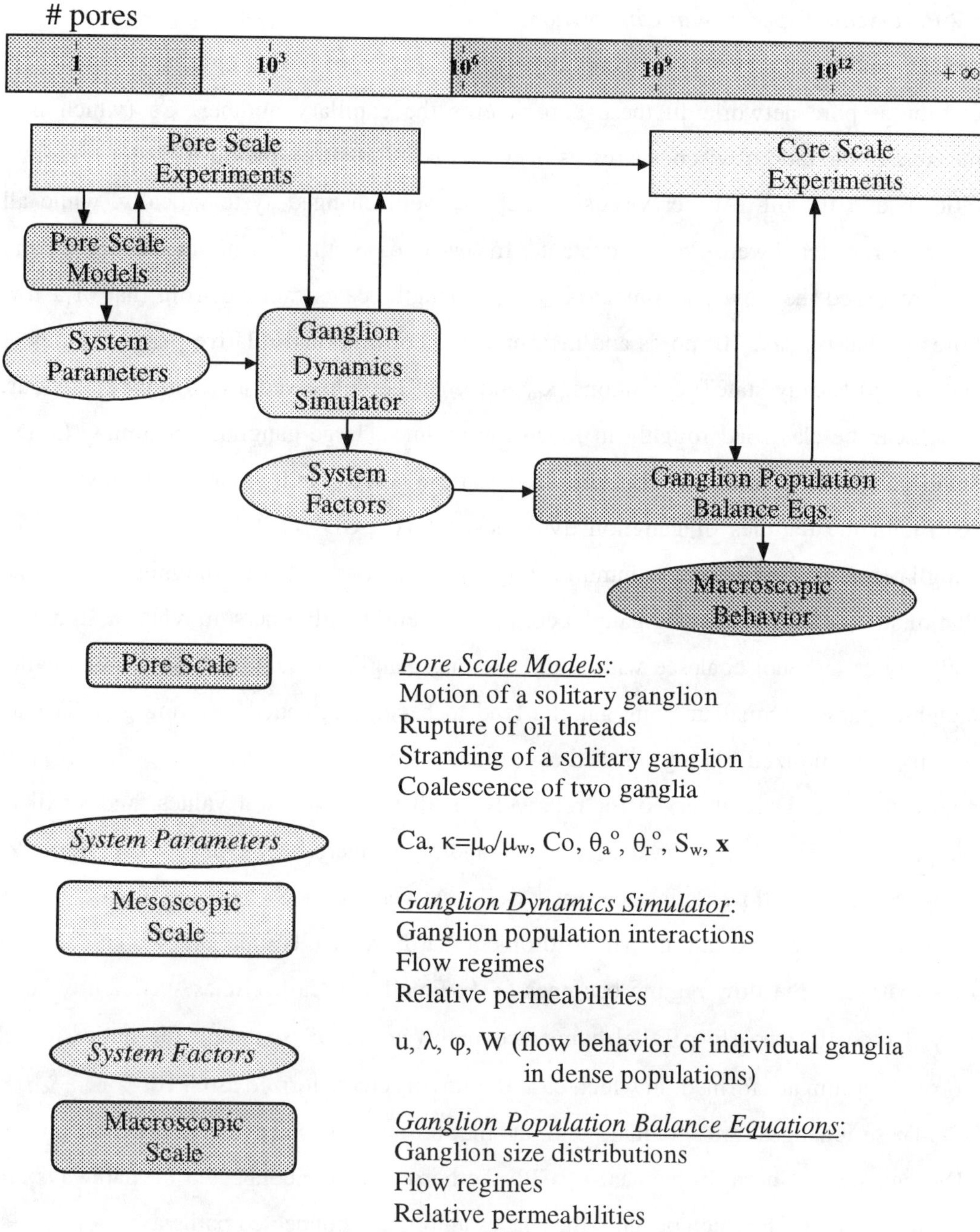

Figure 1: Hierarchical theoretical modeling and simulation of two-phase flow in porous media

3.2 Pore-Scale Experimental Observations

Avraam and Payatakes [1] conducted experiments of SS2φFPM using a model chamber-and-throat pore network. In these experiments the capillary number, Ca (which is a measure of the ratio of viscous stresses over capillary pressure), the oil-to-water flow-rate ratio, r, and the oil-to-water viscosity ratio, κ, were changed systematically, while all other parameters were kept constant. In each case these authors observed and videorecorded the flow phenomena at various length scales ranging from that of a few pores to that of, say, 10^4 pores and measured the conventional relative permeabilities at stationary ("steady state") conditions, k_{ro} and k_{rw}. They observed a rich flow behaviour, which can be classified roughly in four flow regimes: large-ganglion dynamics (LGD), small-ganglion dynamics (SGD), drop-traffic flow (DTF), and connected pathway flow (CPF). In the regimes of ganglion dynamics, GD (=LGD & SGD), the population of ganglia is composed of two communicating subpopulations, that of moving ganglia and that of stranded ones. Moving ganglia collide frequently with others, moving or stranded, and may or may not coalesce with them. Moving ganglia frequently fission into smaller daughter ganglia. Small moving ganglia tend to become stranded. Moving ganglia may become remobilized, being pushed by other moving ones, with or without prior coalescence. LGD is observed for relatively small to medium Ca values, and small to medium oil saturation, S_o, values. As Ca and/or S_o increase the mean ganglion size decreases and the flow regime becomes SGD. As Ca and/or S_o increase further, the oil becomes disconnected into numerous droplets that have size comparable to the diameter of throats, and the flow regime becomes DTF. Finally, if Ca becomes sufficiently large (say, $Ca \geq 5\times10^{-6}$) nearly parallel connected pathways of oil spanning the entire model porous medium are formed, in which case the flow is characterized as CPF.

These changes are gradual, and at the borders between flow regimes appear characteristics of both. In the case of CPF the borders of the connected oil pathways are surrounded by DTF which occupy the spaces among the connected pathways.

3.3 Pore-Scale Theoretical Models

From the above experimental observations certain pore-scale flow events emerge as of central importance, specifically: (a) ganglion stranding and mobilization; (b) ganglion motion; (c) ganglion fission (or breakup); (d) ganglion collision and coalescence. An analytical criterion for ganglion stranding and mobilization is given in [8]. A computer aided simulation predicting the motion of solitary ganglia in porous media was developed in [6]. Analytical criteria for ganglion breakup, Fig. 2(a), are given in [8] and [6]. Finally, a computer-aided simulator of ganglion-ganglion collision and possible coalescence in porous media, Fig. 2(b), was developed in [4]. Fig. 3 shows a typical simulated collision

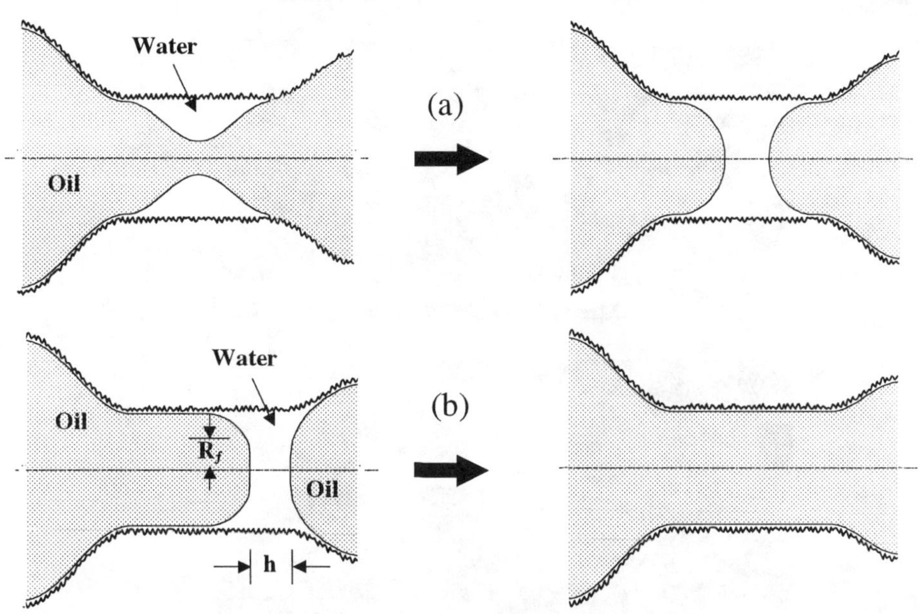

Figure 2: Pore scale mechanisms of two-phase flow in porous media:
a. Disconnection of oil by pinch-off
b. Collision and coalescence of two oil bodies

resulting in the coalescence of two ganglia. It is important to point out that this simulation involves calculations on two different length scales (the "inner problem" of the drainage of the water film that is inteposed between the two colliding ganglia within a throat, and the "outer problem" which consists of the motion of the two ganglia on the scale of several pores), as well as on two different time scales (slow film drainage versus "fast" ganglion motion).

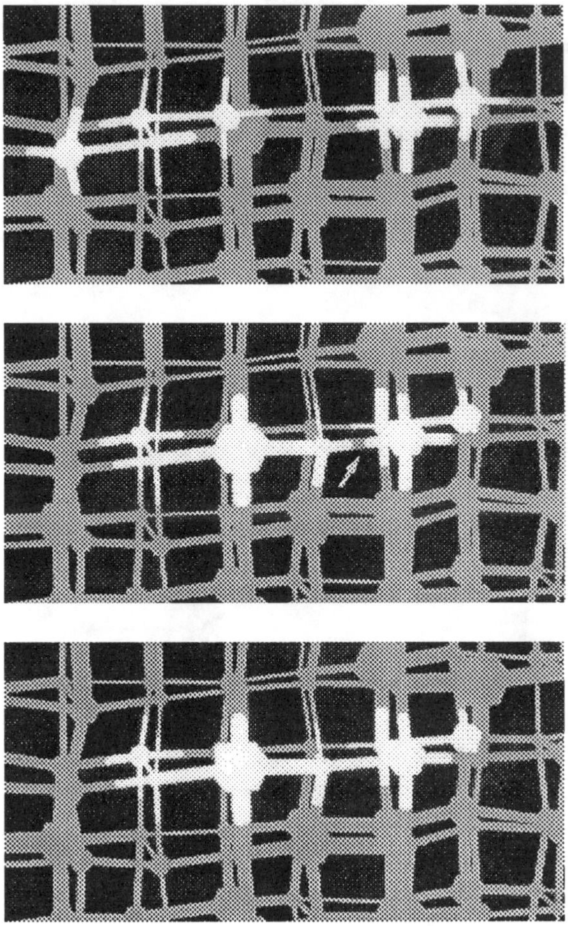

t*=5

t*=6

t*=6.2

Figure 3: Stages of simulation of collision and coalescence of a pair of ganglia for $Ca=10^{-5}$, $\kappa=0.5$, $\theta_a^0=45°$, $\theta_r^0=35°$ in a two-layer pore network representing a typical Berea sandstone.

3.4 Ganglion Dynamics Simulator

The collective behavior of a large population of interacting ganglia (moving and stranded) can be predicted with a network-type simulator [5]. This simulator incorporates all the pore-scale theoretical models cited in Section 3.3, and includes all the pertinent cooperative network-wide effects. Typical close-up snapshots of a ganglion population at two moments separated by a small time difference are shown in Fig. 4.

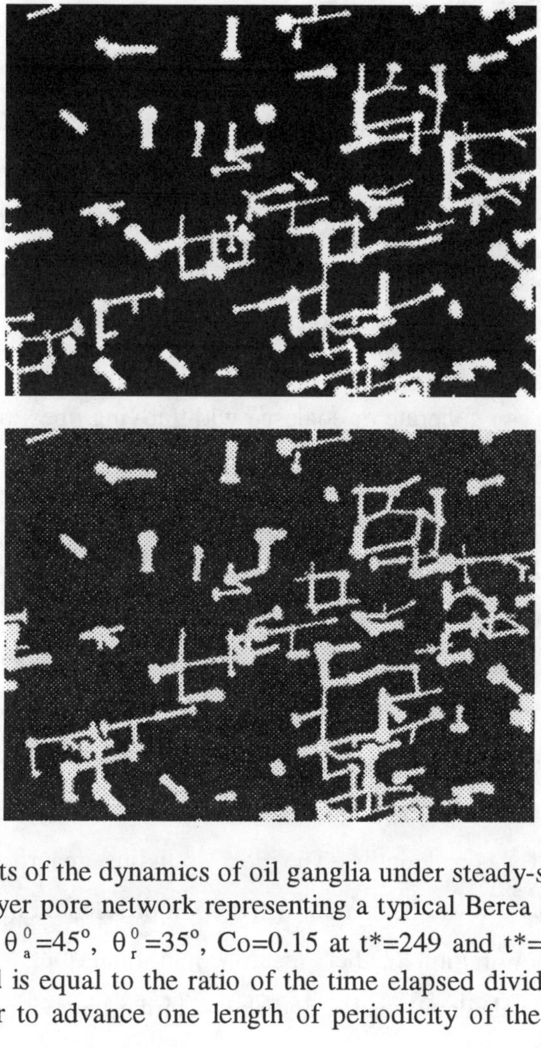

t*=249

t*=250

Figure 4: Snapshots of the dynamics of oil ganglia under steady-state flow conditions in a two-layer pore network representing a typical Berea sandstone for $Ca=10^{-5}$, $\kappa=3.35$, $\theta_a^0=45°$, $\theta_r^0=35°$, Co=0.15 at t*=249 and t*=250 (t* is the reduced time and is equal to the ratio of the time elapsed divided by the time needed by water to advance one length of periodicity of the network). Only oil is shown.

The collective behavior of the entire population can be codified in terms of the so called *system factors*, namely: the average ganglion velocity, u; the ganglion stranding coefficient, λ; the ganglion breakup coefficient, W. These are functions of the system parameters (Fig. 1, and Section 3.1) and can be calculated as such using the aforementioned GD simulator, [5] and [13]. The system factor values for a given set of values of the system parameters is used as input to population balance equations described below.

3.5 Ganglion Population Balance Equations

The two integrodifferential population balance equations that describe the collective flow behavior in the case of ganglion dynamics, steady-state as well as transient, are given in [8] and in slightly improved form in [13]. These equations are not presented here because of space limitations. It suffices to say that: (a) the first of these equations is a balance of moving ganglia of all sizes; (b) the second equation is a balance of stranded ganglia of all sizes; (c) the two equations are coupled because moving ganglia can generate stranded ones and stranded ones can generate or coalesce with moving ones; (d) the system factors that are calculated with the Ganglion Dynamics Simulator, Section 3.4, appear as parameters in the balance equations.

Integration of the ganglion population balance equations, with the appropriate initial and boundary conditions, is done numerically and it produces a wealth of information about the macroscopic behavior of the system. For example, Fig. 5 shows that in SS2φFPM the ganglion size distribution of moving ganglia, n, and that of stranded ganglia, σ, evolve into fully-developed profiles, even if one feeds the fluids into the porous medium with very different inlet size distributions. Fully-developed profiles are achieved within *ca* 10^3 pore-lengths. The corresponding macroscopic quantities are readily calculated; Fig. 6 shows the spatial distribution of selected macroscopic quantities in the developing flow region. Other macroscopic quantities, notably the relative permeabilities, are also calculated readily and show similar behavior.

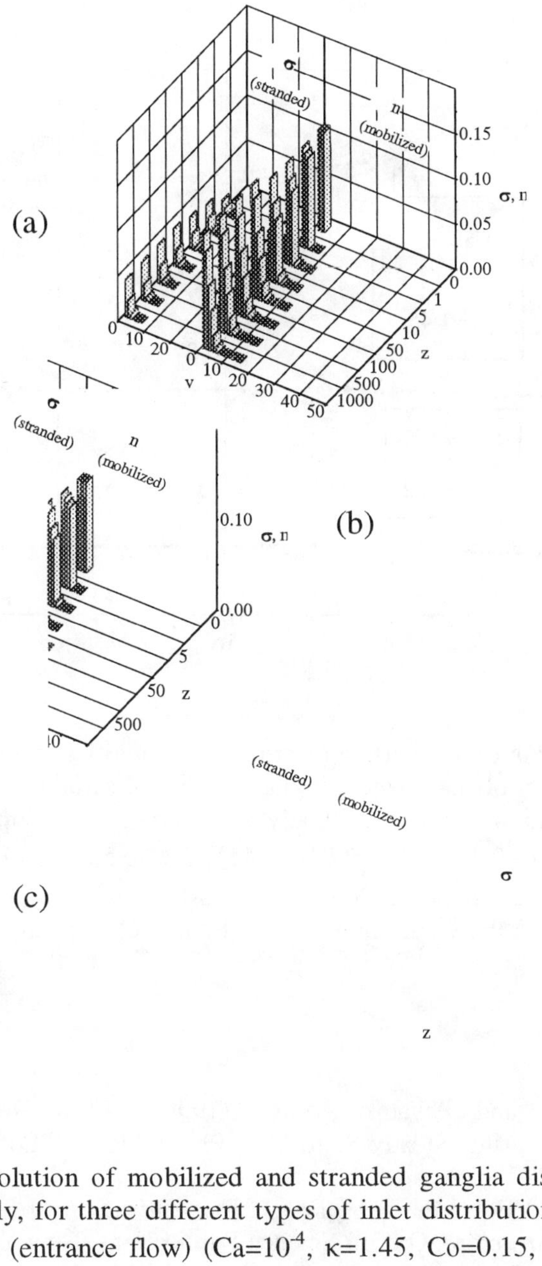

Figure 5: Spatial evolution of mobilized and stranded ganglia distributions, n and σ, respectively, for three different types of inlet distribution, under steady-state conditions (entrance flow) ($Ca=10^{-4}$, κ=1.45, Co=0.15, $\theta_a^0 = 45°$, $\theta_a^0 = 45°$, $S_{o,in}=0.5$ and r=0.2). Theoretical results are shown for: (a) unimodal, (b) uniform, and (c) triagonal, inlet distribution.

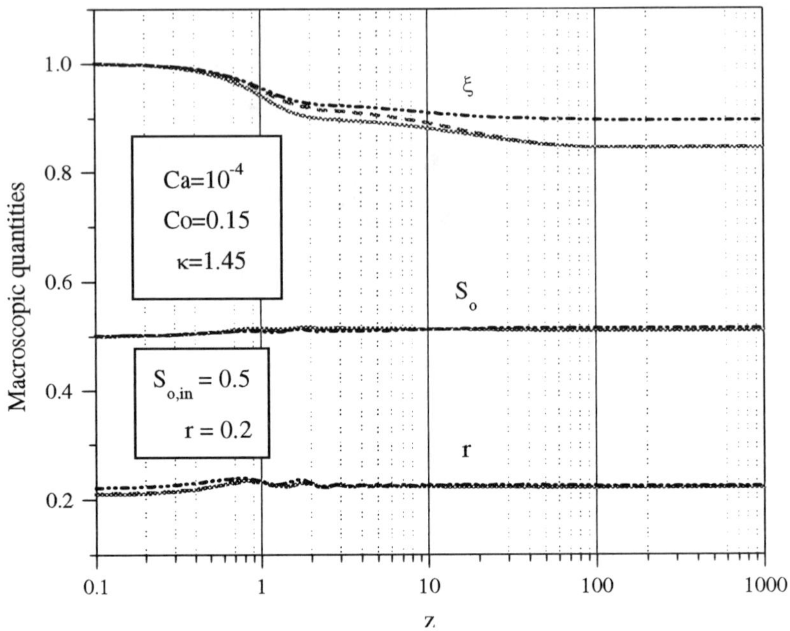

Figure 6: Spatial evolution of certain macroscopic quantities, namely, the fraction of oil in the form of mobilized ganglia, ξ_o, the oil saturation, S_o and the oil/water flowrate ratio, r, for three steady state entrance development flows (Ca=10^{-4}, κ=1.45, Co=0.15, $\theta_a^0 = 45°$, $\theta_a^0 = 45°$, $S_{o,in}$=0.5 and r=0.2) with different inlet ganglion distributions. Continuous, dashed and dash-dotted lines correspond to the flows shown in Figures 5(a), (b) and (c) respectively.

4. References

[1] Avraam, D.G. and Payatakes, A.C. (1995), "Flow Regimes and Relative Permeabilities During Steady-State Two-Phase Flow in Porous Media", *J. Fluid Mech.*, **293**, 207-236.

[2] Beris, A.N., and Edwards, B.J. (1994), *Thermodynamics of Flowing Systems with Internal Microstructure*, Oxford Science Publications, New York.

[3] Constantinides, G.N. and Payatakes, A.C. (1989), "A Three-Dimensional Network Model for Consolidated Porous Media. Basic Studies," *Chem. Eng. Commun.*, **81**, 55-81.

[4] Constantinides G.N. and Payatakes, A.C. (1991), A Theoretical Model of Collision and Coalescence of Ganglia in Porous Media," *J. Colloid and Interface Sci.*, **141**, 486-504.

[5] Constantinides, G.N. and Payatakes, A.C. (1996), "Network Simulation of Steady-State Two-Phase Flow in Consolidated Porous Media", *AIChE J.*, **42**, 369-382.

[6] Dias, M.M. and Payatakes, A.C. (1986), "Network model for two-phase flow in porous media. Part 2. Motion of oil ganglia" , *J. Fluid Mech.* , **164**, 337-358.

[7] Lymberopoulos D.P, and Payatakes, A.C. (1991), "Derivation of Topological, Geometrical and Correlational Properties of Porous Media from Pore-Chart Analysis of Serial Section Data", *J. Colloid Interface Sci.*, **150**, 61-80.

[8] Payatakes, A.C. (1982) , "Dynamics of Oil Ganglia during Immiscible Displacement in Water-Wet Porous Media" , *Ann. Rev. Fluid Mech.* , **14**, 365-393.

[9] Theodorou, D.N. (1994), "Editorial", *Chem. Eng. Sci.*, **49**, 2715-2716 (See also the "Symposium in Print" hosted in the same issue of this journal).

[10] Tsakiroglou, C.D. and Payatakes, A.C. (1990), "A new simulator of mercury porosimetry for the characterization of porous materials," *J. Colloid Interface Sci.*, **137**, 315-339.

[11] Tsakiroglou, C.D. and Payatakes, A.C. (1991), "Effects of pore size correlations on mercury porosimetry curves," *J. Colloid Interface Sci.*, **146**, 479-494.

[12] Tsakiroglou, C.D. and Payatakes, A.C. (1993), , "Pore-Wall Roughness as a Fractal Surface and Theoretical Simulation of Mercury Intrusion/Retraction in Porous Media", *J. Colloid Interface Sci.*, **159**, 287-301.

[13] Valavanides, M.S. , Constantinides, G.N., and Payatakes, A.C. (1998) , "Mechanistic Model of Steady-State Two-Phase Flow in Porous Media Based on Ganglion Dynamics" , *Transp. Porous Media* (in press).

[14] Yao, J., Thovert, J.-F., Adler, P.M., Burganos, V.N., Payatakes, A.C., Moulu, J.-C., and Kalaydjian, F. (1997), "Characterization Reconstruction and Transport Properties of Vosges Sandstones", *Revue de l' Institute Francais du Petrole*, **52**, (No 1), 3-21.

SECTION C:
Mathematical Theory of Multiple Scattering

CHRISTODOULOS ATHANASIADIS AND IOANNIS G STRATIS

On the scattering of electromagnetic waves by a piecewise homogeneous chiral obstacle

Dedicated to Professor Gary F. Roach on the occasion of his retirement

1 Introduction

During much of the 19th century and until quite recently in this century, electromagnetics researchers focussed on either vacuum or metals or certain dielectric media; some attention was paid to magnetic materials as well, mainly at low frequencies. During the 1960s, however, attention began to be sporadically focussed on general electromagnetic media. This picture began to dramatically alter during the mid – 1980s. Chiral media arrived on the scene, with the possibility of being technologically significant at microwave frequencies. This became possible owing to huge advances in polymer sciences (biomimetic materials, extremely long–chain polymers with chiral conformations, the recognition of enantioselectivity by the pharmaceutical industry, thin film technology, electromagnetically controllable smart materials for transduction and actuation, etc.). The electromagnetics of complex media has become a truly multidisciplinary research area spanning the bridge from mathematical, theoretical and experimental research at universities to industrial production of a diverse array of electrical, microwave, infrared and optical materials and devices. This variety of topics is very well represented up–to–date in [12].

Chiral media are examples of media responding with both magnetic and electric polarization to either electric or magnetic excitation. Thus they belong to the general class of bianisotropic media. Chiral media can be characterized by different sets of constitutive relations; in this work we employ the Drude–Born–Fedorov equations. For interesting explanation of these equations and various aspects of propagation problems regarding chiral media, we refer to the monograph [10] by Lakhtakia and to the references therein. The original investigations of the effect of chirality on the polarization of light date back to the 19th century. Arago, Biot, Cauchy, Pasteur and Fresnel all examined optical activity in chiral media.

In [2], [3] the present authors investigated the scattering problems by a chiral and a general bi–isotropic medium, respectively, embedded in achiral free space; they proved existence and uniqueness of the solutions of these problems by weak solutions approaches. Moreover, in [4] they established that the inverse chiral scattering problem is uniquely solvable if the far field patterns coincide for all pairs of incident directions and polarizations.

The purpose of the present work is to establish existence and uniqueness of the solution of the scattering problem arising when a plane, time–harmonic, electromagnetic wave propagating in achiral free space is incident upon a multi–layered chiral scatterer. Moreover, an integral representation for the total exterior field is constructed. The paper is organized as follows: in Section 2 we describe the multi–layered chiral scatterer and formulate the scattering problem. In Section 3 we prove that this scattering problem has a unique solution. In Section 4 we obtain an integral representation for the total exterior electric field in which are incorporated the transmission conditions and the behaviour at infinity. Moreover, we derive the electric far field pattern of the considered problem.

2 Formulation of the problem

The multi–layered chiral scatterer Ω is a bounded, convex and closed subset of $\mathbb{R}^3$, with a C^2–boundary S_0. The interior of Ω is divided by means of closed and nonintersecting C^2–surfaces into subsets (layers) Ω_j, with $\partial\Omega_j \cap \partial\Omega_{j+1} = S_j$, $j = 1, 2, \ldots, N-1$. The surface S_{j-1} surrounds S_j and there is one normal unit vector $\mathbf{n}(\mathbf{r})$ at each point $\mathbf{r}$ of any surface S_j pointing at Ω_j. Each Ω_j, $j = 1, 2, \ldots, N$ is occupied by a homogeneous isotropic chiral medium of electric permittivity ε_j, magnetic permeability μ_j, chirality measure β_j, and vanishing conductivity.

The exterior Ω_0 of the scatterer is an infinite homogeneous isotropic achiral medium of electric permittivity ε_0, magnetic permeability μ_0 and vanishing conductivity.

We shall consider the scattering of time–harmonic electromagnetic plane waves by the multi–layered chiral scatterer Ω. Let $(\mathbf{E}^{\rm inc}, \mathbf{H}^{\rm inc})$ and $(\mathbf{E}^{\rm sc}, \mathbf{H}^{\rm sc})$ be the incident and the scattered waves, respectively. $(\mathbf{E}^{\rm sc}, \mathbf{H}^{\rm sc})$ satisfies the well known Silver–Müller radiation condition [5], where $k_0 = \omega(\varepsilon_0\, \mu_0)^{1/2}$ is the wave number in Ω_0, and ω is the angular frequency,

$$\omega\mu_0\, \mathbf{H}^{\rm sc}(\mathbf{r}) \times \frac{\mathbf{r}}{r} - k_0\, \mathbf{E}^{\rm sc}(\mathbf{r}) = o(r^{-1}), \quad r \to \infty, \tag{2.1}$$

uniformly in all directions $\mathbf{r}/r$. The total exterior field $(\mathbf{E}_0, \mathbf{H}_0)$ in Ω_0 is given by

$$\mathbf{E}_0 = \mathbf{E}^{\rm sc} + \mathbf{E}^{\rm inc}, \quad \mathbf{H}_0 = \mathbf{H}^{\rm sc} + \mathbf{H}^{\rm inc}, \tag{2.2}$$

and satisfies the Maxwell equations

$$\nabla \times \mathbf{E}_0 = i\omega\mu_0\, \mathbf{H}_0, \tag{2.3}$$

$$\nabla \times \mathbf{H}_0 = -i\omega\varepsilon_0\, \mathbf{E}_0. \tag{2.4}$$

The total field $(\mathbf{E}_j, \mathbf{H}_j)$ in each Ω_j, $j = 1, 2, \ldots, N$, satisfies the equations

$$\nabla \times \mathbf{E}_j = i\omega\mu_j\, \frac{\gamma_j^2}{k_j^2}\, \mathbf{H}_j + \beta_j\gamma_j^2\, \mathbf{E}_j, \tag{2.5}$$

$$\nabla \times \mathbf{H}_j = -i\omega\varepsilon_j\, \frac{\gamma_j^2}{k_j^2}\, \mathbf{E}_j + \beta_j\gamma_j^2\, \mathbf{H}_j, \tag{2.6}$$

where

$$\gamma_j^2 = \frac{k_j^2}{1 - k_j^2\beta_j^2}, \quad \text{with} \quad k_j^2 = \omega^2 \varepsilon_j\, \mu_j\,. \tag{2.7}$$

Let us note that for $j = 1, 2. \ldots, N$, k_j is simply a shorthand notation and does not represent a wave number. The equations (2.5) and (2.6) are derived by the source–free Maxwell curl postulates

$$\nabla \times \mathbf{E}_j = i\omega \mathbf{B}_j\,, \;\; \nabla \times \mathbf{H}_j = -i\omega \mathbf{D}_j \tag{2.8}$$

along with the Drude–Born–Fedorov constitutive relations

$$\mathbf{D}_j = \varepsilon_j(\mathbf{E}_j + \beta_j \nabla \times \mathbf{E}_j)\,,\;\; \mathbf{B}_j = \mu_j(\mathbf{H}_j + \beta_j \nabla \times \mathbf{H}_j). \tag{2.9}$$

For the physical meaning of β_j, γ_j, k_j and more details about the constitutive relations for chiral media we refer to [10]. We note that any solution of (2.3)–(2.6) is divergence free.

On the surfaces S_j, $j = 0, 1, \ldots, N-1$ we have the transmission conditions

$$\mathbf{n} \times \mathbf{E}_j = \mathbf{n} \times \mathbf{E}_{j+1}, \quad \mathbf{n} \times \mathbf{H}_j = \mathbf{n} \times \mathbf{H}_{j+1} . \tag{2.10}$$

For subsequent reference let us recall the Bohren transform [10] in Ω_j, $j = 1, 2, \ldots, N$

$$\mathbf{E}_j = \mathbf{Q}_L^{(j)} + i\eta_j \, \mathbf{Q}_R^{(j)}, \quad \mathbf{H}_j = \frac{1}{i\eta_j} \mathbf{Q}_L^{(j)} + \mathbf{Q}_R^{(j)}, \tag{2.11}$$

where $\mathbf{Q}_L^{(j)}$ is a left–handed and $\mathbf{Q}_R^{(j)}$ is a right–handed Beltrami field satisfying

$$\nabla \times \mathbf{Q}_L^{(j)} = \gamma_L^{(j)} \, \mathbf{Q}_L^{(j)} \;\text{ and }\; \nabla \times \mathbf{Q}_R^{(j)} = -\gamma_R^{(j)} \, \mathbf{Q}_R^{(j)} \text{ in } \Omega_j, \tag{2.12}$$

with $\eta_j = (\mu_j/\varepsilon_j)^{\frac{1}{2}}$ being the intrinsic impedance of the chiral medium in Ω_j, and the wave numbers corresponding to $\mathbf{Q}_L^{(j)}$, $\mathbf{Q}_R^{(j)}$ being given, respectively, by

$$\gamma_L^{(j)} = k_j(1 - k_j\beta_j)^{-1}, \quad \gamma_R^{(j)} = k_j(1 + k_j\beta_j)^{-1}. \tag{2.13}$$

As in [10], it is assumed that $\operatorname{Re}\eta_j$, $\operatorname{Re}\gamma_L^{(j)}$, $\operatorname{Re}\gamma_R^{(j)} > 0$, and $\operatorname{Im}\gamma_L^{(j)}$, $\operatorname{Im}\gamma_R^{(j)} \geq 0$.

3 Solvability of the scattering problem

We proceed to study the solvability of the scattering problem described in Section 2, i.e. the following mathematical transmission problem: find $\mathbf{E}_j$, $\mathbf{H}_j \in C^2(\Omega_j) \cap C(\overline{\Omega}_j)$, $j = 0, 1, \ldots, N$ satisfying (2.1) – (2.4) in Ω_0, (2.5), (2.6) in Ω_j, $j = 1, 2, \ldots, N$, and (2.10) on S_j, $j = 0, 1, \ldots, N-1$.

The first step to proving that this transmission problem has a unique solution, is to consider the corresponding homogeneous transmission problem, i.e. take $\mathbf{E}^{\text{inc}} = \mathbf{H}^{\text{inc}} = 0$ in (2.2), and establish the following result

Theorem 3.1 *The homogeneous transmission problem has only the trivial solution.*

The proof is omitted for the sake of brevity; it involves the Bohren decomposition described in Section 2, and may be performed in a fashion analogous to the corresponding result of [2].

To prove existence of solutions for our transmission problem we follow a weak solutions approach. Let $(\mathbf{U},\mathbf{V})_D = \int_D \mathbf{U}\cdot\overline{\mathbf{V}}\,dv(\mathbf{r})$ denote the $L^2(D)$ inner product for scalar– and vector–valued functions, and for $\mathbf{U} \in L^2(\partial D)$ let $<\mathbf{U},\mathbf{V}>_{\partial D} = \int_{\partial D} \mathbf{U}\cdot\overline{\mathbf{V}}\,ds(\mathbf{r})$ denote the dual brackets in $(H^{-\frac{1}{2}}(\partial D), H^{\frac{1}{2}}(\partial D))$. Green's formulae can then be written as

$$(\nabla\times\mathbf{U},\mathbf{V})_D - (\mathbf{U},\nabla\times\mathbf{V})_D = <\mathbf{n}\times\mathbf{U},\mathbf{V}>_{\partial D}, \tag{3.1}$$

$$(\nabla\cdot\mathbf{U},v)_D + (\mathbf{U},\nabla v)_D = <\mathbf{n}\cdot\mathbf{U},v>_{\partial D}\ . \tag{3.2}$$

Our transmission problem may be compactly written as

$$\nabla\times\boldsymbol{\mathcal{E}} - i\tilde{\xi}\boldsymbol{\mathcal{H}} - \tilde{\theta}\boldsymbol{\mathcal{E}} = \mathbf{0} \ \text{ and } \ \nabla\times\boldsymbol{\mathcal{H}} + i\tilde{\lambda}\boldsymbol{\mathcal{E}} - \tilde{\theta}\boldsymbol{\mathcal{H}} = \mathbf{0}, \ \text{ in } \cup_{j=0}^{N}\Omega_j, \tag{3.3}$$

$$[\mathbf{n}\times\boldsymbol{\mathcal{E}}]_-^+ = [\mathbf{n}\times\boldsymbol{\mathcal{H}}]_-^+ = \mathbf{0}, \ \text{ on } S_j,\ j=0,1,\dots,N-1, \tag{3.4}$$

$$\omega\mu_0(\boldsymbol{\mathcal{H}} - \mathbf{H}^{\text{inc}})\times\frac{\mathbf{r}}{r} - k_0(\boldsymbol{\mathcal{E}} - \mathbf{E}^{\text{inc}}) = o(r^{-1}), \ \ r\to\infty, \tag{3.5}$$

where

$$\boldsymbol{\mathcal{E}} = \begin{cases} \mathbf{E}_0 & ,\ \text{in } \Omega_0 \\ \mathbf{E}_1 & ,\ \text{in } \Omega_1 \\ \vdots & \\ \mathbf{E}_N & ,\ \text{in } \Omega_N \end{cases}, \quad \boldsymbol{\mathcal{H}} = \begin{cases} \mathbf{H}_0 & ,\ \text{in } \Omega_0 \\ \mathbf{H}_1 & ,\ \text{in } \Omega_1 \\ \vdots & \\ \mathbf{H}_N & ,\ \text{in } \Omega_N \end{cases}, \quad \tilde{\xi} = \begin{cases} \omega\mu_0 & ,\ \text{in } \Omega_0 \\ \omega\mu_1\frac{\gamma_1^2}{k_1^2} & ,\ \text{in } \Omega_1 \\ \vdots & \\ \omega\mu_N\frac{\gamma_N^2}{k_N^2} & ,\ \text{in } \Omega_N \end{cases},$$

$$\tilde{\lambda} = \begin{cases} \omega\varepsilon_0 & ,\ \text{in } \Omega_0 \\ \omega\varepsilon_1\frac{\gamma_1^2}{k_1^2} & ,\ \text{in } \Omega_1 \\ \vdots & \\ \omega\varepsilon_N\frac{\gamma_N^2}{k_N^2} & ,\ \text{in } \Omega_N \end{cases}, \quad \tilde{\theta} = \begin{cases} 0 & ,\ \text{in } \Omega_0 \\ \beta_1\gamma_1^2 & ,\ \text{in } \Omega_1 \\ \vdots & \\ \beta_N\gamma_N^2 & ,\ \text{in } \Omega_N \end{cases} \tag{3.6}$$

and $[\mathbf{h}]_-^+ := \mathbf{h}^+ - \mathbf{h}^-$, where $\mathbf{h}^+$ ($\mathbf{h}^-$) denotes the limit of $\mathbf{h}$ on ∂D from the exterior (interior) of ∂D. The transmission problem consisting of (3.3)–(3.5) will be denoted by (P) in the sequel. Eliminating $\boldsymbol{\mathcal{H}}$ from (P) we obtain the transmission problem (P_E) consisting of

$$\Delta \boldsymbol{\mathcal{E}} + \tilde{\gamma}^2 \boldsymbol{\mathcal{E}} + 2\tilde{\beta}\tilde{\gamma}^2 \nabla \times \boldsymbol{\mathcal{E}} = \mathbf{0}\,, \quad \text{in } \cup_{j=0}^{N} \Omega_j, \tag{3.7}$$

$$[\mathbf{n} \times \boldsymbol{\mathcal{E}}]_-^+ = \left[\mathbf{n} \times \left(\frac{1}{\tilde{\xi}} \nabla \times \boldsymbol{\mathcal{E}} + \tilde{\theta} \boldsymbol{\mathcal{E}} \right) \right]_-^+ = \mathbf{0}\,, \quad \text{on } S_j,\ j = 0, 1, \ldots, N-1, \tag{3.8}$$

$$(\nabla \times (\boldsymbol{\mathcal{E}} - \mathbf{E}^{\mathrm{inc}})) \times \frac{\mathbf{r}}{r} - ik_0 (\boldsymbol{\mathcal{E}} - \mathbf{E}^{\mathrm{inc}}) = o(1)\,, \quad r \to \infty\,, \tag{3.9}$$

where

$$\tilde{\gamma} = \begin{cases} k_0 & , \text{ in } \Omega_0 \\ \gamma_1 & , \text{ in } \Omega_1 \\ \vdots & \\ \gamma_N & , \text{ in } \Omega_N \end{cases} , \quad \tilde{\beta} = \begin{cases} 0 & , \text{ in } \Omega_0 \\ \beta_1 & , \text{ in } \Omega_1 \\ \vdots & \\ \beta_N & , \text{ in } \Omega_N \end{cases} . \tag{3.10}$$

Let B_R be a ball of radius $R > 0$, centred at $\mathbf{0}$, containing $\overline{\Omega}$. Let $\boldsymbol{\mathcal{E}}$, $\boldsymbol{\mathcal{H}}$ be a classical solution of (P). Then by noting that $(\boldsymbol{\mathcal{E}}, \nabla \times V)_{\mathbb{R}^3} = (\nabla \times \boldsymbol{\mathcal{E}}, V)_{\mathbb{R}^3}$, $\forall V \in H_0^1(\mathbb{R}^3)$, we get that $\nabla \times \boldsymbol{\mathcal{E}} \in L^2(B_R)$. Moreover, one easily gets that $\nabla \cdot \boldsymbol{\mathcal{E}} \in L^2(\mathbb{R}^3)$. By standard regularity results [6], we have that $\boldsymbol{\mathcal{E}} \in H^1(B_R)$.

Using Green's formula in Ω and in $B_R \setminus \overline{\Omega}$, the equation (3.7), and the transmission conditions (3.8) we obtain

$$\begin{aligned} &\left(\frac{1}{\tilde{\xi}} \nabla \times \boldsymbol{\mathcal{E}}, \nabla \times \mathbf{V} \right)_{B_R} - (\tilde{\alpha} \boldsymbol{\mathcal{E}}, \mathbf{V})_{B_R} - (\tilde{\alpha}\tilde{\rho} \nabla \times \boldsymbol{\mathcal{E}}, \mathbf{V})_{B_R} = \\ &= \; - < \mathbf{n} \times \nabla \times \boldsymbol{\mathcal{E}}, \mathbf{V} >_{S_R} - \sum_{j=0}^{N-1} < \tilde{\beta}\tilde{\gamma}^2 \mathbf{n} \times \boldsymbol{\mathcal{E}}, \mathbf{V} >_{S_j}, \forall \mathbf{V} \in H^1(B_R), \end{aligned} \tag{3.11}$$

where $S_R = \partial B_R$, and

$$\tilde{\alpha} = \begin{cases} \omega\varepsilon_0 & , \text{ in } \Omega_0 \\ \omega\varepsilon_1 & , \text{ in } \Omega_1 \\ \vdots & \\ \omega\varepsilon_N & , \text{ in } \Omega_N \end{cases} , \quad \tilde{\rho} = \begin{cases} \frac{1}{k_0^2} & , \text{ in } \Omega_0 \\ 2\beta_1 & , \text{ in } \Omega_1 \\ \vdots & \\ 2\beta_N & , \text{ in } \Omega_N \end{cases} . \tag{3.12}$$

Now consider the Dirichlet–to–Neumann operator [7], [8], defined as

$$\mathbb{T} : H^{\frac{1}{2}}(S_R) \to H^{-\frac{1}{2}}(S_R), \quad \mathbb{T}\mathbf{g} = \frac{\partial \mathbf{u}}{\partial n} \tag{3.13}$$

where $\mathbf{u}$ is the unique solution of the exterior Dirichlet problem

$$\begin{aligned} &\Delta \mathbf{u} + k_0^2\, \mathbf{u} = \mathbf{0}\,, \quad \text{in } \mathbb{R}^3 \setminus \overline{B_R} \\ &\mathbf{u} = \mathbf{g}\,, \quad \text{on } S_R \\ &\nabla \times \mathbf{u} \times \frac{\mathbf{r}}{r} - i\, k_0\, \mathbf{u} = o(1)\,, \quad r \to \infty. \end{aligned} \tag{3.14}$$

Consider, also, the Dirichlet–to–Neumann operator $\mathbb{T}_0$ with the respect to the Laplacian in $\mathbb{R}^3 \setminus \overline{B_R}$, i.e. setting $k_0 = 0$ in (3.14). Then we may summarize the basic properties of these operators in the following

Lemma 3.1 *(i) $\mathbb{T}$ is continuous from $H^{\frac{1}{2}}(S_R)$ into $H^{-\frac{1}{2}}(S_R)$. (ii) $-\mathbb{T}_0$ is strictly coercive from $H^{\frac{1}{2}}(S_R)$ into $H^{-\frac{1}{2}}(S_R)$. (iii) $\mathbb{T} - \mathbb{T}_0$ is bounded from $H^{\frac{1}{2}}(S_R)$ into $H^{\frac{1}{2}}(S_R)$.*

The proof may be found in [8].

Incorporating the radiation condition in (3.11) by the Dirichlet–to–Neumann operator, see [3], [7], [8], we may obtain a variational formulation of the equivalent to (P_E) truncated problem in B_R. To do this we first consider the boundary operator

$$\mathbb{A} : \; H^{\frac{1}{2}}(S_R) \to H^{-\frac{1}{2}}(S_R), \;\; \mathbb{A}\mathbf{u} = \mathbf{n} \times \nabla \times \mathbf{u} - (\nabla \cdot \mathbf{u})\mathbf{n} + \frac{\partial \mathbf{u}}{\partial n} \tag{3.15}$$

which is known [7] to be a bounded operator. Next, we consider the sesquilinear form

$$\begin{aligned} \mathbb{L}(\boldsymbol{\mathcal{E}}, \mathbf{V}) \;&=\; \left(\frac{1}{\overline{\overline{\xi}}} \nabla \times \boldsymbol{\mathcal{E}}, \nabla \times \mathbf{V}\right)_{B_R} + \left(\frac{1}{\overline{\overline{\xi}}} \nabla \cdot \boldsymbol{\mathcal{E}}, \nabla \cdot \mathbf{V}\right)_{B_R} \\ &-\; (\tilde{\alpha}\boldsymbol{\mathcal{E}}, \mathbf{V})_{B_R} - (\tilde{\alpha}\tilde{\rho}\, \nabla \times \boldsymbol{\mathcal{E}}, \mathbf{V})_{B_R} + \sum_{j=0}^{N-1} < \tilde{\beta}\tilde{\gamma}^2 \mathbf{n} \times \boldsymbol{\mathcal{E}}, \mathbf{V} >_{S_j} \\ &+\; < \mathbb{A}\boldsymbol{\mathcal{E}}, \mathbf{V} >_{S_R} - < \mathbb{T}\boldsymbol{\mathcal{E}}, \mathbf{V} >_{S_R} \end{aligned} \tag{3.16}$$

for $\boldsymbol{\mathcal{E}}$, $\mathbf{V} \in H^1(B_R)$.

Then the variational formulation of the truncated problem in B_R has the form

$$\mathbb{L}(\boldsymbol{\mathcal{E}}, \mathbf{V}) = < (\frac{\partial \mathbf{E}^{\mathrm{inc}}}{\partial n} - \mathbb{T}\mathbf{E}^{\mathrm{inc}}), \mathbf{V} >_{S_R}, \;\; \forall \mathbf{V} \in H^1(B_R). \tag{3.17}$$

It is not hard to prove the following

Lemma 3.2 *The variational equation (3.17) is an equivalent weak formulation of* (P_E).

Remark 3.1 In view of Lemma 3.2, Theorem 3.1 establishes that the corresponding to (P_E) homogeneous transmission problem that has the variational formulation $\mathbb{L}(\boldsymbol{\mathcal{E}}, \mathbf{V}) = 0, \ \forall \mathbf{V} \in H^1(B_R)$, has only the trivial solution in B_R.

Now we proceed to the main result of this section

Theorem 3.2 *Assume that*

$$\mathrm{Re}\,\frac{1}{\tilde{\tilde{\xi}}} \geq 1. \tag{3.18}$$

Then for any $\mathbf{F} \in H^1(B_R)$ *there exists a unique solution* $\boldsymbol{\mathcal{E}} \in H^1(B_R)$ *of*

$$\mathbb{L}(\boldsymbol{\mathcal{E}}, \mathbf{V}) = (\mathbf{F}, \mathbf{V})_{H^1(B_R)}, \quad \forall \mathbf{V} \in H^1(B_R). \tag{3.19}$$

Proof. We first prove that $\mathbb{L}(\cdot,\cdot)$ is coercive on $H^1(B_R)$.
Using the vector identity $\nabla \times \nabla \times \mathbf{U} = \nabla(\nabla \cdot \mathbf{U}) - \Delta \mathbf{U}$, and Green's formula we obtain

$$(\nabla \mathbf{U}, \nabla \mathbf{V})_{B_R} = (\nabla \times \mathbf{U}, \nabla \times \mathbf{V})_{B_R} + (\nabla \cdot \mathbf{U}, \nabla \cdot \mathbf{V})_{B_R} + < \mathbb{A}\mathbf{U}, \mathbf{V} >_{S_R} \tag{3.20}$$

where $\nabla \mathbf{U} \cdot \nabla \mathbf{V} = \sum_{m=1}^{3} \nabla U_m \cdot \nabla V_m$, $\mathbf{U} = (U_1, U_2, U_3)$, $\mathbf{V} = (V_1, V_2, V_3)$.

Consider the sesquilinear form

$$\mathbb{M}(\mathbf{U}, \mathbf{V}) = \left(\frac{1}{\tilde{\tilde{\xi}}} \nabla \times \mathbf{U}, \nabla \times \mathbf{V}\right)_{B_R} + \left(\frac{1}{\tilde{\tilde{\xi}}} \nabla \cdot \mathbf{U}, \nabla \cdot \mathbf{V}\right)_{B_R} + < \mathbb{A}\mathbf{U}, \mathbf{V} >_{S_R}. \tag{3.21}$$

By (3.18) and (3.20) it is easily seen that $\mathbb{M}(\mathbf{U}, \mathbf{V})$ is coercive. By known results on boundary sesquilinear forms [11, Th. 3.4.2 in view of Prop. 3.2.1 and 3.2.3], it follows that

$$\mathbb{M}(\mathbf{U}, \mathbf{V}) - \sum_{j=0}^{N-1} < \tilde{\beta}\tilde{\gamma}^2 \mathbf{n} \times \mathbf{U}, \mathbf{V} >_{S_j}$$

is coercive. On the other hand since $\mathbb{T} = (\mathbb{T} - \mathbb{T}_0) + \mathbb{T}_0$, Lemma 3.1 yields that

$$\mathbb{M}(\mathbf{U}, \mathbf{V}) + \sum_{j=0}^{N-1} < \tilde{\beta}\tilde{\gamma}^2 \mathbf{n} \times \mathbf{U}, \mathbf{V} >_{S_j} - < \mathbb{T}\mathbf{U}, \mathbf{V} >_{S_R}$$

is coercive, too. Noting that $- \mid (\nabla \times \mathbf{U}, \mathbf{U})_{B_R} \mid \geq -c \parallel \mathbf{U} \parallel^2_{H^1(B_R)}$ where c is a positive constant, we conclude that $\mathbb{L}(\cdot,\cdot)$ is a coercive sesquilinear form on $H^1(B_R)$.

Now, it is well–known that $H^1(B_R)$ is compactly embedded in $L^2(B_R)$, and hence $\mathbb{L}$ is a compact perturbation of a strictly coercive sesquilinear form. Therefore, existence and uniqueness of solutions of (3.19) follows (by the Lax–Milgram Theorem) from Fredholm's Alternative; but the latter holds in view of Remark 3.1. □

Remark 3.2 Standard regularity theory (e.g. see [6], [11]) establishes that the initial transmission problem has a unique classical solution.

4 The exterior field

In this section we construct an integral representation of the total exterior electric field, in which the transmission conditions are incorporated. In order to do this, we begin with the following integral representation for the electric field, [2],

$$\begin{aligned}\mathbf{E}^{\mathrm{sc}}(\mathbf{r}) &= \int_{S_0} \Big[(\nabla \times \mathbf{E}^{\mathrm{sc}}(\mathbf{r}')) \cdot (\mathbf{n} \times \tilde{\Gamma}(\mathbf{r},\mathbf{r}')) - \\ &\quad -(\mathbf{n} \times \mathbf{E}^{\mathrm{sc}}(\mathbf{r}')) \cdot (\nabla_{\mathbf{r}'} \times \tilde{\Gamma}(\mathbf{r},\mathbf{r}'))\Big]\, ds(\mathbf{r}'), \mathbf{r} \in \Omega_0, \end{aligned} \tag{4.1}$$

where $\tilde{\Gamma}(\mathbf{r},\mathbf{r}')$ is the free–space dyadic Green's function

$$\tilde{\Gamma}(\mathbf{r},\mathbf{r}') = -\left(\tilde{I} + k_0^{-2}\nabla \otimes \nabla\right) \frac{e^{ik_0|\mathbf{r}-\mathbf{r}'|}}{4\pi \mid \mathbf{r}-\mathbf{r}' \mid}, \tag{4.2}$$

with $\tilde{I} = \hat{\mathbf{e}}_1 \otimes \hat{\mathbf{e}}_1 + \hat{\mathbf{e}}_2 \otimes \hat{\mathbf{e}}_2 + \hat{\mathbf{e}}_3 \otimes \hat{\mathbf{e}}_3$ the identity dyadic ($\hat{\mathbf{e}}_j$, $j = 1, 2, 3$, are the cartesian unit vectors). Inserting (2.2) in (4.1) and taking into account that $\mathbf{E}^{\mathrm{inc}}$ is a solution of the equation $\nabla \times \nabla \times \mathbf{U} = k_0^2\, \mathbf{U}$ in Ω_0, we obtain

$$\begin{aligned}\mathbf{E}_0(\mathbf{r}) &= \mathbf{E}^{\mathrm{inc}}(\mathbf{r}) + \int_{S_0} \Big[(\nabla \times \mathbf{E}_0(\mathbf{r}')) \cdot (\mathbf{n} \times \tilde{\Gamma}(\mathbf{r},\mathbf{r}')) - \\ &\quad -(\mathbf{n} \times \mathbf{E}_0(\mathbf{r}')) \cdot (\nabla_{\mathbf{r}'} \times \tilde{\Gamma}(\mathbf{r},\mathbf{r}'))\Big]\, ds(\mathbf{r}'). \end{aligned} \tag{4.3}$$

Using the transmission conditions on S_0, the relation (4.3) becomes

$$\mathbf{E}_0(\mathbf{r}) = \mathbf{E}^{\mathrm{inc}}(\mathbf{r}) \quad - \quad \frac{\varepsilon_1 \gamma_0^2}{\varepsilon_0 \gamma_1^2} \int_{S_0} \mathbf{n} \cdot \Big[(\nabla \times \mathbf{E}_1(\mathbf{r}')) \times \tilde{\Gamma}(\mathbf{r},\mathbf{r}')\Big]\, ds(\mathbf{r}')$$

$$
\begin{aligned}
&+ \gamma_0^2 \frac{\varepsilon_1}{\varepsilon_0}\beta_1 \int_{S_0} \mathbf{n} \cdot \left[\mathbf{E}_1(\mathbf{r}') \times \tilde{\Gamma}(\mathbf{r},\mathbf{r}')\right] ds(\mathbf{r}') \\
&- \int_{S_0} \mathbf{n} \cdot \left[\mathbf{E}_1(\mathbf{r}') \times (\nabla_{\mathbf{r}'} \times \tilde{\Gamma}(\mathbf{r},\mathbf{r}'))\right] ds(\mathbf{r}') .
\end{aligned} \tag{4.4}
$$

Now, applying successively the dyadic form of the divergence theorem, taking into account that $\mathbf{E}_j$ and $\tilde{\Gamma}$ are solutions of $\nabla \times \nabla \times \mathbf{U} = \gamma_j^2 \mathbf{U} + 2\beta_j\gamma_j\nabla \times \mathbf{U}$ in Ω_j, $j = 1, 2, \ldots, N$, and Ω_0 respectively, and using the transmission conditions

$$
\mathbf{n} \times \mathbf{E}_j = \mathbf{n} \times \mathbf{E}_{j+1}, \tag{4.5}
$$

$$
\mathbf{n} \times \nabla \times \mathbf{E}_j = \frac{\varepsilon_{j+1}}{\varepsilon_j}\frac{\gamma_j^2}{\gamma_{j+1}^2}\, \mathbf{n} \times \nabla \times \mathbf{E}_{j+1} + \gamma_j^2 \left(\beta_j - \frac{\varepsilon_{j+1}}{\varepsilon_j}\beta_{j+1}\right) \mathbf{n} \times \mathbf{E}_{j+1} , \tag{4.6}
$$

we obtain that

$$
\mathbf{E}_0(\mathbf{r}) = \mathbf{E}^{\mathrm{inc}}(\mathbf{r}) + \mathbf{E}^{\mathrm{sc(c)}}(\mathbf{r}) + \mathbf{E}^{\mathrm{sc(a)}}(\mathbf{r}), \tag{4.7}
$$

where $\mathbf{E}^{\mathrm{sc(c)}}$ and $\mathbf{E}^{\mathrm{sc(a)}}$ denote, respectively, the chiral and the achiral parts of the electric scattered field $\mathbf{E}^{\mathrm{sc}}$, and are given by

$$
\begin{aligned}
\mathbf{E}^{\mathrm{sc(c)}}(\mathbf{r}) &= k_0^2 \sum_{j=1}^{N} \frac{\varepsilon_j}{\varepsilon_0}\beta_j^2 \int_{\Omega_j} (\nabla \times \mathbf{E}_j(\mathbf{r}')) \cdot (\nabla_{\mathbf{r}'} \times \tilde{\Gamma}(\mathbf{r},\mathbf{r}'))\, dv(\mathbf{r}') \\
&- k_0^2 \sum_{j=1}^{N} \frac{\varepsilon_j}{\varepsilon_0}\beta_j^2 \int_{\Omega_j} (\nabla \times \mathbf{E}_j(\mathbf{r}')) \cdot \tilde{\Gamma}(\mathbf{r},\mathbf{r}')\, dv(\mathbf{r}') \\
&+ k_0^2 \sum_{j=1}^{N} \frac{\varepsilon_j}{\varepsilon_0}\beta_j^2 \int_{\Omega_j} \mathbf{E}_j(\mathbf{r}') \cdot (\nabla_{\mathbf{r}'} \times \tilde{\Gamma}(\mathbf{r},\mathbf{r}'))\, dv(\mathbf{r}'), \quad \mathbf{r} \in \Omega_0,
\end{aligned} \tag{4.8}
$$

and

$$
\begin{aligned}
\mathbf{E}^{\mathrm{sc(a)}}(\mathbf{r}) &= k_0^2 \sum_{j=1}^{N} \left(1 - \frac{\varepsilon_j}{\varepsilon_0}\right) \int_{\Omega_j} \mathbf{E}_j(\mathbf{r}') \cdot \tilde{\Gamma}(\mathbf{r},\mathbf{r}')\, dv(\mathbf{r}') \\
&+ \sum_{j=1}^{N} \left(\frac{\mu_0}{\mu_j} - 1\right) \int_{\Omega_j} (\nabla \times \mathbf{E}_j(\mathbf{r}')) \cdot (\nabla_{\mathbf{r}'} \times \tilde{\Gamma}(\mathbf{r},\mathbf{r}'))\, dv(\mathbf{r}'), \quad \mathbf{r} \in \Omega_0.
\end{aligned} \tag{4.9}
$$

Remark 4.1 The relation (4.6) arises from the standard dielectric condition $\mathbf{n} \times \mathbf{H}_j = \mathbf{n} \times \mathbf{H}_{j+1}$ on S_j, by substituting the magnetic field from (2.5).

The behaviour of the scattered electric wave in the region of radiation is described by the electric far field pattern $\mathbf{E}_\infty(\hat{\mathbf{r}})$, which is defined [5] by the relation

$$
\mathbf{E}^{\mathrm{sc}}(\mathbf{r}) = \mathbf{E}_\infty(\hat{\mathbf{r}})\, h(k_0 r) + O(r^{-2}), \quad r \to \infty , \tag{4.10}
$$

uniformly in all directions $\hat{\mathbf{r}} = \mathbf{r}/r$, where $h(x) = e^{ix}/(ix)$ is the zeroth–order spherical Hankel function of the first kind. Substituting the asymptotic forms, [1],

$$\begin{aligned} \tilde{\Gamma}(\mathbf{r},\mathbf{r}') &= -\frac{ik_0}{4\pi}(\tilde{I} - \hat{\mathbf{r}}\otimes\hat{\mathbf{r}})\, h(k_0 r)\, e^{-ik_0\hat{\mathbf{r}}\cdot\mathbf{r}'} + O(r^{-2})\,, \quad r\to\infty && (4.11)\\ \nabla_{\mathbf{r}'}\times\tilde{\Gamma}(\mathbf{r},\mathbf{r}') &= -\frac{k_0^2}{4\pi}(\tilde{I}\times\hat{\mathbf{r}})\, h(k_0 r)\, e^{-ik_0\hat{\mathbf{r}}\cdot\mathbf{r}'} + O(r^{-2}),\,, \quad r\to\infty && (4.12)\end{aligned}$$

in (4.8), (4.9) and in view of (4.10) we obtain

$$\mathbf{E}_\infty(\hat{\mathbf{r}}) = \mathbf{E}_\infty^{(c)}(\hat{\mathbf{r}}) + \mathbf{E}_\infty^{(a)}(\hat{\mathbf{r}})\,, \tag{4.13}$$

where

$$\begin{aligned} \mathbf{E}_\infty^{(c)}(\hat{\mathbf{r}}) &= k_0^4\sum_{j=1}^N \frac{\varepsilon_j}{\varepsilon_0}\beta_j\int_{\Omega_j}(\nabla\times\mathbf{E}_j(\mathbf{r}'))\cdot(\tilde{I}\times\hat{\mathbf{r}})\, e^{-ik_0\hat{\mathbf{r}}\cdot\mathbf{r}'}dv(\mathbf{r}')\\ &+ ik_0^3\sum_{j=1}^N \frac{\varepsilon_j}{\varepsilon_0}\beta_j\int_{\Omega_j}(\nabla\times\mathbf{E}_j(\mathbf{r}'))\cdot(\tilde{I} - \hat{\mathbf{r}}\otimes\hat{\mathbf{r}})\, e^{-ik_0\hat{\mathbf{r}}\cdot\mathbf{r}'}dv(\mathbf{r}')\\ &+ k_0^4\sum_{j=1}^N \frac{\varepsilon_j}{\varepsilon_0}\beta_j\int_{\Omega_j}\mathbf{E}_j(\mathbf{r}')\cdot(\tilde{I}\times\hat{\mathbf{r}})\, e^{-ik_0\hat{\mathbf{r}}\cdot\mathbf{r}'}dv(\mathbf{r}') && (4.14)\end{aligned}$$

and

$$\begin{aligned} \mathbf{E}_\infty^{(a)}(\hat{\mathbf{r}}) &= -ik_0^3\sum_{j=1}^N\left(1-\frac{\varepsilon_j}{\varepsilon_0}\right)\int_{\Omega_j}\mathbf{E}_j(\mathbf{r}')\cdot(\tilde{I} - \hat{\mathbf{r}}\otimes\hat{\mathbf{r}})\, e^{-ik_0\hat{\mathbf{r}}\cdot\mathbf{r}'}dv(\mathbf{r}')\\ &- k_0^2\sum_{j=1}^N\left(\frac{\mu_0}{\mu_j}-1\right)\int_{\Omega_j}(\nabla\times\mathbf{E}_j(\mathbf{r}'))\cdot(\tilde{I}\times\hat{\mathbf{r}})\, e^{-ik_0\hat{\mathbf{r}}\cdot\mathbf{r}'}dv(\mathbf{r}')\,. && (4.15)\end{aligned}$$

Remark 4.2 (i) If we let the radius of the core go to zero, and reduce the number of layers to one ($N = 1$), from (4.13), (4.14), (4.15) we obtain that

$$\begin{aligned} 4\pi\mathbf{E}_\infty(\hat{\mathbf{r}}) &= ik_0^3\frac{\varepsilon}{\varepsilon_0}\int_\Omega\left[\left(1-\frac{\varepsilon}{\varepsilon_0}\right)\mathbf{E}(\mathbf{r}') + \beta(\nabla\times\mathbf{E}(\mathbf{r}'))\right]\cdot\\ &\quad\cdot(\tilde{I} - \hat{\mathbf{r}}\otimes\hat{\mathbf{r}})e^{-ik_0\hat{\mathbf{r}}\cdot\mathbf{r}'}\, dv(\mathbf{r}')\\ &+ k_0^2\int_\Omega\left[k_0^2\beta\frac{\varepsilon}{\varepsilon_0}\mathbf{E}(\mathbf{r}') + \left(\left(1-\frac{\mu_0}{\mu}\right) + k_0^2\beta\frac{\varepsilon}{\varepsilon_0}\right)(\nabla\times\mathbf{E}(\mathbf{r}'))\right]\cdot\\ &\quad\cdot(\tilde{I}\times\hat{\mathbf{r}})e^{-ik_0\hat{\mathbf{r}}\cdot\mathbf{r}'}\, dv(\mathbf{r}'), && (4.16)\end{aligned}$$

where ε, μ, β, $\mathbf{E}$ denote the quantities in the (non–layered now) scatterer Ω. This expression is completely analogous to the one given in [9] after elimination of the terms containing the magnetic field there.

(ii) If the chirality measures β_j, $j = 1, 2, \ldots, N$, are all equal to zero, then Ω is an achiral multi–layered scatterer and the results of this section coincide with the corresponding ones in [1]. In particular, $\mathbf{E}^{sc(c)}(\mathbf{r})$ and $\mathbf{E}_{\infty}^{(c)}(\hat{\mathbf{r}})$ are both equal to zero in this case.

References

[1] C. Athanasiadis, "Low–frequency electromagnetic scattering theory for a multi–layered scatterer", *Quart. J. Mech. Appl. Math.*, 44, 1991, 55–67.

[2] C. Athanasiadis and I.G. Stratis, "Electromagnetic scattering by a chiral object", *IMA J. Appl. Math.*, 58, 1997, 83–91.

[3] C. Athanasiadis and I.G. Stratis, "A transmission problem for bi–isotropic media", *Appl. Anal.*, to appear.

[4] C. Athanasiadis and I.G. Stratis, "Uniqueness of the inverse scattering problem by a chiral obstacle", *Intern. J. Appl. Electromagn. Mech.*, in press.

[5] D.L. Colton and R. Kress, *Inverse Acoustic and Electromagnetic Scattering Theory*, Springer, Berlin, 1992.

[6] R. Dautray and J.L. Lions, *Mathematical Analysis and Numerical Methods for Science and Technology*, vol. 3: *Spectral Theory and Applications*, Springer, Berlin, 1990.

[7] F. Hettlich, "Uniqueness of the inverse conductive scattering problem for time–harmonic electromagnetic waves", *SIAM J. Appl. Math.*, 56, 1996, 588–601.

[8] A. Kirsch, "The domain derivative and two applications in inverse scattering theory", *Inverse Problems*, 9, 1993, 81–96.

[9] A. Lakhtakia, V.K. Varadan and V.V. Varadan, "Dilute random distribution of small chiral spheres", *Appl. Optics*, 29, 1990, 3627–3632.

[10] A. Lakhtakia, *Beltrami Fields in Chiral Media*, World Scientific, Singapore, 1994.

[11] J. Nečas, *Les Méthodes Directes en Théorie des Équations Elliptiques*, Masson, Paris, 1967.

[12] W.S. Weiglhofer (ed.), *Proceedings of the International Conference "Bianisotropics 97"*, Glasgow University, Glasgow, 1997.

Authors' address: University of Athens,
Department of Mathematics,
Panepistimiopolis,
GR 157 84 Athens, Greece.

e–mail address: istratis@eudoxos.uoa.gr

A CHARALAMBOPOULOS, G DASSIOS AND M HADJINICOLAOU

Bispherical geometry in multiple scattering

1. Introduction

Many diagnostic methods in physics, chemistry and medical science such as spectroscopy, velocimetry, anemometry, ultrasounds, tomography, etc. are based on scattering theory. Accorrding to this, useful results can be extracted by studying the effects upon an obstacle (scatterer) of a propagating wave. Whenever the scatterer is more than one, multiple scattering occurs since the scattered field from the one becomes incident field for the other [13]. A plethora of published works concern the single scatterer problem [4,5]. Analytic solutions are available in low and high frequencies, for spherical and ellipsoidal shaped scatteres and for plane wave excitation [2,9]. Recently, [1,6,7] point generated fields were considered for studying scattering problems in low frequencies and analytic expansions for all quantities of interest were obtained for the case of a spherical scatterer with different boundary conditions. Although the scattering problem with one single scatterer has been extensively investigated, not much has been done towards the direction of many scatterers.

In the present work we study a multiple scattering problem of first order. The incident field is an acoustic plane wave and the scatterer consists of two small spheres, having their centers at a distance d apart, along the surfaces of which, the total field (incident and scattered) assumes zero value.

The standard technique for treating such a problem is to use spherical coordinate system, with origin at the center of the one sphere. The analytic solution is then obtained by employing the classical addition theorems for spherical harmonics [10,13,14]. This technique requires long manipulations and the series form of the resulting expansions does not help in their physical interpretation.

A different approach is presented in this work. We employ the bispherical coordinate system [10,11,12] which is the most appropriate for the geometry of the problem, since the surfaces of the two given spheres can be described by particular values of one of the coordinate variables. It has been shown [3] that such a system exists and is unique, whenever the radii of the two spheres and their distance are known. Bispherical system of coordinates has been also employed in [9] for obtaining analytic solution for the scattering by a small sphere embedded in a half space.

Since in bispherical coordinate system Helmholtz operator does not separate variables while Laplace's allows R- separation [11,12], low frequency theory [5] is utilized for obtaining analytic solution for the problem at hand. The formulation of the problem and the description of the bispherical system is presented in Section 2. The leading two non vanishing low frequency coefficients for the near field are presented in Section 3. Then in Section 4, the relevant approximations for both, the scattering amplitude and the cross section are obtained by calculating the corresponding surface integrals. Extensive discussion of the obtained results is the subject of Section 5. Some 2-D and 3-D plots are also provided demonstrating noteworthy features, useful for the above mentioned applications.

2. Formulation of the Problem

2.1 Bispherical system of coordinates

Given the radii a_1 and a_2 of two spheres and the distance d of their centers, it has been shown in [3] that there exist exactly one Bispherical system of coordinates (ζ,θ,φ) that fits to this geometry. The surfaces of the two spheres are then described by two specific values of the ζ coordinate variable.

The bispherical system of coordinates Fig. 1 is orthogonal and axisymmetric [10,11,12] and is related with the Cartesian (x_1,x_2,x_3) through the formulae:

$$x_1 = a\frac{\sin\theta\cos\varphi}{\cosh\zeta - \cos\theta} \tag{1}$$

$$x_2 = a\frac{\sin\theta\sin\varphi}{\cosh\zeta - \cos\theta} \tag{2}$$

$$x_3 = a\frac{\sinh\zeta}{\cosh\zeta - \cos\theta} \tag{3}$$

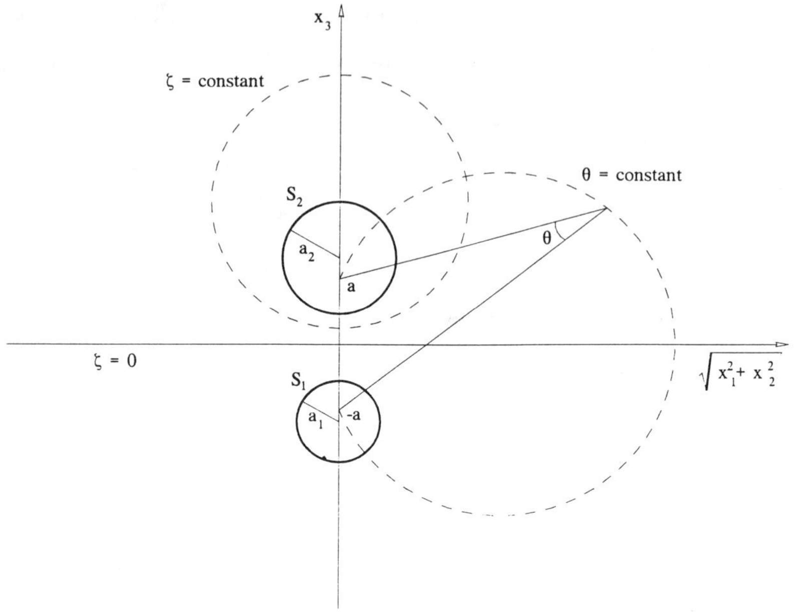

Figure 1. The two-sphere scattering system in bispherical geometry. S_1, S_2 correspond to $\zeta=-\zeta_1$, $\zeta=\zeta_2$ respectively.

where a is the semifocii distance, $\zeta \in R$, $\theta \in [0,\pi]$, $\varphi \in [0,2\pi)$

Positive values of ζ are associated with spheres whose centers are in the upper x_1x_2 plane, while negative values with spheres whose centers are in the lower x_1x_2 plane. Zero ζ specifies spheres of infinite radius, since $r = \frac{a}{|\sinh\zeta|}$. The case $\zeta=\pm\infty$, characterizes points on the x_3 axis at $x_3=\pm a$.

Surfaces θ=const. are apple-shaped if $\theta<\pi/2$, and spindle-shaped if $\theta>\pi/2$. $\theta=\pi/2$ describes a sphere of radius a, and $\theta=\pi/2$ the focii distance of length 2a.

The φ=const surface is, each time, a particular meridian plane.

2.2 Mathematical formulation

We asssume ζ_1, ζ_2 positive real numbers. Then the sphere described by the value ζ=-ζ_1 is situated in the lower half space and with the value ζ_2 in the upper half space. Knowing the geometrical magnitudes a_1, a_2 and d, the corresponding bispherical ones: a, ζ_1 and ζ_2 are identified through the solution of the non-linear system [3]:

$$\frac{a}{\sinh\zeta_1} = a_1 \tag{4}$$

$$\frac{a}{\sinh\zeta_2} = a_2 \tag{5}$$

$$\frac{a}{\tanh\zeta_1} + \frac{a}{\tanh\zeta_2} = d \tag{6}$$

which is

$$a = \frac{\sqrt{d^2-(a_1-a_2)^2}\sqrt{d^2-(a_1+a_2)^2}}{2d} \tag{7}$$

$$\zeta_1 = \ell n\frac{d^2+(a_1-a_2)(a_1+a_2)+\sqrt{d^2-(a_1-a_2)^2}\sqrt{d^2-(a_1+a_2)^2}}{2da_1} \tag{8}$$

$$\zeta_2 = \ell n\frac{d^2-(a_1-a_2)(a_1+a_2)+\sqrt{d^2-(a_1-a_2)^2}\sqrt{d^2-(a_1+a_2)^2}}{2da_2} \tag{9}$$

The uniqueness of this solution is a consequence of the positiveness of a, ζ_1 and ζ_2 [3].

In this particular scattering problem the scattered wave propagates in the exterior space, V, of the two spheres defined by:

$$\mathbf{V} = \{(\zeta,\theta,\varphi) \mid \zeta \in (-\zeta_1,\zeta_2),\ \theta \in [0,\pi],\ \varphi \in [0,2\pi)\} \tag{10}$$

The spherical radial distance is given by

$$r = a\sqrt{\frac{\cosh\zeta+\cos\theta}{\cosh\zeta-\cos\theta}} \tag{11}$$

which implies that the far-field region corresponds to a small neighborhood of $(\zeta,\theta)=(0,0)$ in the bispherical domain.

We work in the frequency domain, where the harmonic time dependence exp{-iωt} is suppressed through out. We assume now the plane wave incidence

$$u^i(\mathbf{r}) = e^{i\mathbf{k}\cdot\mathbf{r}} \tag{12}$$

with $\mathbf{k} = k\hat{\mathbf{k}}$ the propagation vector and k the wave number.

We define the total field

$$u(\mathbf{r}) = u^i(\mathbf{r}) + u^s(\mathbf{r}) \quad , \ \mathbf{r}\in\mathbf{V} \tag{13}$$

where u^i and u^s stand for the incident and the scattered field respectively. Then the scalar scattering problem for a pair of acoustically soft spheres is dictated as follows: We seek a wave field u(**r**) that satisfies the Helmholtz equation

$$(\Delta + k^2)u(\mathbf{r}) = 0 \quad , \ \mathbf{r}\in\mathbf{V} \tag{14}$$

the boundary conditions

$$u(\mathbf{r}) = 0 \quad , \quad \mathbf{r}\in S_1 \cup S_2 \tag{15}$$

and the scattered field u^s satisfies the Sommerfeld Radiation Condition

$$\lim_{r\to\infty} r\left(\frac{\partial u}{\partial r} - iku\right) = 0 \tag{16}$$

uniformly over the unit sphere S^2.

Assuming very small scatterers compared with the wave number, k, i.e. for scatterers with ka_1, $ka_2 << 1$, the well known low-frequency theory [5] can be applied where the field u(**r**) is expanded as

$$u(\mathbf{r}) = \sum_{n=0}^{\infty} \frac{(ik)^n}{n!} u_n(\mathbf{r}) \quad , \ \mathbf{r}\in\mathbf{V} \tag{17}$$

and the low-frequency approximations satisfy the subsequent problems in potential theory [8].

In what follows we restrict ourselves in the leading two order approximations. The coefficient of the zeroth order approximation, the so called Rayleigh approximation, solves the boundary value problem

$$\Delta u_0(\mathbf{r}) = 0 \quad , \ \mathbf{r}\in\mathbf{V} \tag{18}$$

$$u_0(\mathbf{r}) = 0 \quad , \quad \mathbf{r}\in S_1\cup S_2 \tag{19}$$

$$u_0(\mathbf{r}) = 1 + 0\left(\frac{1}{r}\right) \quad , \quad r\to\infty \tag{20}$$

The first order low-frequency approximation solves the following problem

$$\Delta u_1(\mathbf{r}) = 0 \quad , \ \mathbf{r}\in\mathbf{V} \tag{21}$$

$$u_1(\mathbf{r}) = 0 \quad , \quad \mathbf{r}\in S_1\cup S_2 \tag{22}$$

$$u_1(\mathbf{r}) = \hat{\mathbf{k}}\cdot\mathbf{r} - \frac{1}{4\pi}\int_{S_1\cup S_2}\frac{\partial u_0(\mathbf{r})}{\partial n}\,ds(r) + 0\left(\frac{1}{r}\right) \quad , \quad r\to\infty \tag{23}$$

3. The Near Field

In bispherical coordinates, Laplace equation assumes the form:

$$(\cosh\zeta - \cos\theta)\left[\frac{\partial}{\partial\zeta}\left(\frac{1}{\cosh\zeta-\cos\theta}\frac{\partial}{\partial\zeta}\right)g(\zeta,\theta,\varphi)\right.$$

$$\left.+\frac{1}{\sin\theta}\frac{\partial}{\partial\theta}\left(\frac{\sin\theta}{\cosh\zeta-\cos\theta}\frac{\partial}{\partial\theta}\right)g(\zeta,\theta,\varphi)\right] + \frac{1}{\sin^2\theta}\frac{\partial^2}{\partial\varphi^2}g(\zeta,\theta,\varphi) = 0 \tag{24}$$

It accepts R-separation [10] and a typical eigenfunction expansion, regular in the axis of symmetry has the form:

$$f(\zeta,\theta,\varphi) = \sqrt{\cosh\zeta-\cos\theta}\sum_{n=0}^{\infty}\sum_{m=0}^{n}\left[D_n^m e^{\left(n+\frac{1}{2}\right)\zeta} + E_n^m e^{-\left(n+\frac{1}{2}\right)\zeta}\right]$$

$$P_n^m(\cos\theta)\left[F_n^m\cos(m\varphi) + G_n^m\sin(m\varphi)\right] \tag{25}$$

where P_n are the Legendre Polynomials.

In the following calculations we use the uniformly convergent expansion given in [12]:

$$\frac{1}{\sqrt{\cosh\zeta-\cos\theta}} = \sqrt{2}\sum_{n=0}^{\infty} e^{-\left(n+\frac{1}{2}\right)|\zeta|}P_n(\cos\theta) \tag{26}$$

3.1 The Rayleigh approximation

The boundary value problem (18-20) leads to the axially symmetric zeroth order low frequency approximation

$$u_0(\zeta,\theta) = \sqrt{\cosh\zeta - \cos\theta}\sum_{n=0}^{\infty}\left[\sqrt{2}\, e^{-\left(n+\frac{1}{2}\right)|\zeta|} + A_n\, e^{\left(n+\frac{1}{2}\right)\zeta} + B_n\, e^{-\left(n+\frac{1}{2}\right)\zeta}\right] P_n(\cos\theta) \qquad (27)$$

where the coefficients A_n and B_n, Fig. 2, evaluated from the B.C' s are given by

$$A_n = -\sqrt{2}\,\Gamma(\zeta_1) \qquad (28)$$

$$B_n = -\sqrt{2}\,\Gamma(\zeta_2) \quad , \quad n=0,1,2,\ldots \qquad (29)$$

where

$$\Gamma(\zeta) = \sum_{n=0}^{\infty} \frac{e^{(2n+1)\zeta}}{e^{(2n+1)(\zeta_1+\zeta_2)} - 1} \qquad (30)$$

Using (11), the asymptotic behaviour of u_0 given in (20), is written as

$$u_0(\zeta,\theta) = 1 + 0\left(\sqrt{\frac{\cosh\zeta - \cos\theta}{\cosh\zeta + \cos\theta}}\right) \quad , \quad (\zeta,\theta)\to(0,0) \qquad (31)$$

and

$$\lim_{r\to\infty} r\,[u_0 - 1] = \lim_{(\zeta,\theta)\to(0,0)} a\sqrt{\frac{\cosh + \cos\theta}{\cosh\zeta - \cos\theta}}\,[u_0(\zeta,\theta) - 1] = -2a\big(\Gamma(\zeta_1) + \Gamma(\zeta_2)\big) \qquad (32)$$

which confirms that, when $r\to\infty$ i.e $(\zeta,\theta)\to(0,0)$, $u_0(\zeta,\theta)$ satisfies the asymptotic condition (31). It is observed that u_0 is azimouthal independent since the azimouthal angle φ does not appear in (27).

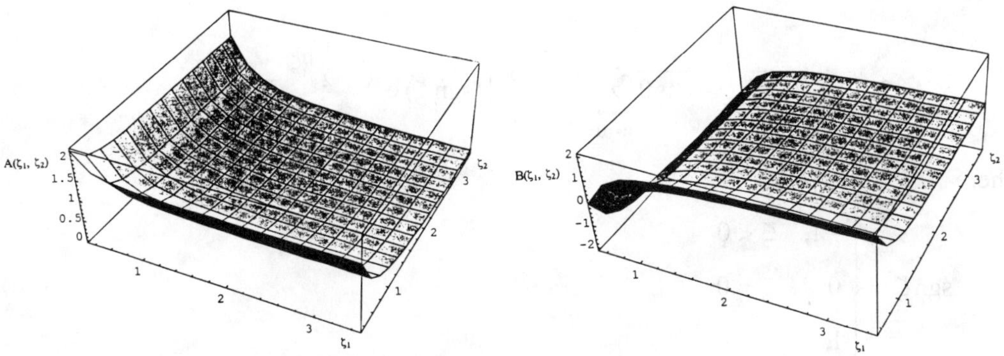

Figure 2a,b: 3-D plot of series A. B approximated with n=5 terms.

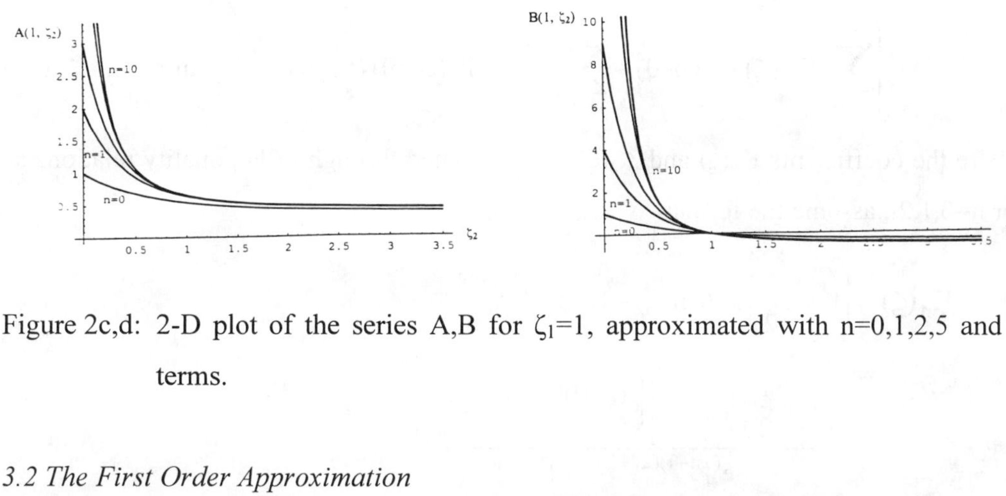

Figure 2c,d: 2-D plot of the series A,B for $\zeta_1=1$, approximated with n=0,1,2,5 and 10 terms.

3.2 The First Order Approximation

For the calculation of the first order approximation u_1, we need to transform in bispherical coordinates the boundary value problem (21-23). To achieve this, the linear harmonic functions x_1, x_2, x_3 that appear when analyzing the $\hat{\mathbf{k}}\cdot\mathbf{r}$ term in (23) ($\hat{\mathbf{k}}\cdot\mathbf{r}=k_1x_1+k_2x_2+k_3x_3$), must be expressed in R-separable form, which is:

$$x_1 = 2\sqrt{2}\,a\sqrt{\cosh\zeta-\cos\theta}\sum_{n=1}^{\infty} e^{-\left(n+\frac{1}{2}\right)|\zeta|}\, P_n^1(\cos\theta)\cos\varphi \tag{33}$$

$$x_2 = 2\sqrt{2}\, a\sqrt{\cosh\zeta - \cos\theta}\sum_{n=1}^{\infty} e^{-\left(n+\frac{1}{2}\right)|\zeta|} P_n^1(\cos\theta)\sin\varphi \tag{34}$$

$$x_3 = 2\sqrt{2}\, a\sqrt{\cosh\zeta - \cos\theta}\sum_{n=0}^{\infty}\left(n+\frac{1}{2}\right)(\operatorname{sgn}\zeta)\, e^{-\left(n+\frac{1}{2}\right)|\zeta|} P_n(\cos\theta) \tag{35}$$

where

$$\operatorname{sgn}\zeta = \begin{cases} -1, & \zeta < 0 \\ 0 & \zeta = 0 \\ 1, & \zeta > 0 \end{cases} \tag{36}$$

Then u_1 that solves the boundary value problem (21)-(23) has the form

$$u_1(\zeta,\theta,\varphi) = 2\sqrt{2}\, a\sqrt{\cosh\zeta - \cos\theta}$$

$$\left\{\sum_{n=0}^{\infty} K_n(\zeta)\, P_n(\cos\theta) + \sum_{n=1}^{\infty} \Lambda_n(\zeta)\, P_n^1(\cos\theta)\,(k_1\cos\varphi + k_2\sin\varphi)\right\} \tag{37}$$

where the coefficients $K_n(\zeta)$ and $\Lambda_n(\zeta)$, are evaluated through orthogonality relations and for n=0,1,2...assume the forms:

$$\begin{aligned} K_n(\zeta) &= \left[\left(n+\frac{1}{2}\right)k_3(\operatorname{sgn}\zeta) - C\right] e^{-\left(n+\frac{1}{2}\right)|\zeta|} \\ &- \frac{\left[\left(n+\frac{1}{2}\right)k_3 - C\right]\left[e^{(2n+1)\zeta_1}+1\right] + 2C}{e^{(2n+1)(\zeta_1+\zeta_2)} - 1}\, e^{\left(n+\frac{1}{2}\right)\zeta} \\ &+ \frac{\left[\left(n+\frac{1}{2}\right)k_3 + C\right]\left[e^{(2n+1)\zeta_2}+1\right] - 2C}{e^{(2n+1)(\zeta_1+\zeta_2)} - 1}\, e^{-\left(n+\frac{1}{2}\right)\zeta} \end{aligned} \tag{38}$$

and

$$\Lambda_n(\zeta) = e^{-\left(n+\frac{1}{2}\right)|\zeta|} - \frac{e^{(2n+1)\zeta_1} - 1}{e^{(2n+1)(\zeta_1+\zeta_2)} - 1}\, e^{\left(n+\frac{1}{2}\right)\zeta} - \frac{e^{(2n+1)\zeta_2} - 1}{e^{(2n+1)(\zeta_1+\zeta_2)} - 1}\, e^{-\left(n+\frac{1}{2}\right)\zeta} \tag{39}$$

where

$$C = 2a\left(\Gamma(\zeta_1) + \Gamma(\zeta_2)\right) \tag{40}$$

These expansions involve the azimouthal angle φ as an azimouthal monopole ($P_n(\cos\theta)$) as well as an azimouthal dipole term ($P_n^1(\cos\theta)\cos\varphi$)) and therefore u_1 is azimouthal dependent. Azimouthal independence is allowed when the expression ($k_1 \cos\varphi + k_2 \sin\varphi$) in the RHS of (37) vanishes. Physically, this expression denotes the inner product of the horizontal projections of the direction of propagation $\hat{\mathbf{k}}$ and the direction of observation $\hat{\mathbf{r}}$, on a plane perpendicular to the axis of symmetry. Thus, whenever these directions are perpendicular, u_1 turns out to be azimouthal independent.

4. The Far Field

The scattered field enjoys the integral representation

$$u^s(\mathbf{r}) = -\frac{ik}{4\pi} \int_{S_1 \cup S_2} h(k\,|\,\mathbf{r} - \mathbf{r}'\,|)\frac{\partial}{\partial n'} u(\mathbf{r}')\, ds(\mathbf{r}') \quad , \quad \mathbf{r} \in V \tag{41}$$

where $h(x) = \dfrac{e^{ix}}{ix}$ is the fundamental solution of the Helmholtz operator in R^3.

Then we obtain the following asymptotic form

$$u^s(\mathbf{r}) = g(\hat{\mathbf{r}}, \hat{\mathbf{k}})\, h(kr) + 0\left(\frac{1}{r^2}\right) \quad , \quad \mathbf{r} \to \infty \tag{42}$$

where

$$g(\hat{\mathbf{r}}, \hat{\mathbf{k}}) = -\frac{ik}{4\pi} \int_{S_1 \cup S_2} e^{-ik\hat{\mathbf{r}}\cdot\mathbf{r}'} \frac{\partial u(\mathbf{r}')}{\partial n'} ds(\mathbf{r}') \tag{43}$$

According to scattering theory [5], the scattering amplitude which is direction depended, incorporates all the information about the geometrical and the physical characteristics of the scatterer. From the representation of u^s in the near field (41) and of the far field (42), it is observed that they differ in the kernel function which in the near field denotes a point source located at $\mathbf{r}'$, while in the far field located at the origin but modified by the phase factor of a plane wave. Consequently the scattering amplitude (43) describes the far field response of an obstacle when it is excited by a plane wave propagating in the direction $\hat{\mathbf{k}}$. Low frequency analysis of the scattering amplitude [8] leads, after long manipulations, to the low frequency approximation

$$g(\hat{\mathbf{r}},\hat{\mathbf{k}}) = -2\,i\,k\,a\,A - 2(ka)^2\left\lfloor(0_3 - k_3)B + 2A^2\right\rfloor + 0\left((ka)^3\right)\ ,\quad ka\to 0 \tag{44}$$

where

$$\hat{\mathbf{r}} = (0_1, 0_2, 0_3) \tag{45}$$

$$A = \Gamma(\zeta_1) + \Gamma(\zeta_2) \tag{46}$$

and

$$B = \Gamma'(\zeta_1) - \Gamma'(\zeta_2) \tag{47}$$

where the prime denotes the first derivate.

It is observed that while the first order approximation of g is a monopole term with respect to both directions of incidence and observation, the second order approximation consists of a monopole and a dipole term as well. However this dipole term vanishes whenever the two spheres are equal ($\zeta_1=\zeta_2$) allowing isotropic behaviour of g up to the second order approximation. This result can be useful when studying inverse scattering problems in the sense that it provides kriteria for an estimation on whether the two spheres are equal or not. This is performed by varying either the direction of incidence or the direction of observation and checking whether this variation causes changes in the second order approximation of g.

To the same conclusions we can arrive by using instead of g the scattering cross section σ, given by

$$\sigma = \frac{1}{k^2}\int_{S^2}\left|g(\hat{\mathbf{r}},\hat{\mathbf{k}})\right|^2 ds(\hat{\mathbf{r}}) \tag{48}$$

which is a global measure of the interaction due the local energy distribution. Substituting (44) into (48), we obtain

$$\sigma = 16\pi\alpha^2 \left\{ A^2 + (ka)^2 \left[(2A^2 - k_3 B)^2 + \frac{B^2}{3} \right] \right\} + 0\left((ka)^4\right) \quad , \quad ka \to 0 \tag{49}$$

Figures 3,4 presented below, show σ/α^2 for different k_3 and for ka=0.1 and 0.01.

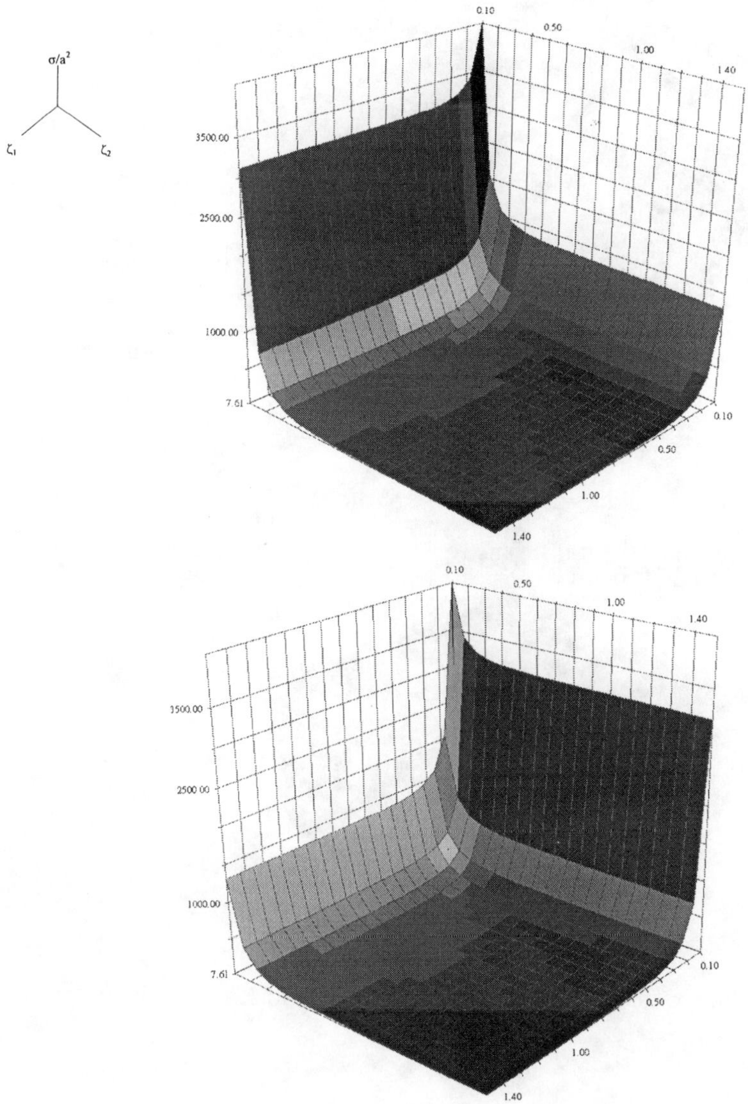

(a)

(b)

Figure 3a,b: σ/a^2 for ka=0.1, $k_3 = -\frac{\sqrt{2}}{2}$ and k_3=1 respectively

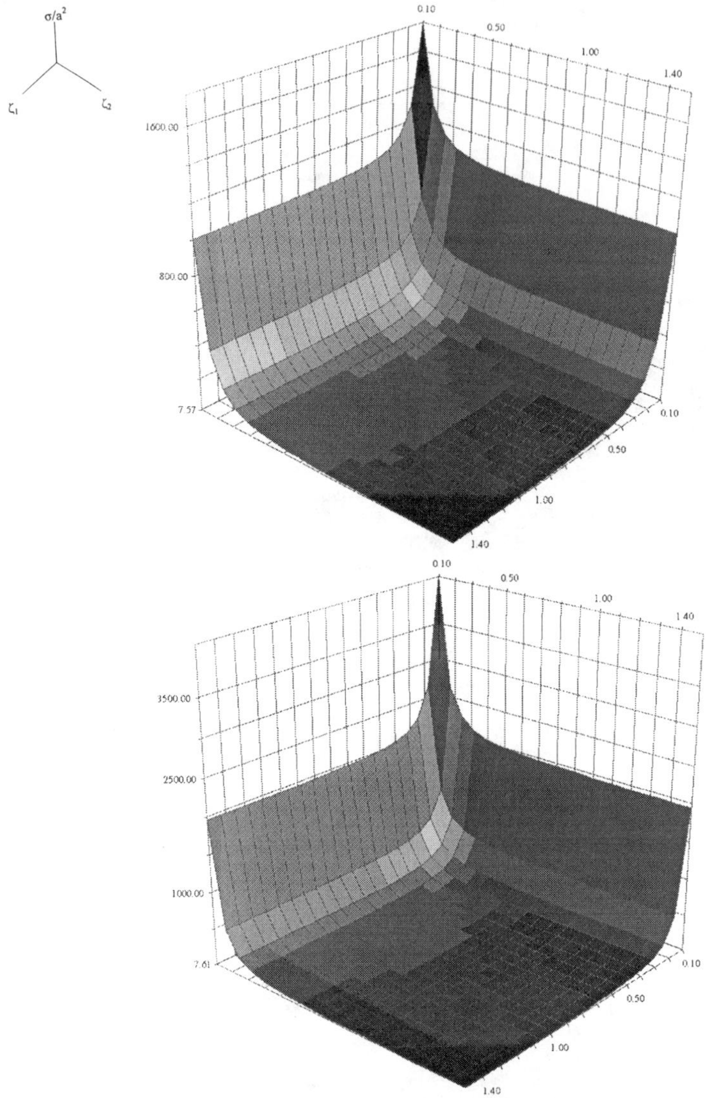

Figure 4a,b: σ/a^2 for $k_3=0$ and $ka=0.01$ and $ka=0.1$, respectively

5. Conclusions

Analytical solution for the multiple scattering problem by two spheres is obtained by employing the bispherical system of coordinates. It is an axisymmetric coordinate system and the surfaces of the two spheres are described by two different values of one of the coordinate variables. This system fits perfectly to the given geometry and allows R-separability for the Laplace equation but not for the Helmholtz [12]. Therefore low frequency assumptions are made and the near field as well as the far field expressions are derived through series expansions. The series form reveals the characteristic of multiple scattering [13,14] which is the interaction between the spheres and the scattered field.

In the present study, we consider multiple scattering of first order [13]. Consequently, we restrict ourselves only to the leading two order of approximation since these are sufficient for practical applications. This statement holds true because of the fast convergence of the series defining A (28) and B (29) and (30) as it is shown in Fig. 2.

It is also shown that the coefficient of the first term of the expansion of the scattering amplitude is a monopole term which is equivalent to the result obtained for the single sphere. The second coefficient consists of a monopole term and a dipole term which vanishes whenever the two spheres are equal. In the case of equal spheres, the scattering cross section remains unchangeable under the variation of the incident direction. In addition, no difference in the size of the two spheres is noticeable whenever the angle formed by the direction of incidence and the focii axis is equal to the observation angle. These results are useful for studying inverse scattering problems and also for credibility assessment of numerical codes.

References

1. Arnaoudov, Y., Dassios, G., and Hadjinicolaou, M., *The Resistive Coated Sphere in the Presence of a Point Generated wave Field* (submitted).
2. Bowman, J.J. Senior, T.B.A., and Uslenghi, P.L.E. (1969), *Electromagnetic and Acoustic Scattering by Simple Shapes*, North-Holland, Amsterdam.
3. Charalambopoulos, A., Dassios, G., and Hadjinicolaou M., *An Analytic Solution for Low-Frequency Scattering by Two Soft Spheres*, SIAM Journal of Applied Mathematics (in press).

4. Colton, D., and Kress, R. (1983), *Integral Equation Methods in Scattering Theory*, John Wiley and Sons, New York.

5. Colton, D., and Kress, R. (1992), *Inverse Acoustic and Electromagnetic Scattering Theory*, Springer-Verlag, Berlin.

6. Dassios, G., M. Hadjinicolaou, and G. Kamvyssas, "The Penetrable Coated Sphere Embeded in a Point Source Excitation Field" (submitted).

7. Dassios, G., and Kamvyssas, G. (1995), *Point Source Excitation in Direct and Inverse Scattering. The Soft and the Hard Sphere*, IMA J. Appl. Math., **55**, pp. 67-84.

8. Dassios, G., and Kleinman, R.E. (1989), *On Kelvin Inversion and Low-Frequency Scattering*, SIAM Review, **31**, pp. 565-585.

9. Dassios, G., and Kleinman, R.E. (1996), *Low-Frequency Scattering by Targets Above a Ground Plane*, Proceedings of the AGARD Meeting on Radar Singature Analysis of Imaging of Military Targets, Ankara.

10. Moon, P., and Spencer, D.E. (1961), *Field Theory Handbook*, Springer-Verlag, Berlin.

11. Moon, P., and Spencer, D.E. (1961), *Field Theory for Engineers*, Van Nostrand, Princeton.

12. Morse, P.M., and Feshbach, H. (1953), *Methods of Theoretical Physics*, Vol. I, II, McGraw-Hill, New York.

13. Twersky, V. (1962), *Multiple Scattering by Arbitrary Configuration in Three Dimensions*, J. Math. Phys., **3**, pp. 83-91.

14. Twersky, V. (1967), *Multiple Scattering of Electromagnetic Waves by Arbitrary Configurations*, J. Math. Phys., **8**, pp. 589-610.

DROSSOS GINTIDES AND KIRIAKIE KIRIAKI

Multiple scattering of elastic waves by sphere configurations

1. Introduction

The problem of multiple scattering has been treated by many scientists because of its significant importance to many areas of science such as atmospheric, chemistry, biology, astrophysics, photonic crystals and various other technological applications.

Many approaches have been adopted in order to handle this very complicated scattering process in acoustics and electromagnetism. Between them one was devised by Zaviska [15]. This approach exploits the results of single scattering and uses addition theorems. The method is very useful for numerical computations. Many researchers based on this idea have solved the problem for scatterers of known shape. Among the recent papers we refer to [8]. One of the pioneers of the mathematical formulation of the multiple scattering problem is Twersky. In [13] he proposed a method of constructing the solution in a step-by-step fashion as follows : Calculate the wave scattered by each body separately when it is excited by the incident wave. This is called the "first order of scattering". The wave scattered by the first (second) body is then rescattered by the second (first) body ; this is the "second order of scattering" and this procedure is infinitely continued. This approach is very complicated for numerical implementations. Another approach reduces the problem to finding an integral equation. Theoretical results using the integral formalism are presented in [9]. A recent paper with results of light scattering based on this method is [10]. Results for multiple scattering in the low-frequency region are presented in [2] where the authors exploit a bispherical coordinate system in order to find an analytical solution for the scattering by two soft spheres in acoustics. Modinos et al, in a series of papers have investigated arrays of particles in studying photonic crystals. We refer to their last review article [11].

The problem of elastic wave scattering by a single scatterer has also been investigated by many scientists. It is a very complicated problem. Its difficulty is due to the nature of the wave propagation in the elastic medium. The mathematical theory of linear elasticity is very well exposed in [7]. For low-frequency elastic scattering we refer to [5]. It is obvious that the problem of multiple elastic scattering is much more complicated than the single one. For this reason, despite the many results existing for the case of acoustics and electromagnetics, there are not many results for the elastic case. Among them we distinguish Phanord and Berger's work [12] where the multiple scattering of elastic waves by a distribution of identical spheres is presented. Their formulation of the problem is based on Twersky's approach and their numerical results are for the Rayleigh region.

In this work we examine the multiple scattering in an elastic medium. In Section 2 we consider the mathematical formulation of the scattering problem. In Section 3, in order to handle the scattering process for $N-$spheres randomly oriented in $\mathbf{R}^3$, we exploit Zaviska's ideas. Through eigenvector expansions of the displacement field in terms of Hansen eigenvectors and using addition theorems we conclude to an infinite linear system which gives as solution the coefficients in the Fourier analysis of the displacement field. In Section 4 we consider the scattering amplitudes and cross section and we establish the relations that calculate the scattering amplitudes in terms of the coefficients of the scattered field. Numerical results for two spheres are presented in Section 5. The infinite system which gives the Fourier coefficients is truncated to a finite one according to a suitable criterion. The case of the single sphere is examined as a degenerate case and comparison with known results is given. Plots for scattering cross section are also presented.

2. Statement of the problem

Let D_i, $i = 1,....,N$ denote bounded, convex and closed subsets of $\mathbf{R}^3$ having smooth boundaries ∂D_i, $\partial D = \partial D_1 \cup \partial D_2 \cup ... \partial D_N$. Let D_+ be the complement of D_- in $\mathbf{R}^3$ where $D_- = D_1 \cup D_2 \cup ... \cup D_N$ and assume that an isotropic, homogeneous elastic medium fills up D_+, with Lamé constants λ, μ and mass density ρ. We choose N-origins

$\mathrm{O_i}$ with $\mathrm{O_i} \in D_i$ and $\mathbf{r}_p^i$ denotes the position vector of a point P with respect to $\mathrm{O_i}$. We also assume a coordinate system which has its origin O outside D_- and $\mathbf{r}_p$ denotes the position vector of the point P with respect to this system. We will refer to it as the reference system.

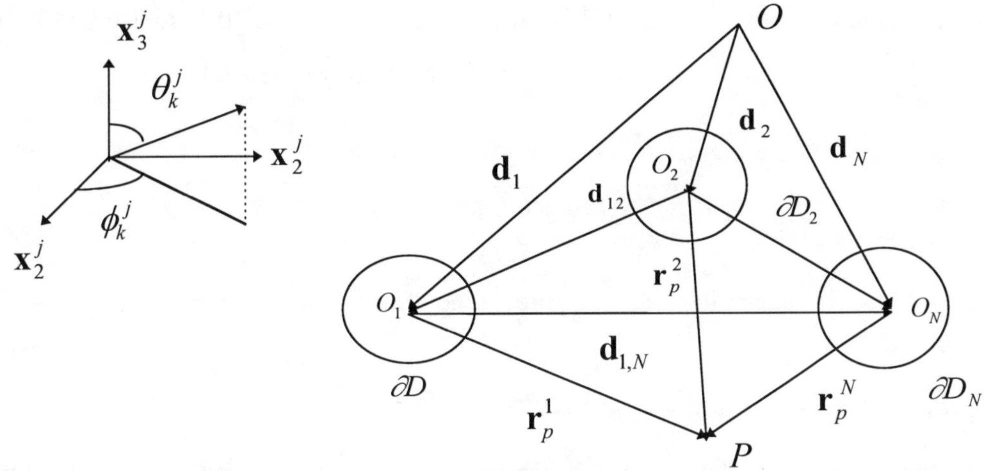

Figure 1

In the absence of body forces the displacement field $\mathbf{u}$ satisfies the stationary Navier equation

$$\Delta^{*}\mathbf{u}(P) + \omega^{2}\mathbf{u}(P) = \mathbf{0} \text{ for } P \in D_{+} \tag{2.1}$$

where Δ^{*} is the Lamé operator. We assume that a plane wave propagates in the infinite medium, which could be either a longitudinal or a transverse time independent plane wave of the form

$$\mathbf{u}_{in}^{a}(\mathbf{r}_p) = \hat{\mathbf{a}}\, e^{ik_a \hat{\mathbf{k}} \cdot \mathbf{r}} \tag{2.2}$$

where a = p or s and $\hat{\mathbf{a}}$ the unit polarization vector which is equal to $\hat{\mathbf{k}}$ for longitudinal incidence and to $\hat{\mathbf{b}}$ for a transverse wave, $\hat{\mathbf{b}} \cdot \hat{\mathbf{k}} = 0$, k_a are the corresponding wavenumbers.

The total field is the superposition of the incident $\mathbf{u}_{in}$ and the scattered field $\mathbf{u}_{sc}$ that is

$$\mathbf{u}_{tot}(\mathbf{r}_p) = \mathbf{u}_{in}^a(\mathbf{r}_p) + \mathbf{u}_{sc}(\mathbf{r}_p), \quad \mathbf{r}_p \in D_+, \; a = p, s \tag{2.3}$$

and it satisfies Eq (2.1). We assume that the scatterers are rigid bodies, so the following boundary conditions must be satisfied on the boundary ∂D

$$\mathbf{u}_{tot}(\hat{\mathbf{r}}_p) = \mathbf{0}, \qquad P \in \partial D \tag{2.4}$$

For the scattered field $\mathbf{u}_{sc}$ the well-known radiation conditions due to Kupradze are satisfied by its longitudinal and transverse parts [7] uniformly over all directions

$$\begin{aligned} & r_p \left\{ \frac{\partial \mathbf{u}_{sc}^a(\hat{\mathbf{r}}_p)}{\partial r_p} - ik_a \mathbf{u}_{sc}^a(\hat{\mathbf{r}}_p) \right\} = o\,(1), \\ & \mathbf{u}_{sc}^a(\hat{\mathbf{r}}_p) = o\,(1) \quad , \quad r_p = |\mathbf{r}_p| \to \infty \end{aligned} \tag{2.5}$$

where $a = p, s$. These conditions are required to hold for all directions $\mathbf{r}_p$, uniformly.

3. The multiple scattering problem

In the sequel, we will follow Zaviska's approach [15] in order to examine the multiple scattering problem for N rigid spheres of radius R_j, $j = 1, \ldots, N$ arbitrarily distributed in $\mathbf{R}^3$. The total field at a point $\mathbf{r}_p$ can be represented as

$$\mathbf{u}_{tot}(P) = \mathbf{u}_{in}(P) + \sum_{j=1}^{N} \mathbf{u}_{sc}^j(\mathbf{r}_p^j), \quad P \in D_+ \tag{3.1}$$

where the displacement field $\mathbf{u}_{sc}^j$ is the scattered field emanated from the $j-$th sphere, and the total scattered field is the superposition of all the scattered fields.

In order to deal with the scattering problem we will express every component of the total scattered field in terms of Hansen eigenvectors as

$$\mathbf{u}_{sc}^j(\mathbf{r}_p^j) = \sum_{n=1,1,0}^{\infty} \sum_{m=-n}^{n} \left\{ a_{mn}^{(j)} \mathbf{M}_{mn}^e(k_s r_p^j) + b_{mn}^{(j)} \mathbf{N}_{mn}^e(k_s r_p^j) + c_{mn}^j \mathbf{L}_{mn}^e(k_p r_p^j) \right\}, \quad j = 1, \ldots, N \tag{3.2}$$

where the summation over n for the first two vectors starts from $n = 1$, while for the third one starts from $n = 0$.

As it is well-known the Hansen vectors are expressed via the mutually orthogonal surface vector harmonics $\mathbf{P}_{mn}$, $\mathbf{B}_{mn}$ and $\mathbf{C}_{mn}$ [1] via the relations

$$\mathbf{M}_{mn}^{e,i}(k_a r_p^j) = \left(n(n+1)\right)^{1/2} \mathbf{C}_{mn}\, f_n^{e,i}(k_a r_p^j) \tag{3.3}$$

$$\mathbf{N}_{mn}^{e,i}(k_a r_p^j) = n(n+1)\,\mathbf{P}_{mn}\,\frac{f_n^{e,i}(k_a r_p^j)}{k_a r_p^j} + \left(n(n+1)\right)^{1/2} \mathbf{B}_{mn}\left[f_n^{e,i\,'}(k_a r_p^j) + \frac{f_n^{e,i}(k_a r_p^j)}{k_a r_p^j}\right] \tag{3.4}$$

$$\mathbf{L}_{mn}^{e,i}(k_a r_p^j) = \mathbf{P}_{mn}\, f_n^{e,i\,'}(k_a r_p^j) + \left(n(n+1)\right)^{1/2} \mathbf{B}_{mn}\,\frac{f_n^{e,i}(k_a r_p^j)}{k_a r_p^j} \tag{3.5}$$

where $a = p,s$ and e,i stand for exterior and interior Hansen eigenvectors, while $f_n^{e,i}$ denote the spherical Bessel functions. More precisely the superscript i indicates an internal solution, finite at the origin, and the superscript e indicates an external solution, finite at infinity. The prime denotes differentiation with respect to the argument.

The incident waves p or s can also be expanded in terms of Hansen vectors as [1]

$$\mathbf{u}_{in}^{p}(P) = 4\pi \sum_{n=0}^{\infty}\sum_{m=-n}^{n} \frac{i^{n-1}}{\Omega_{mn}}\, \hat{\mathbf{k}}\cdot \mathbf{P}_{mn}(\theta_k^j, \varphi_k^j)\, \overset{*}{\mathbf{L}}{}^{i}_{mn}(k_p r_p^j) \tag{3.6}$$

$$\begin{aligned}\mathbf{u}_{in}^{s}(P) = 4\pi \sum_{n=1,1}^{\infty}\sum_{m=-n}^{n} \frac{i^{n-1}}{\Omega_{mn}\sqrt{n(n+1)}}\, &\hat{\mathbf{b}}\cdot \mathbf{B}_{mn}(\theta_k^j, \varphi_k^j)\, \overset{*}{\mathbf{N}}{}^{j}_{mn}(k_s r_p^j)\\ &+ \hat{\mathbf{b}}\cdot \mathbf{C}_{mn}(\theta_k^j, \varphi_k^j)\, \overset{*}{\mathbf{M}}{}^{i}_{mn}(k_s r_p^j)\end{aligned} \tag{3.7}$$

where the starred quantities are obtained by changing $\exp(im\varphi)$ into $\exp(-im\varphi)$ and $\Omega_{mn} = \dfrac{4\pi\,(m+n)!}{(2n+1)\,(n-m)!}$.

Now, we are in the position to apply the boundary conditions in order to calculate the coefficients $a_{mn}^{(j)}$, $b_{mn}^{(j)}$, $c_{mn}^{(j)}$. So, we have on ∂D

$$\mathbf{u}_{in}^{a}(\mathbf{r}_p^i) + \sum_{j=1}^{N} \mathbf{u}_{sc}^{j}(\mathbf{r}_p^j) = \mathbf{0}, \quad \mathbf{r}_p^i \in \partial D_i,\, i = 1,\ldots,N,\ a = p,s \tag{3.8}$$

or, if we express all the scattered fields emanated from the j spheres in terms of the coordinate system centered on O_i, we have

$$\mathbf{u}_{in}^{a}(\mathbf{r}_p^i) + \mathbf{u}_{sc}^{(i)}(\mathbf{r}_p^i) + \sum_{\substack{j=1\\ j\neq i}}^{N} \mathbf{u}_{sc}^{i}(\mathbf{r}_p^i + \operatorname{sgn}(i-j)\,\mathbf{d}_{ij}) = \mathbf{0}, \quad \mathbf{r}_p^i \in \partial D_i,\, i = 1,\ldots,N,\ a = p,s \tag{3.9}$$

where $\mathbf{d}_{ij}$ is the vector connecting the centers of the spheres O_i, O_j (See Fig.1).

We will exploit now, addition theorems for Hansen vectors [1] and the orthogonality relations which hold for the surface vector spherical harmonics $\mathbf{P}_{mn}, \mathbf{B}_{mn}, \mathbf{C}_{mn}$. So we conclude to the linear system which the coefficients of the scattered fields have to satisfy.

First we formulate the system for $P-$incidence and we conclude to

$$(2n+1)\frac{(n+m)!}{(n-m)!}i^{n-1}\hat{\mathbf{k}}\cdot\mathbf{P}_{(-m)n}(\theta_k^i,\varphi_k^i)\, j_n'(k_pR_i)+b_{mn}^{(i)}\, n(n+1)\frac{h_n^{(1)}(k_sR_i)}{k_sR_i}+c_{mn}^{(i)}\, h_n^{(1)'}(k_pR_i)$$

$$+\sum_{\substack{j=1\\ j\neq i}}^{N}\sum_{\nu=0}^{\infty}\sum_{\mu=-\nu}^{\nu}\left\{c_{\mu\nu}^{(j)}\, B_{mn}^{(ij)\mu\nu}\, j_n'(k_pR_i)+n(n+1)\left(a_{\mu\nu}^{(j)}\, D_{mn}^{(ij)\mu\nu}+b_{\mu\nu}^{(j)}\, C_{mn}^{(ij)\mu\nu}\right)\frac{j_n(k_sR_i)}{k_sR_i}\right\}=0$$

$$n=0,1,\ldots,\quad -n\le m\le n \tag{3.10}$$

$$(2n+1)\frac{(n+m)!}{(n-m)!}i^{n-1}\hat{\mathbf{k}}\cdot\mathbf{P}_{(-m)n}(\theta_k^i,\varphi_k^i)\frac{j_n(k_pR_i)}{k_pR_i}+b_{mn}^{(i)}\left(n(n+1)\right)^{1/2}\left[h_n^{(1)'}(k_sR_i)+\frac{h_n^{(1)}(k_sR_i)}{k_sR_i}\right]$$

$$+c_{mn}^{(i)}\left(n(n+1)\right)^{1/2}\frac{h_n^{(1)}(k_pR_i)}{k_pR_i}+\sum_{\substack{j=1\\ j\neq i}}^{N}\sum_{\nu=0}^{\infty}\sum_{\mu=-\nu}^{\nu}\left\{\left(a_{\mu\nu}^{(j)}\, D_{mn}^{(ij)\mu\nu}+b_{\mu\nu}^{(j)}\, C_{mn}^{(ij)\mu\nu}\right)\cdot\left(j_n'(k_sR_i)+\frac{j_n(k_sR_i)}{k_pR_i}\right)\right.$$

$$\left.+c_{\mu\nu}^{(j)}B_{mn}^{(ij)\mu\nu}(n(n+1))^{1/2}\frac{j_n(k_pR_i)}{k_pR_i}\right\}=0$$

$$n=0,1,\ldots,-n\le m\le n \tag{3.11}$$

$$a_{mn}^{(i)}\left(n(n+1)\right)^{1/2}h_n^{(1)}(k_sR_i)+\sum_{\substack{j=1\\ j\neq i}}^{N}\sum_{\nu=0}^{\infty}\sum_{\mu=-\nu}^{\nu}\left(a_{mn}^{(j)}\, C_{mn}^{(ij)\mu\nu}+b_{mn}^{(j)}\, D_{mn}^{(ij)\mu\nu}\right)\left(n(n+1)\right)^{1/2}j_n(k_sR_i)=0$$

$$n=0,1,\ldots,-n\le m\le n \tag{3.12}$$

where the quantities appearing in (3.10) - (3.12) are given by the relations

$$B_{mn}^{(ij)\mu\nu}=\sum_{q=|\nu-n|}^{\nu+n}(-1)^q\, B(\nu,\mu/n,m/q)\, h_q^{(1)}(k_pd_{ij})\, Y_{\mu-m,q}(\theta_k^j,\varphi_k^j) \tag{3.13}$$

$$C_{mn}^{(j)\mu\nu}=\sum_{q=|\nu-n|}^{\nu+n}(-1)^q\, B(\nu,\mu/n,m/q)\, a(\nu,n,q)\, h_q^{(1)}(k_sd_{ij})\, Y_{\mu-m,q}(\theta_k^j,\varphi_k^j) \tag{3.14}$$

$$D_{mn}^{(ij)\mu\nu}=\sum_{q=|\nu-n|}^{\nu+n}(-1)^q\, G(\nu,\mu/n,m/q)\, b(\nu,n,q)\, h_q^{(1)}(k_sd_{ij})\, Y_{\mu-m,q}(\theta_k^j,\varphi_k^j) \tag{3.15}$$

and the quantities appearing in (3.13) - (3.15) are

$$A\left(\nu,\mu/n,m/q\right) = (-1)^{\mu} i^{\nu-n-q}\ (2n+1)\ (2q+1) \begin{pmatrix} \nu & n & q \\ \mu & -m & m-\mu \end{pmatrix} \begin{pmatrix} \nu & n & q \\ 0 & 0 & 0 \end{pmatrix} \tag{3.16}$$

$A\left(\nu,\mu/n,m/q\right)$ is non zero only when $\nu+n+q=\text{even}$,

$$B\left(\nu,\mu/n,m/q\right) = \left[\frac{(\nu+\mu)\ !\,(n-m)!\,(q-\mu+m)!}{(\nu-\mu)\ !\,(n+m)!\,(q+\mu-m)!}\right]^{1/2} \cdot A\left(\nu,\mu/n,m/q\right) \tag{3.17}$$

$$a\,(\nu,n,q) = 1 + \frac{(n+1)\,(\nu-n+q+1)\,(\nu+n-q) - n\,(n-\nu+q+1)\,(\nu+n+q+2)}{2n\,(n+1)\,(2n+1)} \tag{3.18}$$

$$b\,(\nu,n,q) = \frac{\left[(\nu+n+q+1)\,(n-\nu+q)\,(\nu-n+q)\,(\nu+n-q+1)\right]^{1/2}}{2n\,(n+1)} \tag{3.19}$$

$$G\left(\nu,\mu/n,m/q\right) = (-1)^{\mu}\, i^{\nu-n-q}\ (2n+1)\ (2q+1) \left[\frac{(\nu+\mu)\ !\,(n-m)!\,(q-\mu+m)!}{(\nu-\mu)\ !\,(n+m)!\,(q+\mu-m)!}\right]^{1/2}$$

$$\begin{pmatrix} \nu & n & q \\ \mu & -m & m-\mu \end{pmatrix} \begin{pmatrix} \nu & n & q-1 \\ 0 & 0 & 0 \end{pmatrix} \tag{3.20}$$

The Wigner's symbol which appears in all the previous equations is given by the relation

$$\begin{pmatrix} j_1 & j_2 & j_3 \\ m_1 & m_2 & m_3 \end{pmatrix} = \left[\frac{(j_1+j_2-j_3)\ !\,(j_1-j_2+j_3)!\,(-j_1+j_2-j_3)!}{(j_1+j_2+j_3+1)!}\right]^{1/2}$$

$$\sum_{z\in Z} \frac{(-1)^{z+j_1-j_2-m_3}}{z!\,(j_1+j_2-j_3-z)!(j_1-m_1-z)!(j_2+m_2-z)!(j_3-j_2+m_1+z)!(j_3-j_1-m_2+z)!} \tag{3.21}$$

The summation in (3.21) is over all integers z but since the factorial of a negative number is infinite, the sum contains only a finite number of terms. We mention that in all the above relations we follow closely the notation of [1].

Solving the system (3.10)-(3.12) we have the coefficients $a_{mn}^{(i)}$, $b_{mn}^{(i)}$, $c_{mn}^{(i)}$ in (3.2), so by (3.1) we have evaluated the total field at a point $\mathbf{r}_p$. Similar results hold for $S-$incidence. At this point we want to mention that the results of Einspruch, Witterholt, Truell [6] for a single rigid sphere are deduced as a degenerate case for $N=1$.

4.The scattering amplitudes and the scattering cross section

In what follows, we will exploit the evaluation of the displacement field in order to find the scattering amplitudes and the scattering cross section. As it is well known the relation connecting the radiation patterns or scattering amplitudes, which are the vector functions $\left(\mathbf{g}_p(\hat{\mathbf{r}}), \mathbf{g}_s(\hat{\mathbf{r}})\right)$, with the scattered field is given as follows [5]

$$\mathbf{u}_s(\mathbf{r}_p) = \frac{\exp\,(ik_p r_p)}{r_p}\,\mathbf{g}_p(\hat{\mathbf{r}}_p) + \frac{\exp\,(ik_s r_p)}{r_p}\,\mathbf{g}_s(\hat{\mathbf{r}}_p) + 0\,(\frac{1}{r_p^2}), \tag{4.1}$$

where $\mathbf{g}_p(\hat{\mathbf{r}}_p)$ has the outgoing radial direction $\hat{\mathbf{r}}_p$ and $\mathbf{g}_s(\hat{\mathbf{r}}_p)$ has a tangential direction. The scattering cross-sections σ_p and σ_s corresponding to an incident P and S wave respectively are expressed as [5]

$$\sigma_a = k_a \int_{|\hat{\mathbf{r}}_p|=1} \left[k_p^{-3} \left| \mathbf{g}_p(\hat{\mathbf{r}}_p) \right|^2 + k_s^{-3} \left| \mathbf{g}_r(\hat{\mathbf{r}}_p) \right|^2 \right] d\Omega\,(\hat{\mathbf{r}}_p) \qquad a = p,\, s \tag{4.2}$$

In order to continue with the evaluation of the scattering amplitudes, first, we need some theoretical results, expressing the scattering amplitudes, as a convergent series. Similar results hold in acoustics [4]. So, we establish the following proposition.

Proposition (4.1) The far-field patterns $\mathbf{g}_p$ and $\mathbf{g}_s$ are given by the uniformly convergent series

$$\mathbf{g}_p(\hat{\mathbf{r}}_p) = \frac{1}{k_p} \sum_{n=0}^{+\infty} \sum_{m=n}^{n} \frac{1}{i^n} c_{mn}\, \mathbf{P}_{mn}(\theta, \varphi) \tag{4.3}$$

and

$$\mathbf{g}_s(\hat{\mathbf{r}}_p) = \frac{1}{k_s} \sum_{n=1}^{+\infty} \sum_{m=-n}^{n} \frac{1}{i^n \sqrt{n(n+1)}} \left\{ b_{mn}\, \mathbf{B}_{mn}(\theta,\varphi) + a_{mn}\, \mathbf{C}_{mn}(\theta,\varphi) \right\} \tag{4.4}$$

where a_{mn}, b_{mn}, c_{mn} are given by the relations

$$a_{mn} = \sum_{j=1}^{N} \sum_{\nu=1}^{+\infty} \sum_{\mu=-\nu}^{\nu} \left\{ a_{\mu\nu}^{(j)}\, C_{mn}^{(j)\mu\nu} + b_{\mu\nu}^{(j)}\, D_{mn}^{(j)\mu\nu} \right\} \tag{4.5}$$

$$b_{mn} = \sum_{j=1}^{N} \sum_{\nu=1}^{+\infty} \sum_{\mu=-\nu}^{\nu} \left\{ a_{\mu\nu}^{(j)}\, D_{mn}^{(j)\mu\nu} + b_{\mu\nu}^{(j)}\, C_{mn}^{(j)\mu\nu} \right\} \tag{4.6}$$

$$c_{mn} = \sum_{j=1}^{N} \sum_{\nu=1}^{+\infty} \sum_{\mu=-\nu}^{\nu} c_{\mu\nu}^{(j)} \, B_{mn}^{(j)\mu\nu} \tag{4.7}$$

where

$$B_{\mu\nu}^{(j)mn} = \sum_{q=|n-\nu|}^{n+\nu} (-1)^q \, B\,(n,m/\nu,\mu/q) \, j_q(k_p d_j) \, Y_{m-\mu,q}\,(\theta_k\,,\varphi_k) \tag{4.8}$$

$$C_{\mu\nu}^{(j)mn} = \sum_{q=|n-\nu|}^{n+\nu} (-1)^q \, B\,(n,m/\nu,\mu/q) \, a\,(n,\nu,q) \, j_q(k_s d_j) \, Y_{m-\mu,q}\,(\theta_k\,,\varphi_k) \tag{4.9}$$

$$D_{\mu\nu}^{(j)mn} = \sum_{q=|n-\nu|}^{n+\nu} (-1)^q \, G\,(n,m/\nu,\mu/q) \, b\,(n,\nu,q) \, j_q(k_s d_j) \, Y_{m-\mu,q}\,(\theta_k\,,\varphi_k) \tag{4.10}$$

$a(n,\nu,q), b\,(n,\nu,q), B\,(n,m/\nu,\mu/q), G\,(n,m/\nu,\mu/q)$ are given by the relations (3.17) - (3.20) and the distance d_j is the distance between the origin O of the reference coordinate system and the origins O_j, $j = 1, 2, \ldots, N$ of the other systems (see Fig. 1).

Proof

The total scattered field, as we have already assumed is the superposition of all the scattered fields emanated by the $N-$spheres. If we assume the reference coordinate system, with origin O, taking into account Eq. (3.2) and using addition theorems we conclude that

$$u_{sc}(\mathbf{r}_p) = \sum_{n=1,1,0}^{\infty} \sum_{m=-n}^{n} \left\{ a_{mn} \, \mathbf{M}_{mn}^{e}(k_s r_p) + b_{mn} \, \mathbf{N}_{mn}^{e}(k_s r_p) + c_{mn} \, \mathbf{L}_{mn}^{e}(k_p r_p) \right\} \tag{4.11}$$

where a_{mn}, b_{mn}, c_{mn} are given by Eqs (4.5) - (4.7). This expansion holds outside a sphere centered at O and surrounding all the scatterers.

In order to take the scattering amplitudes we have to assume that $r_p \to \infty$. But we cannot pass to the limit $r_p \to \infty$ in (3.2) using the asymptotic behaviour of Hankel functions, because the latter does not hold uniformly in n. So, in order to overcome this difficulty we exploit a Müller's type theorem which holds in linear elasticity [3]. It states that the far-field patterns $\mathbf{g}_p(\hat{\mathbf{r}}_p), \mathbf{g}_s(\hat{\mathbf{r}}_p)$ can be expanded in a uniformly convergent series of vector spherical harmonics. Exploiting this result we deduce

$$\mathbf{g}_p(\hat{\mathbf{r}}_p) = \sum_{n=1}^{+\infty} \sum_{m=-n}^{n} d_{mn} \mathbf{P}_{mn}(\theta,\varphi) \tag{4.12}$$

$$\mathbf{g}_s(\hat{\mathbf{r}}_p) = \sum_{n=0}^{+\infty} \sum_{m=-n}^{n} \{e_{mn} \mathbf{B}_{mn}(\theta,\varphi) + f_{mn} \mathbf{C}_{mn}(\theta,\varphi)\} \tag{4.13}$$

with Fourier coefficients

$$d_{mn} = \Omega_{mn}^{-1} \int_{\partial\Omega} \mathbf{g}_p(\hat{\mathbf{r}}_p) \cdot \overset{*}{\mathbf{P}}_{mn}(\theta,\varphi)\, d\,\Omega\,(\hat{\mathbf{r}}_p) \tag{4.14}$$

$$e_{mn} = \Omega_{mn}^{-1} \int_{\partial\Omega} \mathbf{g}_s(\hat{\mathbf{r}}_p) \cdot \overset{*}{\mathbf{B}}_{mn}(\theta,\varphi)\, d\,\Omega\,(\hat{\mathbf{r}}_p) \tag{4.15}$$

$$f_{mn} = \Omega_{mn}^{-1} \int_{\partial\Omega} \mathbf{g}_s(\hat{\mathbf{r}}_p) \cdot \overset{*}{\mathbf{C}}_{mn}(\theta,\varphi)\, d\,\Omega\,(\hat{\mathbf{r}}_p) \tag{4.16}$$

From the definition of the scattering amplitudes that is from relation (4.1) we can write Eqs (4.14) - (4.16) in the form

$$d_{mn} = \lim_{r_p \to \infty} r_p\, e^{-ik_p r_p} \int_{\partial\Omega} \mathbf{u}_{sc}(r_p\, \hat{\mathbf{r}}_p) \cdot \overset{*}{\mathbf{P}}_{mn}\, (\theta,\varphi)\, d\,\Omega\,(\hat{\mathbf{r}}_p) \tag{4.17}$$

$$e_{mn} = \lim_{r_p \to \infty} r_p\, e^{-ik_s r_p} \int_{\partial\Omega} \mathbf{u}_{sc}(r_p\, \hat{\mathbf{r}}_p) \cdot \overset{*}{\mathbf{B}}_{mn}\, (\theta,\varphi)\, d\,\Omega\,(\hat{\mathbf{r}}_p) \tag{4.18}$$

$$f_{mn} = \lim_{r_p \to \infty} r_p\, e^{-ik_s r_p} \int_{\partial\Omega} \mathbf{u}_{sc}(r_p\, \hat{\mathbf{r}}_p) \cdot \overset{*}{\mathbf{C}}_{mn}\, (\theta,\varphi)\, d\,\Omega\,(\hat{\mathbf{r}}_p) \tag{4.19}$$

From (4.11) we have that the scattered field is given as Fourier series of vector Hansen functions, or in terms or vector spherical harmonics as

$$\begin{aligned} \mathbf{u}_{sc}(\mathbf{r}_p) = \sum_{n=1,1,0}^{\infty} \sum_{m=-n}^{n} \Bigg\{ & \left\{ n(n+1) \frac{h_n^{(1)}(k_s r_p)}{k_s r_p}\, b_{mn} + h_n^{(1)\prime}\, (k_p r_p)\, c_{mn} \right\} \mathbf{P}_{mn}\, (\theta,\varphi) \\ & + \sqrt{n(n+1)} \left\{ \left(h_n'^{(1)\prime}\, (k_s r) + \frac{h_n^{(1)}(k_s r_p)}{k_s r_p} \right) b_{mn} + \frac{h_n^{(1)}(k_p r_p)}{k_p r_p}\, c_{mn} \right\} \mathbf{B}_{mn}\, (\theta,\varphi) \\ & + \sqrt{n(n+1)}\, h_n^{(1)}(k_s r_p)\, a_{mn}\, \mathbf{C}_{mn}\, (\theta,\varphi) \Bigg\} \end{aligned} \tag{4.20}$$

If we substitute (4.20) in (4.17) - (4.19) and use the asymptotic behaviour of Hankel functions we conclude that

$$d_{mn} = \frac{c_{mn}}{k_p i^n} \quad , \; e_{mn} = \frac{b_{mn}}{k_s i^n} \frac{1}{\sqrt{n(n+1)}} \text{ and } \quad f_{mn} = \frac{a_{mn}}{k_s i^n} \frac{1}{\sqrt{n(n+1)}} \tag{4.21}$$

So, the Proposition (4.1) is established.

Now, we are in the position to evaluate the scattering amplitudes in view of the relations (4.3), (4.4) and the scattering cross section with a substitution in Eq. (4.2).

5. Numerical results

In order to derive numerical results we consider the multiple scattering process by two spheres. We consider that the two spheres are centered at the positions $\mathbf{d}_1 = d\,\hat{\mathbf{x}}$ and $\mathbf{d}_2 = -d\,\hat{\mathbf{x}}$ to the reference system O, and that a plane transverse wave propagates in the positive z – direction and polarized in the $\hat{\mathbf{x}}$ direction, that is of the form

$$\mathbf{u}_{in}(P) = \hat{\mathbf{x}}_x e^{ikz} \tag{5.1}$$

To handle the infinite system given by Eqs. (3.10) - (3.12) for N=2, we have to conclude to a finite linear system. The two things which we have to take in mind to do the truncation are the following : First, the characteristic quantity which usually is taken into account $k\alpha$, where k is the wave number and α the characteristic dimension of the scatterer and second the distance between the two spheres. When the distance between the two spheres is long enough we use the same criterion for truncation as in Mie scattering calculation [14].. After tests we have estimated that it holds for $d_{12} \geq 5 \max(R_1,R_2)$. The truncation order N_j of the series for every j =1,2 can be taken as $N_j \sim \left[(kR_j)+4\sqrt[3]{kR_j}+2\right]$, j =1,2. This approximation holds for $0.02 < kR_j < 8$.

In Table 1 we give for different d_{12} and $R_1 = R_2 = R$ the ratio σ_s^s/σ_s of multiple scattering cross section σ_s over the single scattering cross section σ_s^1 for $R_1 = R$. In Fig.1 (left) the scattering cross section σ_s^1 versus size parameter $k_s R$ for a single sphere is plotted. and in Fig.1 (right) σ_s versus size parameter $k_s R$ for two equal spheres of radius R =0.1 with distance d_{12}= 4, k_s = 1.1 and k_p = 0.8 As the distance increases the ratio tends to 2. This means that each sphere scatters independently and the total cross section is the superposition of the single scattering cross sections. To compute the scattering cross section we have used relation (4.2) where the scattering amplitudes are given by Eqs (4.3), (4.4). In Fig. 2 (left) we plot the ratio $\frac{\sigma_s}{\sigma_s^1}$ that is the multiple cross section for two spheres divided by the scattering cross section by an isolated sphere. In Fig.2 (right) we plot the ratio $\frac{\sigma_s}{\sigma_s^1}$ versus size distance d_{12} for two spheres with $R_1 = R_2 = 0.1$ and $k_s = 1.1$, $k_p = 1$.

$R_1 = R_2$	d_{12}	k_s	k_p	$\frac{\sigma_s}{\sigma_s^1}$
0.2	10	1.1	1	1.96727
0.2	5	1.1	1	0.00878046
0.2	0.4	1.1	1	0.00109047
0.4	5	1.1	1	1.76597
0.5	2	1.1	1	0.0120838
0.7	2	1.1	1	5.17565 10

Table 1

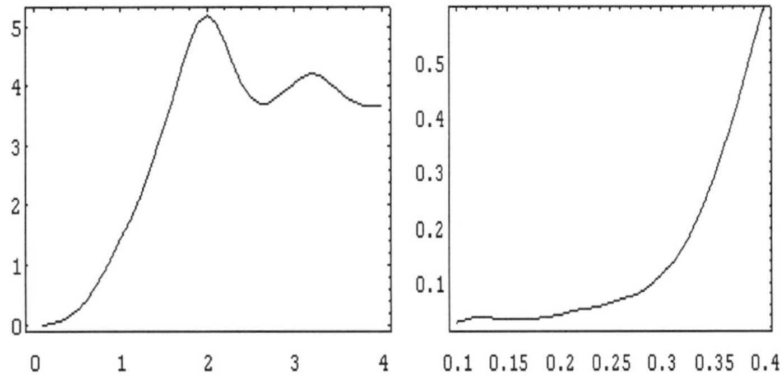

Figure 1: Single scattering cross section σ_s^1 versus size parameter $k_s R$ for a sphere (left) and multiple scattering cross section σ_s versus size parameter $k_s R$ for two equal spheres $d_{12} = 4$, and $k_s = 1.1$, $k_p = 0.8$ (right).

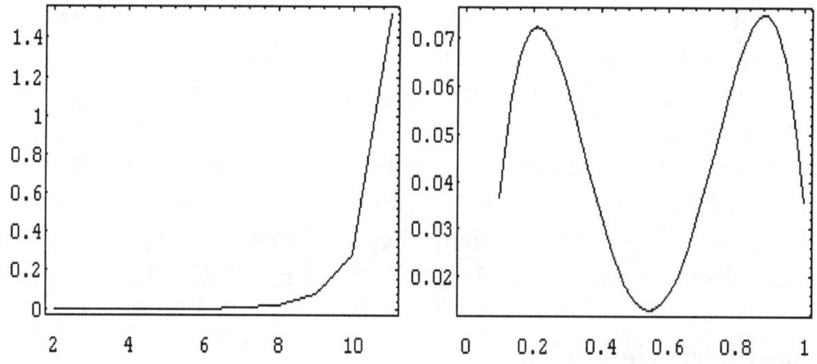

Figure 2: Ratio $\frac{\sigma_s}{\sigma_s^1}$ versus the distance d_{12} for two equal spheres $R_1 = R_2 = 0.1$ and $k_s =$ 1.1 , $k_p = 0.8$ (left) and ratio $\frac{\sigma_s}{\sigma_s^1}$ versus size distance d_{12} for two spheres with $R_1 = R_2 = 0.1$ and $k_s = 1.1$, $k_p = 1$ (right).

REFERENCES

1. Ari Ben - Menahem and S. J. Singh: "Seismic waves and sources" Springer-Verlag, New York (1981).
2. A. Charalambopoulos, G. Dassios and M. Hadjinicolaou: "An analytic solution for low-frequency scattering by two soft spheres" SIAM (to appear).
3. A. Charalambopoulos and K. Kiriaki: "Characterization of functions as radiation patterns in linear elasticity" Math. Methods in Appl. Sc., V. 15 (1992) pp. 547-558.
4. D. Colton and R. Kress: "Inverse acoustic and electromagnetic scattering theory" Springer - Verlag, New York (1992).
5. G. Dassios and K. Kiriaki: "The low-frequency theory of elastic wave scattering" Quart. Appl. Math., 42, (1984) pp. 225-248.
6. N. Einspruch, E Witterholt and R Truell.: "Scattering of a plane transverse wave by a spherical, obstacle in an elastic medium" J. of Applied Physics, Vol. 31, 5 (1960) pp. 806-818.
7. Y. D. Kupradze: "Progress in Solid Mechanics (V. III) North - Holland (1963).
8. C. M. Linton and P. Mc Iver: "The scattering of water waves by an array of circular cylinders in a channel" J. Eng. Math., Vol. 30, (1996) pp. 661-682.
9. P. A. Martin: "Integral - equation methods for multiple - scattering problems - I. Acoustics" Q. J. Mech. Appl. Math, Vol 38, (1985), pp. 105-118.
10. M. I. Mishchenko, D. W. Mackowski and L. D. Travis: "Scattering of light by bispheres with touching and separated components" Applied Optics, Vol. 34, No. 21, (1995) pp. 4589 - 4599.
11. A. Modinos, N. Stefanou and V. Karathanos: "Layer - by - layer methods in the study of photonic crystals and related problems" Photonic Band Cap. Mat., Kluwer Ac. Publ., (1996) pp. 229-251.

12. D. Phanord and N. E. Berger: “Multiple Scattering of elastic waves by a distribution of identical spheres” J. Math. Phys., 31 (11) (1990) pp. 2619 - 2625.
13. V. Twersky: “Multiple Scattering by arbitrary configuration in three dimensions” J. Math. Phys., 3, (1962) pp. 83-91.
14. W. J. Wiscombe: “Improved Mie scattering algorithms” Appl. Opt, 19 (1980) pp. 1505-1509.
15. F. Zaviska: “Uber die Beugung elektromagnetischer Wellen an parallelen, unendlich langen Kreiszylinder” Annalen der Physik, Vol. 40 (1913) pp. 1023 - 1056.

Drossos Gintides and Kiriakie Kiriaki,
National Technical University, Department of Mathematics,
Zografou Campus, Athens157 80,
Greece

ERDOĞAN S ŞUHUBI

Equivalence transformations for one-dimensional wave equations

1. Introduction

If a given system of equations depends on arbitrary functions or parameters we then have in essence a family of equations of the same type. For instance the field equations of hyperelastic solids depend heavily on the stress potentials which reflect the properties of the constitution of materials although the derivation rule of these equations is the same for all the hyperelastic materials. The equivalence transformations are defined as transformations which leave the family of equations invariant, namely, which map one member of the family to another member of the same family [7]. Furthermore it is desired that these transformations should possess a group structure. It is obvious that the classical symmetry groups, or Lie groups, are none other than the special subgroups of equivalence group of transformations which map a member of the family to itself.

In this work we deal with finding equivalence transformations for the following quite general family of one – dimensional nonlinear wave equations of balance form

$$u_{tt} = [f(x,t,u,u_x,u_t)]_x + g(x,t,u,u_x,u_t) \tag{1.1}$$

where f and g are arbitrary functions of their arguments. An equivalence transformation maps (1.1) to an equation of the same type but with different functions f and g than the original ones. We attack the problem of determining equivalence transformations within the framework of exterior calculus. To this end we first reduce (1.1) to a system of first order partial differential equations by introducing new dependent variables as derivatives of u with respect to x and t and then we increase the number of the variables by adding f and g and their derivatives with respect to their arguments. We then consider the exterior algebra over this extended manifold, define appropriate contact 1- forms which yield eventually the relations between original and additional variables and a balance 2- form which gives rise to Eq. (1.1) on the solution manifold. We further construct the closure of the ideal of the exterior algebra generated by these forms. Afterwards we look for vector fields in the tangent space of the extended manifold with respect to which Lie derivatives of the generators of the ideal remain in the ideal. These vector fields are called isovector fields. Their orbits provide transformations which leave the ideal invariant and it can be shown that they possess group properties [1,2]. These are equivalence transformations corresponding to the family of equations given by (1.1).

The strategy described above is quite general and can be applied to an arbitrary system of balance equations. Symmetry transformations associated with general balance equations of finite order involving arbitrary but finite number of dependent and independent variables are thoroughly studied in [8]. By employing a generalised version of the classical approach to determine symmetry groups, equivalence transformations are found for a special nonlinear one – dimensional wave equation which is not necessarily of balance form [3] and for a model equation of detonation of balance form [4]. A somewhat more general case than the latter is treated in [5]. Some rather special solutions can also be found in [6].

2. Fundamental equations

If we define new variables by

$$p = u_x, v = u_t \tag{2.1}$$

Eq (1.1) in terms of new variables becomes

$$v_t = [f(x,t,u,p,v)]_x + g(x,t,u,p,v) \tag{2.2}$$

(1.1) and the relations (2.1) constitute a first order system of equations. In order to study equivalence transformations we introduce an extended 17 – dimensional manifold M with a Cartesian coordinate cover (x, t, u, p, v, f, g, $\{f_i, g_i, i=1,..,5\}$) where f_i and g_i denote derivatives of f and g with respect to their arguments and they will be thought henceforth as independent variables. Explicitly they are

$$\begin{aligned} &f_1 = f_x, f_2 = f_t, f_3 = f_u, f_4 = f_p, f_5 = f_v \\ &g_1 = g_x, g_2 = g_t, g_3 = g_u, g_4 = g_p, g_5 = g_v \end{aligned} \tag{2.3}$$

We then define the following contact 1- forms of the exterior algebra over the manifold M

$$
\begin{aligned}
\sigma_1 &= du - pdx - \upsilon dt \\
\sigma_2 &= df - f_1dx - f_2dt - f_3du - f_4dp - f_5d\upsilon \\
\sigma_3 &= dg - g_1dx - g_2dt - g_3du - g_4dp - g_5d\upsilon
\end{aligned} \tag{2.4}
$$

and a balance 2 – form

$$
\alpha = d\upsilon \wedge dx + df \wedge dt + gdx \wedge dt \tag{2.5}
$$

These forms are so chosen that they are annihilated by solutions of Eqs (2.1-3). Let I be an ideal of $\Lambda(M)$ generated by the forms $\alpha, \sigma_1, \sigma_2, \sigma_3, d\sigma_1, d\sigma_2$ and $d\sigma_3$ where

$$
\begin{aligned}
d\sigma_1 &= -dp \wedge dx - d\upsilon \wedge dt \\
d\sigma_2 &= -df_1 \wedge dx - df_2 \wedge dt - df_3 \wedge du - df_4 \wedge dp - df_5 \wedge d\upsilon \\
d\sigma_3 &= -dg_1 \wedge dx - dg_2 \wedge dt - dg_3 \wedge d\upsilon - dg_4 \wedge dp - dg_5 \wedge d\upsilon
\end{aligned} \tag{2.6}
$$

This ideal is closed. In fact we immediately see that

$$
d\alpha = dg \wedge dx \wedge dt = dx \wedge dt \wedge (\sigma_3 + g_3\sigma_1) - (g_4dt + g_5dx) \wedge d\sigma_1 \in I
$$

As it well – known a solution of the forms (2.4) and (2.5) annihilates every form in the ideal I.

3. Isovector fields of the ideal $I(\alpha, \sigma_1, \sigma_2, \sigma_3, d\sigma_1, d\sigma_2, d\sigma_3)$

If V is a vector field in the tangent space $T(M)$ it is expressible as

$$
\mathcal{V} = X\frac{\partial}{\partial x} + T\frac{\partial}{\partial t} + U\frac{\partial}{\partial u} + P\frac{\partial}{\partial p} + V\frac{\partial}{\partial \upsilon} + F\frac{\partial}{\partial f} + G\frac{\partial}{\partial g} + \sum_{i=1}^{5}\left(F_i\frac{\partial}{\partial f_i} + G_i\frac{\partial}{\partial g_i}\right) \tag{3.1}
$$

where the coefficients, namely, the components of the vector field $\mathcal{V}$ are smooth functions of coordinates of M. We wish now to determine $\mathcal{V}$ in such a way that Lie derivative of any form in the ideal I remains in the ideal, i. e., we require that $\forall \alpha \in I, \pounds_{\mathcal{V}}\alpha \in I$. It can be shown that the necessary and sufficient conditions for the existence of such vector fields are that Lie derivatives of all generators of I with respect to $\mathcal{V}$ should remain in the ideal I [2]. This is tantamount to demand that Lie derivatives of the generators are expressible as forms in the ideal. This in turn means that we have to find appropriate smooth functions and 1- forms so that only the following relations are to be satisfied

$$
\pounds_{\mathcal{V}}\sigma_1 = \lambda_1\sigma_1 + \lambda_2\sigma_2 + \lambda_3\sigma_3
$$

$$\pounds_{\mathscr{v}}\sigma_2 = \mu_1\sigma_1 + \mu_2\sigma_2 + \mu_3\sigma_3 \tag{3.2}$$

$$\pounds_{\mathscr{v}}\sigma_3 = \nu_1\sigma_1 + \nu_2\sigma_2 + \nu_3\sigma_3$$

$$\pounds_{\mathscr{v}}\alpha = \lambda\alpha + \kappa_1 d\sigma_1 + \kappa_2 d\sigma_2 + \kappa_3 d\sigma_3 + \Lambda_1 \wedge \sigma_1 + \Lambda_2 \wedge \sigma_2 + \Lambda_3 \wedge \sigma_3$$

if we note the identity $\pounds_{\mathscr{v}} d\omega = d\pounds_{\mathscr{v}}\omega$ for every $\omega \in \Lambda(M)$.

Let us also recall that the Lie derivative of an exterior form can be evaluated as follows

$$\pounds_{\mathscr{v}}\omega = d(\mathscr{v}\rfloor\omega) + \mathscr{v}\rfloor d\omega \tag{3.3}$$

where operator $\rfloor$: $\Lambda^k(M) \rightarrow \Lambda^{k-1}(M)$ denotes the interior product of a vector field with a form field.

We first start with determining isovector field of the contact ideal $I(\sigma_1, \sigma_2, \sigma_3)$[2]. Noting that

$$\mathscr{v}\rfloor\sigma_1 = U - pX - \upsilon T = \omega \in \Lambda^0(M) \tag{3.4}$$

$$\mathscr{v}\rfloor d\sigma_1 = -Pdx + Xdp - Vdt + Td\upsilon$$

we obtain from (3.4), (3.3) and $(3.2)_1$ by comparing both sides that

$$\lambda_1 = \omega_u + \omega_f f_3 + \omega_g g_3, \quad \lambda_2 = \omega_f, \; \lambda_3 = \omega_g$$

$$\frac{\partial\omega}{\partial f_i} = 0, \; \frac{\partial\omega}{\partial g_i} = 0, \; i = 1,\ldots,5$$

and

$$X = -\omega_p - \omega_f f_4 - \omega_g g_4, \; T = -\omega_\upsilon - \omega_f f_5 - \omega_g g_5$$

$$P = \omega_x + (\omega_u + \omega_f f_3 + \omega_g g_3)p + \omega_f f_1 + \omega_g g_1$$

$$V = \omega_t + (\omega_u + \omega_f f_3 + \omega_g g_3)\upsilon + \omega_f f_2 + \omega_g g_2 \tag{3.5}$$

$$U = \omega - (\omega_p + \omega_f f_4 + \omega_g g_4)p - (\omega_\upsilon + \omega_f f_5 + \omega_g g_5)\upsilon$$

where it is clear that $\omega = \omega(x, t, u, p, \upsilon, f, g)$.

Similarly we obtain

$$\mathcal{V}\rfloor\sigma_2 = F - f_1X - f_2T - f_3U - f_4P - f_5V = \Omega \in \Lambda^0(M)$$
$$\mathcal{V}\rfloor d\sigma_1 = -F_1dx + Xdf_1 - F_2dt + Tdf_2 - F_3du$$
$$+Udf_3 - F_4dp + Pdf_4 - F_5dv + Vdf_5 \in \Lambda^1(M) \tag{3.6}$$

and it follows from $(3.2)_2$ that

$$\mu_1 = \Omega_u + \Omega_f f_3 + \Omega_g g_3 - F_3, \mu_2 = \Omega_f, \mu_3 = \Omega_g$$

$$\frac{\partial\Omega}{\partial g_i} = 0, \; i = 1,\ldots,5$$

$$F_1 = \overline{F} - pF_3, \quad \overline{F}_1 = \Omega_x + (\Omega_u + \Omega_f f_3 + \Omega_g g_3)p + \Omega_f f_1 + \Omega_g g_1$$

$$F_2 = \overline{F}_2 - vF_3, \quad \overline{F}_2 = \Omega_t + (\Omega_u + \Omega_f f_3 + \Omega_g g_3)v + \Omega_f f_2 + \Omega_g g_2 \tag{3.7}$$

$$F_4 = \Omega_p + \Omega_f f_4 + \Omega_g g_4$$

$$F_5 = \Omega_v + \Omega_f f_5 + \Omega_g g_5$$

and

$$X + \frac{\partial\Omega}{\partial f_1} = 0, \quad T + \frac{\partial\Omega}{\partial f_2} = 0, \quad U + \frac{\partial\Omega}{\partial f_3} = 0, \quad P + \frac{\partial\Omega}{\partial f_4} = 0, \quad V + \frac{\partial\Omega}{\partial f_5} = 0 \tag{3.8}$$

Evaluating the derivatives in (3.8) by noting $(3.7)_4$ we arrive at the following results

$$\frac{\partial F}{\partial f_1} = \omega_f f_4, \quad \frac{\partial F}{\partial f_2} = \omega_f f_5, \quad \frac{\partial F}{\partial f_3} = \omega_f (pf_4 + vf_5), \quad \frac{\partial F}{\partial f_4} = -\omega_f (f_1 + pf_3)$$

$$\frac{\partial F}{\partial f_5} = -\omega_f (f_2 + vf_3), \quad \frac{\partial F}{\partial g_i} = 0, \quad i = 1,\ldots,5 \tag{3.9}$$

It is straightforward to see that the foregoing system of equations is solvable if and only if $\omega_f = 0$. Hence we conclude that

$$\frac{\partial F}{\partial f_i} = 0, \quad i = 1, \ldots, 5 \tag{3.10}$$

Proceeding in the same fashion we rapidly obtain from $(3.2)_3$ that

$$\nu_1 = \Delta_u + \Delta_f f_3 + \Delta_g g_3 - G_3, \; \nu_2 = \Delta_f, \; \nu_3 = \Delta_g$$

$$\frac{\partial\Delta}{\partial f_i} = 0, \; i = 1,\ldots,5$$

$$X+\frac{\partial\Delta}{\partial g_1}=0,\ T+\frac{\partial\Delta}{\partial g_2}=0,\ U+\frac{\partial\Delta}{\partial g_3}=0,\ P+\frac{\partial\Delta}{\partial g_4}=0,\ V+\frac{\partial\Delta}{\partial g_5}=0$$

where

$$\Delta=G-g_1X-g_2T-g_3U-g_4P-g_5V\in\Lambda^0(M) \tag{3.11}$$

and

$$\begin{aligned}
&G_1=\overline{G}_1-pG_3,\ \overline{G}_1=\Delta_x+(\Delta_u+\Delta_f f_3+\Delta_g g_3)p+\Delta_f f_1+\Delta_g g_1\\
&G_2=\overline{G}_2-vG_3,\ \overline{G}_2=\Delta_t+(\Delta_u+\Delta_f f_3+\Delta_g g_3)v+\Delta_f f_2+\Delta_g g_2\\
&G_4=\Delta_p+\Delta_f f_4+\Delta_g g_4\\
&G_5=\Delta_v+\Delta_f f_5+\Delta_g g_5
\end{aligned} \tag{3.12}$$

Similarly we find that

$$\begin{aligned}
&\frac{\partial G}{\partial g_1}=\omega_g g_4,\ \frac{\partial G}{\partial g_2}=\omega_g g_5,\ \frac{\partial G}{\partial g_3}=\omega_g(pg_4+vg_5),\ \frac{\partial G}{\partial g_4}=-\omega_g(g_1+pg_3)\\
&\frac{\partial G}{\partial g_5}=-\omega_g(g_2+vg_3),\ \frac{\partial G}{\partial f_i}=0,\ i=1,\ \ldots,\ 5
\end{aligned} \tag{3.13}$$

The above equations are solvable if and only if $\omega_g=0$ and we thus obtain

$$\frac{\partial G}{\partial g_i}=0,\ i=1,\ \ldots,\ 5 \tag{3.14}$$

Collecting what we have found so far we conclude that the functions ω, F and G must have the forms given below

$$\omega=\omega(x,t,u,p,v),\qquad F=F(x,t,u,p,v,f,g),\qquad G=G(x,t,u,p,v,f,g) \tag{3.15}$$

It then follows from (3.5) that

$$\begin{aligned}
&X=-\omega_p,\ T=-\omega_v,\ U=\omega-\omega_p p-\omega_v v,\\
&P=\omega_x+\omega_u p,\qquad V=\omega_t+\omega_u v
\end{aligned} \tag{3.16}$$

The functions $\overline{F}_1$, $\overline{F}_2$, F_4 and F_5 are determined by the relations (3.7) and $\overline{G}_1$, $\overline{G}_2$, $\overline{G}_4$ and $\overline{G}_5$ by (3.12). F_3 and G_3 are left as arbitrary smooth functions over the manifold M. They generate vector fields

$$\mathfrak{I}=F_3\left(\frac{\partial}{\partial f_3}-p\frac{\partial}{\partial f_1}-\upsilon\frac{\partial}{\partial f_2}\right), \qquad \mathcal{G}=G_3\left(\frac{\partial}{\partial g_3}-p\frac{\partial}{\partial g_1}-\upsilon\frac{\partial}{\partial g_2}\right)$$

which satisfy the following relations $\mathcal{G}$

$$\mathfrak{I}\rfloor\sigma_1=0,\ \mathfrak{I}\rfloor d\sigma_1=0,\ \mathfrak{I}\rfloor\sigma_2=0,\ \mathfrak{I}\rfloor d\sigma_2=-F_3\sigma_1,\ \mathfrak{I}\rfloor\sigma_3=0,\ \mathfrak{I}\rfloor d\sigma_3=0$$

$$\mathcal{G}\rfloor\sigma_1=0,\ \mathcal{G}\rfloor d\sigma_1=0,\ \mathcal{G}\rfloor\sigma_2=0,\ \mathcal{G}\rfloor d\sigma_2=0,\ \mathcal{G}\rfloor\sigma_3=0,\ \mathcal{G}\rfloor d\sigma_3=-G_3\sigma_1$$

so we have

$$£_{\mathfrak{I}}\sigma_1=0, £_{\mathfrak{I}}\sigma_2=-F_3\sigma_1, £_{\mathfrak{I}}\sigma_3=0, £_{\mathcal{G}}\rfloor d\sigma_2=0, £_{\mathcal{G}}\sigma_3=0, £_{\mathcal{G}}\rfloor d\sigma_3=-G_3\sigma_1$$

Therefore $\mathfrak{I}$ and $\mathcal{G}$ are trivial isovector fields of the contact ideal and they can be discarded.

We thus find that an arbitrary vector field (3.1) becomes an isovector field of the contact ideal if and only if its components are determined through (3.15), (3.7) and (3.12) depending on three arbitrary functions ω, F and G of the form given in (3.15). We now consider the relation $(3.2)_4$. Since

$$\mathcal{V}\rfloor\alpha=Vdx-Xd\upsilon+Fdt-Tdf+g(Xdt-Tdx)$$

$$\mathcal{V}\rfloor d\alpha=Gdx\wedge dt-Xdg\wedge dt+Tdg\wedge dx$$

This relation can be cast into the form

$$£_{\mathcal{V}}\alpha=dV\wedge dx-dX\wedge d\upsilon+dF\wedge dt-dT\wedge df+gdX\wedge dt-gdT\wedge dx+Gdx\wedge dt$$
$$=\lambda\alpha+\kappa_1 d\sigma_1+\kappa_2 d\sigma_2+\kappa_3 d\sigma_3+\Lambda_1\wedge\sigma_1+\Lambda_2\wedge\sigma_2+\Lambda_3\wedge\sigma_3 \qquad (3.17)$$

where $\lambda,\ldots\ldots,\ \kappa_3$ are arbitrary smooth functions and Λ_1,Λ_2 and Λ_3 are arbitrary 1-forms. Since the left hand side of the foregoing expression is free from forms df_i and dg_1, $i=1,\ldots..,5$ we can assume without loss of generality that

$$\kappa_2=0,\ \kappa_3=0,$$

$$\Lambda_i=\alpha_i dx+b_i dt+c_i du+d_i dp+e_i d\upsilon+\phi_i df+\gamma_i dg,\ i=1,2,3$$

where the coefficients $\alpha_{i,},\ldots\ldots,\gamma_i$ are smooth functions over M. After comparing both sides of (3.17) we deduce the following relations

$$
\begin{aligned}
&F_x - V_t + gX_x - gT_t + G = \lambda g + b_1 p - \alpha_1 v + b_2 f_1 - \alpha_2 f_2 + b_3 g_1 - \alpha_3 - g_2 \\
&F_u + gX_u = -b_1 - c_1 v - c_2 f_2 + b_2 f_3 - c_3 g_2 + b_3 g_3 \\
&F_p + gX_p = -d_1 v - d_2 f_2 + b_2 f_4 - d_3 g_2 + b_3 g_4 \\
&F_v + X_t + gX_v = -\kappa_1 - e_1 v - e_2 f_2 + b_2 f_5 - e_3 g_2 + b_3 g_5 \\
&F_f + T_t = \lambda - b_2 - \phi_1 v - \phi_2 f_2 - \phi_3 g_2 \\
&F_g = -b_3 - \gamma_1 v - \gamma_2 f_2 - \gamma_3 g_2 \\
&T_x = -\alpha_2 - \phi_1 p - \phi_2 f_1 - \phi_3 g_1 \\
&T_u = -c_2 + \phi_1 - \phi_2 f_3 - \phi_3 g_3 \\
&T_p = -d_2 - \phi_2 f_4 - \phi_3 g_4 \\
&T_v = -e_2 - \phi_2 f_5 - \phi_3 g_5 \\
&V_u - gT_u = -\alpha_1 - c_1 p - c_2 f_1 + \alpha_2 f_3 - c_3 g_1 + \alpha_3 g_3 \\
&V_p - gT_p = -\kappa_1 - d_1 p - d_2 f_1 + \alpha_2 f_4 - d_3 g_1 + \alpha_3 g_4 \\
&V_v + X_x - gT_v = \lambda - e_1 p - e_2 f_1 + \alpha_2 f_5 - e_3 g_1 + \alpha_3 g_5 \\
&X_u = e_1 - e_2 f_3 + c_2 f_5 - e_3 g_3 + c_3 g_5 \\
&X_p = -e_2 f_4 + d_2 f_5 - e_3 g_4 + d_3 g_5 \\
&d_1 - d_2 f_3 + c_2 f_4 + d_3 g_3 + c_3 g_4 = 0 \\
&\gamma_1 - \gamma_2 f_3 - \gamma_3 g_3 - c_3 = 0 \\
&\gamma_1 p + \gamma_2 f_1 + \gamma_3 g_1 + \alpha_3 = 0 \\
&\gamma_2 - \phi_3 = 0 \\
&\gamma_2 f_4 + \gamma_3 g_4 + d_3 = 0 \\
&\gamma_2 f_5 + \gamma_3 g_5 + e_3 = 0
\end{aligned}
\tag{3.18}
$$

it is relatively easy to find the solution of the set of equations (3.18) and the result is reported below

$$\alpha_1 = \overline{\alpha}_1 - pc_1 - (f_1 + pf_3)\phi_1 - (g_1 + pg_3)\gamma_1$$

$$\alpha_2 = \overline{\alpha}_2 - p\phi_1 - f_1\phi_2 - g_1\phi_3$$

$$\alpha_3 = -f_1\phi_3 - p\gamma_1 - g_1\gamma_3$$

$$b_1 = \bar{b}_1 - vc_1 - (f_2 + vf_3)\phi_1 - (g_2 + vg_3)\gamma_1$$

$$b_2 = \bar{b}_2 - v\phi_1 - f_2\phi_2 - g_2\phi_3$$

$$b_3 = \bar{b}_3 - f_2\phi_3 - v\gamma_1 - g_2\gamma_3$$

$$c_2 = \bar{c}_2 + \phi_1 - f_3\phi_2 - g_3\phi_3$$

$$c_3 = -f_3\phi_3 + \gamma_1 - g_3\gamma_3 \tag{3.19}$$

$$d_1 = \bar{d}_1 - f_4\phi_1 - g_4\gamma_1$$

$$d_2 = \bar{d}_2 - f_4\phi_2 - g_4\phi_3$$

$$d_3 = -f_4\phi_3 - g_4\gamma_3$$

$$e_1 = \bar{e}_1 - f_5\phi_1 - g_5\gamma_1$$

$$e_2 = \bar{e}_2 - f_5\phi_2 - g_5\phi_3$$

$$e_3 = -f_5\phi_3 - g_5\gamma_3$$

$$\gamma_2 = \phi_3$$

where we have defined

$$\bar{\alpha}_1 = -V_u + (g + f_1)T_u - f_3T_x, \quad \bar{\alpha}_2 = -T_x$$

$$\bar{b}_1 = -F_u - gX_u + f_2T_u + f_3\bar{b}_2 + g_3\bar{b}_3, \quad \bar{b}_2 = \lambda - F_f - T_t, \quad \bar{b}_3 = -F_g$$

$$\bar{c}_2 = -T_u \tag{3.20}$$

$$\bar{d}_1 = -f_3T_p + f_4T_u, \quad \bar{d}_2 = -T_p$$

$$\bar{e}_1 = X_u - f_3T_v + f_5T_u, \quad \bar{e}_2 = -T_v$$

The functions λ and κ_1 are given by

$$\lambda = V_v + X_x + pX_u - (g + f_1 + pf_3)T_v + f_5(pT_u + T_x)$$

$$\kappa_1 = -V_p + (g + f_1 + pf_3)T_p - f_4(pT_u + T_x) \tag{3.21}$$

There are 23 unknown functions in the 21 relations in the set (3.17). Expressions (3.19) and (3.21), depending on 6 arbitrary functions c_1, ϕ_1, ϕ_2, ϕ_3, γ_1 and γ_3 determine 17 of them. In order that the set (3.18) would be solvable we have to impose four compatibility conditions. It is straightforward to see that they given by the following relations

$$X_p - f_4 T_\upsilon + f_5 T_p = 0,$$

$$F_p + gX_p + f_4\left(F_f + T_t - V_\upsilon - X_x - pX_u + \upsilon T_u + gT_\upsilon\right) - \left(f_2 + \upsilon f_3\right)T_p + g_4 F_g$$

$$+ f_4\left(f_1 + pf_3\right)T_\upsilon - f_4 f_5\left(pT_u + T_x\right) = 0,$$

$$F_\upsilon + X_t + g\left(X_\upsilon + T_p\right) + \upsilon X_u - V_p + \left(f_1 + pf_3\right)T_p + f_5\left(F_f + T_t - V_\upsilon - X_x - pX_u\right) + g_5 F_g$$

$$-\left[f_2 + \upsilon f_3 - f_5\left(g + f_1 + pf_3\right)\right]T_\upsilon - \left(f_4 + f_5^2\right)T_x + \left(\upsilon f_5 - pf_4 - pf_5^2\right)T_\upsilon = 0$$

$$G = -F_x + V_t + \upsilon V_u - pF_u + g\left(V_\upsilon - T_t - \upsilon T_u\right) - g^2 T_\upsilon \tag{3.22}$$

$$+\left(f_1 + pf_3\right)\left(V_\upsilon + X_x + pX_u - F_f - T_t - 2gT_\upsilon\right) - \left(g_1 + pg_3\right)F_g - \left(f_1 + pf_3\right)^2 T_\upsilon$$

$$+\left[pf_2 - \upsilon f_1 + pf_5\left(g + f_1 + pf_3\right)\right]T_u + \left[f_2 + \upsilon f_3 + f_5\left(g + f_1 + pf_3\right)\right]T_x$$

Let us now define 1-forms $\overline{\Lambda}_1, \overline{\Lambda}_2$ and $\overline{\Lambda}_3$ by

$$\overline{\Lambda}_1 = \overline{\alpha}_1 dx + \overline{b}_1 dt + \overline{d}_1 dp + \overline{e}_1 d\upsilon,\ \overline{\Lambda}_2 = \overline{\alpha}_2 dx + \overline{b}_2 dt + \overline{c}_2 du + \overline{d}_2 dp + \overline{e}_2 d\upsilon,\ \overline{\Lambda}_3 = \overline{b}_3 dt$$

so that Λ_1, Λ_2 and Λ_3 can be expressed as

$$\Lambda_1 = \overline{\Lambda}_1 + \left(c_1 + f_3\phi_1 + g_3\gamma_1\right)\sigma_1 + \phi_1\sigma_2 + \gamma_1\sigma_3,$$

$$\Lambda_2 = \overline{\Lambda}_2 + \phi_1\sigma_1 + \phi_2\sigma_2 + \phi_3\sigma_3,\quad \Lambda_3 = \overline{\Lambda}_3 + \gamma_1\sigma_1 + \phi_3\sigma_2 + \gamma_3\sigma_3,$$

Consequently we have

$$\Lambda_1 \wedge \sigma_1 + \Lambda_2 \wedge \sigma_2 + \Lambda_3 \wedge \sigma_3 = \overline{\Lambda}_1 \wedge \sigma_1 + \overline{\Lambda}_2 \wedge \sigma_2 + \overline{\Lambda}_3 \wedge \sigma_3$$

Hence without loss of generality we may assume that the functions $c_1, \phi_1, \phi_2, \phi_3, \gamma_1$ and γ_3 vanish.

Since the functions X, T, V, F and G given by (3.15) do not depend on $f_1, \ldots, g_5$ starting from the first equation of (3.22) we find sequentially

$$T_\upsilon = 0,\ T_p = 0,\ X_p = 0$$

$$F_g - 0,\ T_u - 0,\ T_x = 0,\ F_f + T_t - V_\upsilon - X_x - pX_u = 0,\ F_p = 0$$

$$F_\upsilon + X_t + gX_\upsilon + \upsilon X_u - V_p = 0 \tag{3.23}$$

$$G = -F_x + V_t + \upsilon V_u - pF_u + g\left(V_\upsilon - T_t\right)$$

Introducing (3.16) into (3.23) we obtain the following linear partial differential equations to determine the functions ω, F and G

$$\omega_{\upsilon\upsilon}=0,\ \omega_{\upsilon p}=0,\ \omega_{\upsilon u}=0,\ \omega_{\upsilon x}=0,\ \omega_{pp}=0,$$

$$F_g=0,\ F_p=0,\ F_f-\omega_u-2\omega_{\upsilon t}+\omega_{px}+p\omega_{pu}=0,$$

$$F_\upsilon-2\omega_{pt}-2\upsilon\omega_{pu}=0 \tag{3.24}$$

$$G=-F_x-pF_u+\omega_{tt}+2\upsilon\omega_{ut}+\upsilon^2\omega_{uu}+g(\omega_u+2\omega_{\upsilon t}+\upsilon\omega_{uu})$$

Since ω does not depend on f and g equations $(3.24)_{1\text{-}5}$ and $(3.24)_{6,7}$ yield immediately

$$\omega=\alpha(x,t,u)p+\beta(t)\upsilon+\gamma(x,t,u),\ \ F=F(x,t,u,\upsilon,f)$$

so that we deduce from (3.16) that

$$X=-\alpha,\ T=-\beta,\ U=\gamma$$

$$P=\alpha_u p^2+(\alpha_x+\gamma_u)p+\gamma_x,\ V=\alpha_u p\upsilon+(\dot\beta+\gamma_u)\upsilon+\alpha_t p+\gamma_t \tag{3.25}$$

where dots denote ordinary derivatives with respect to t.

In view of these results Eq.$(3.24)_9$ reduces to

$$F_\upsilon=2(\alpha_u\upsilon+\alpha_t)$$

whence we obtain

$$F=\alpha_u\upsilon^2+2\alpha_t\upsilon+\Phi(x,t,u,f)$$

Inserting this relation into $(3.24)_8$ we get

$$\Phi_f=2\dot\beta-\alpha_x+\gamma_u$$

or

$$\Phi=(2\dot\beta-\alpha_x+\gamma_u)f+\Psi(x,t,u)$$

Therefore the function F is constructed as

$$F=\alpha_u\upsilon^2+2\alpha_t\upsilon+(2\dot\beta-\alpha_x+\gamma_u)f+\Psi(x,t,u) \tag{3.26}$$

and the function G follows from the last relation of (3.24) as

$$G=(\gamma_{uu}-\alpha_{ux})\upsilon^2+\left[\ddot\beta+2(\gamma_{ut}-\alpha_{xt})\right]\upsilon-\left[(\gamma_{uu}-\alpha_{ux})p+\gamma_{ux}-\alpha_{xx}\right]f$$

$$+(2\dot\beta+\alpha_{uu}p\upsilon+\gamma_{uu}\upsilon+\alpha_u p+\gamma_u)g+(\alpha_{tt}-\Psi_u)p+\gamma_{tt}-\Psi_x \tag{3.27}$$

In view of (3.7), (3.12), (3.25), (3.26) and (3.27) we see that we have managed to determine all components of the isovector fields in terms of four arbitrary functions $\alpha(x,t,u), \beta(t), \gamma(x,t,u)$ and $\Psi(x,t,u)$. These components are easily identified as infinitesimal generators and their prolongation of the equivalence groups. The equivalence groups of transformations are explicity found by integrating a first order system of ordinary differential equations in the group parameter ε :

$$\frac{d\bar{x}}{d\varepsilon}=-\alpha(\bar{x},\bar{t},\bar{u}),\ \frac{d\bar{t}}{d\varepsilon}=-\beta(\bar{t}),\ \frac{d\bar{u}}{d\varepsilon}=\gamma(\bar{x},\bar{t},\bar{u}),$$

$$\frac{d\bar{p}}{d\varepsilon}=P(\bar{x},\bar{t},\bar{u},\bar{p}),\ \frac{d\bar{\upsilon}}{d\varepsilon}=V(\bar{x},\bar{t},\bar{u},\bar{p},\bar{\upsilon}),$$

$$\frac{d\bar{f}}{d\varepsilon}=F(\bar{x},\bar{t},\bar{u},\bar{\upsilon},\bar{f}),\ \frac{d\bar{g}}{d\eta}=G(\bar{x},\bar{t},\bar{u},\bar{p},\bar{\upsilon},\bar{f},\bar{g})$$

where P, V, F, and G are given in (3.25), (3,26), and (3.27). These equations should be solved under the initial conditions

$$\bar{x}(0)=x,\ \bar{t}(0)=t,\ \bar{u}(0)=u,\ \bar{p}(0)=p,\ \bar{\upsilon}(0)=\upsilon,\ \bar{f}(0)=f,\ \bar{g}(0)=g$$

The ivariance of the contact ideal under the equivalence transformations secures the property that $p=\frac{\partial u}{\partial x}$ is mapped to $\bar{p}=\frac{\partial \bar{u}}{\partial \bar{x}}$, etc. The relevant infinitesimal generators of the infinite dimensional equivalence groups are then classified as follows

$$\mathcal{V}_a=-\alpha\frac{\partial}{\partial x}+p(\alpha_u p+\alpha_x)\frac{\partial}{\partial p}+p(\alpha_u\upsilon+\alpha_t)\frac{\partial}{\partial \upsilon}+(\alpha_u\upsilon^2+2\alpha_t\upsilon-\alpha_x f)\frac{\partial}{\partial f}$$

$$+\left[-\alpha_{ux}\upsilon^2-2\alpha_{xt}\upsilon+(\beta_{ux}p+\alpha_{xx})f+p(\alpha_{uu}\upsilon+\alpha_u)g+\alpha_{tt}\right]\frac{\partial}{\partial g}$$

$$\mathcal{V}_\beta=-\beta\frac{\partial}{\partial t}+\dot{\beta}\upsilon\frac{\partial}{\partial \upsilon}+2\dot{\beta}f\frac{\partial}{\partial f}+(\ddot{\beta}\upsilon+2\dot{\beta}g)\frac{\partial}{\partial g}$$

$$\mathcal{V}_\gamma=\gamma\frac{\partial}{\partial u}+(\gamma_u p+\gamma_x)\frac{\partial}{\partial p}+(\gamma_u\upsilon+\gamma_t)\frac{\partial}{\partial \upsilon}+\gamma_u f\frac{\partial}{\partial f}$$

$$+\left[\gamma_{uu}\upsilon^2+2\gamma_{ut}\upsilon-(\gamma_{uu}p+\gamma_{ux})f+(\gamma_{uu}\upsilon+\gamma_u)g+\gamma_{tt}\right]\frac{\partial}{\partial g}$$

$$V_\Psi = \Psi \frac{\partial}{\partial f} - (\Psi_u p + \Psi_x)\frac{\partial}{\partial g}$$

as a very simple example let us consider the isovector field V_β with $\beta(t) = t^2$. Hence the vector is

$$\mathcal{V} = -t^2 \frac{\partial}{\partial t} + 2tv\frac{\partial}{\partial v} + 4tf\frac{\partial}{\partial f} + (2v + 4tg)\frac{\partial}{\partial g}$$

Hence the transformation group is determined through the differential equations

$$\frac{d\bar{x}}{d\varepsilon} = 0,\ \frac{d\bar{t}}{d\varepsilon} = -t^{-2},\ \frac{d\bar{u}}{d\varepsilon} = 0,\ \frac{d\bar{p}}{d\varepsilon} = 0,\ \frac{d\bar{v}}{d\varepsilon} = 2\bar{t}\bar{v},\ \frac{d\bar{f}}{d\varepsilon} = 4\bar{t}\bar{f},\ \frac{d\bar{g}}{d\varepsilon} = 2\bar{v} + \bar{t}\bar{g}$$

whose integrals are easily obtained as follows

$$\bar{x} = x,\ \bar{t} = \frac{t}{1+\varepsilon t},\ \bar{u} = u,\ \bar{p} = p,\ \bar{v} = (1+\varepsilon t)^2 v,$$

$$\bar{f}(\bar{x},\bar{t},\bar{u},\bar{p},\bar{v}) = (1-\varepsilon\bar{t})^{-4} f\left(x, \bar{t}/1-\varepsilon\bar{t}, p, (1-\varepsilon\bar{t})^2\bar{v}\right),$$

$$\bar{g}(\bar{x},\bar{t},\bar{u},\bar{p},\bar{v}) = (1-\varepsilon\bar{t})^{-4} g\left(x, \bar{t}/1-\varepsilon\bar{t}, p, (1-\varepsilon\bar{t})^2\bar{v}\right) + 2\varepsilon(1-e\bar{t})^{-1}\bar{v},$$

This means that if $u = u(x,t)$ is a solution of the differential equation $u_{tt} = f_x + g$ then $\bar{u} = u(x, \bar{t}/1-\varepsilon\bar{t})$ is the solution of the differential equation $\bar{u}_{\bar{t}\bar{t}} = \bar{f}_{\bar{x}} + \bar{g}$.

4. Special cases

We consider now the case $g = 0$, i. e., the nonlinear wave equation of the form

$$u_{tt} = [f(x,t,,u_x,u_t)]_x \tag{4.1}$$

This implies that the function G defined by (3.27) should vanish identically with $g = 0$. Hence we see that we have to impose the following relations on the functions α, β, γ and Ψ

$$(\gamma_u - \alpha_x)_u = 0,\ (\gamma_u - \alpha_x)_t = -\frac{1}{2}\ddot{\beta},\ (\gamma_u - \alpha_x)_x = 0, \tag{4.2}$$

$$\Psi_u = \alpha_{tt},\qquad \Psi_x = \gamma_{tt}$$

The integration of the first three equations of (4.2) leads to

$$\gamma_u - \alpha_x = -\left(c_1 + \frac{1}{2}\dot{\beta}\right) \tag{4.3}$$

where c_1 is an arbitrary constant. Let us now assume for convenience that γ is represented by

$$\gamma = \Pi_x,\ \Pi = \Pi(x,t,u) \tag{4.4}$$

so that (4.3) can easily be integrated to

$$\alpha = \Pi_u + (c_1 + 1/2\,\dot{\beta})x \tag{4.5}$$

where an arbitrary function of t and u is absorbed into Π without loss of generality. Then the last two equations of (4.2) yield

$$\Psi = \Pi_{tt},\ \dddot{\beta} = 0 \tag{4.6}$$

we again absorb an arbitrary function of t and u into Π. We thus obtain

$$\beta(t) = c_2 t^2 + c_3 t + c_4 \tag{4.7}$$

where c_2, c_3 and c_4 arbitrary constants. Therefore (4.5) now becomes

$$\alpha = \Pi_u + (c_2 t + c_1 + c_3)x \tag{4.8}$$

Let us next define

$$\Gamma(x,t,u) = \Pi(x,t,u) + (c_2 t + c_1 + c_3)xu \tag{4.9}$$

so that we finally have the result

$$\begin{aligned}
&\alpha = \Gamma_u,\ \beta(t) = c_2 t^2 + c_3 t + c_4,\ \gamma = \Gamma_x + (c_2 t + c_1 + c_3)u,\ \Psi = \Gamma_{tt} \\
&P = \Gamma_{uu} p^2 + 2\Gamma_{ux} p + \Gamma_{xx} - (c_2 t + c_1 + c_3)p \\
&V = \Gamma_{uu} p\upsilon + \Gamma_{ux}\upsilon + \Gamma_{ut} p + \Gamma_{xt} - c_2 u + (c_2 t - c_1 + c_3)\upsilon \\
&F = \Gamma_{uu}\upsilon^2 + 2\Gamma_{ut}\upsilon + \Gamma_{ttt} + (3c_2 t - c_1 + 3c_3)f
\end{aligned} \tag{4.10}$$

Therefore transformations the most general isovector field generating equivalence transformations is represented by

$$\begin{aligned}
\mathcal{V} = &-\Gamma_u \frac{\partial}{\partial x} + \Gamma_x \frac{\partial}{\partial u} + \left(\Gamma_{uu} p^2 + 2\Gamma_{ux} p + \Gamma_{xx}\right)\frac{\partial}{\partial p} \\
&+ \left(\Gamma_{uu} p\upsilon + \Gamma_{ux}\upsilon + \Gamma_{ut} p + \Gamma_{xt}\right)\frac{\partial}{\partial \upsilon} + \left(\Gamma_{uu}\upsilon^2 + 2\Gamma_{ut}\upsilon + \Gamma_{tt}\right)\frac{\partial}{\partial f}
\end{aligned} \tag{4.11}$$

$$-c_1\left(u\frac{\partial}{\partial u}+p\frac{\partial}{\partial p}+\upsilon\frac{\partial}{\partial \upsilon}+f\frac{\partial}{\partial f}\right)-c_2\left(t^2\frac{\partial}{\partial t}+tu\frac{\partial}{\partial u}+tp\frac{\partial}{\partial p}+(u-\upsilon t)\frac{\partial}{\partial \upsilon}-3tf\frac{\partial}{\partial f}\right)$$

$$-c_3\left(2t\frac{\partial}{\partial t}+u\frac{\partial}{\partial u}+p\frac{\partial}{\partial p}-\upsilon\frac{\partial}{\partial \upsilon}-3f\frac{\partial}{\partial f}\right)-c_4\frac{\partial}{\partial t}$$

Several other special cases can easily be obtainable from the general scheme developed here. For instance if some arguments were missing in either the function f or g or in both of them then the coordinates f_i or/and g_i should be taken as zero. Consequently F_i or/and G_i associated with these coordinates vanish. These conditions impose of course additional restrictions on the form of the functions $\omega, F,$ and G. For instance if we assume that $f=f(x,p)$ and $g=g(x,p)$ then $f_2, f_3, f_5, g_2, g_3, g_5$ obviously vanish. We then have to impose that F_2, F_3, F_5, G_2, G_3, G_5, should also vanish. It then follows from (3.7) and (3.12) that these conditions are equivalent to

$$\Omega_\upsilon=0,\ \Omega_t+\upsilon\Omega_u=0,\ \Delta_\upsilon=0,\ \Delta_t+\upsilon\Delta_u=0 \tag{4.12}$$

where the functions Ω and Δ are now defined as

$$\Omega=F-f_1X-f_4P,\ \Delta=G-g_1X-g_4P, \tag{4.13}$$

Next we introduce the relations (3.25), (3.26) and (3.27) into (4.13) and the resulting expressions into (4.12) whose first equations yields

$$2(\alpha\upsilon+\alpha_t)=0$$

whence we deduce that $\alpha_u=0, \alpha_t=0$ or $\alpha=\alpha(x)$ while the second equation of (4.12) results in

$$(2\ddot{\beta}+\gamma_{ut}+\upsilon\gamma_{uu})f\Psi_t+\upsilon\Psi_u-f_4(\gamma_{uu}p\upsilon+\gamma_{ut}p+\gamma_{ux}\upsilon+\gamma_{xt})=0.$$

This in turn gives

$$\gamma_{uu}=0,\ \gamma_{ut}=0,\ \gamma_{ux}=0,\ \gamma_{xt}=0,\ \ddot{\beta}=0,\ \Psi_u=0,\ \Psi_t=0.$$

Hence we obtain the following relations

$$\alpha=\alpha(x),\ \beta=c_1t+c_2,\ \gamma=c_3u+\lambda(x)+\mu(t),\ \Psi=\Psi(x) \tag{4.14}$$

where c_1, c_2 and c_3 are arbitrary constants. In view of (4.14) the function G is expressed as

$$G=\alpha''f+(2c_1+c_3)g+\ddot{\mu}-\Psi'$$

Primes denote derivatives with respect to x. The remaining two equations of (4.12) yield solely $\ddot{\mu}=0$. Thus we have

$$\mu = c_4 t^2 + 2c_5 t + c_6$$

Therefore isovector components are given by

$$X=-\alpha(x),\ T=-c_1 t - c_2,\ \gamma = c_3 u + c_4 t^2 + 2c_5 t + c_6 + \lambda(x),\ P=(\alpha' + c_3)p + \lambda'$$

$$V=(c_1+c_3)\upsilon + 2c_4 t + 2c_5,\ F=-(a' - 2c_1 - c_3)f + \Psi,\ G=\alpha'' f + (2c_1+c_3)g + 2c_4 - \Psi'$$

Hence isovectors generating equivalence groups can be classified as follows

$$\mathcal{V}_a = -\alpha\frac{\partial}{\partial x} + \alpha' p\frac{\partial}{\partial p} - \alpha' f\frac{\partial}{\partial f} + \alpha'' f\frac{\partial}{\partial g},\ \mathcal{V}_\lambda = \lambda\frac{\partial}{\partial u} + \lambda'\frac{\partial}{\partial p},\ \mathcal{V}_\Psi = \Psi\frac{\partial}{\partial f} - \Psi'\frac{\partial}{\partial g}$$

$$\mathcal{V}_1 = -t\frac{\partial}{\partial t} + \upsilon\frac{\partial}{\partial \upsilon} + 2f\frac{\partial}{\partial f} + 2g\frac{\partial}{\partial g},\ \mathcal{V}_2 = \frac{\partial}{\partial t},\ \mathcal{V} = u\frac{\partial}{\partial u} + p\frac{\partial}{\partial p} + \upsilon\frac{\partial}{\partial \upsilon} + f\frac{\partial}{\partial f} + g\frac{\partial}{\partial g}$$

$$V_4 = t^2\frac{\partial}{\partial u} + 2t\frac{\partial}{\partial \upsilon} + 2\frac{\partial}{\partial g},\ V_5 = t\frac{\partial}{\partial u} + \frac{\partial}{\partial \upsilon},\ V_6 = \frac{\partial}{\partial u}. \qquad (4.15)$$

Acknowledgement

The author gratefully acknowledgs the support provided by the Turkish Academy of Sciences.

References

[1]. Edelen, D. G. B., (1980). *Isovector Methods for Equations of Balance,* Sijthoff Noordhoff, Alphen aan den Rijn, The Netherlands.

[2]. Edelen, D.G.B., (1985). *Applied Exterior Calculus*, Wiley, New York.

[3]. Ibragimov, N. H., M. Torrisi and A. Valenti, (1991). Preliminary Group Classification of Equations $\upsilon_{tt} = f(x,\upsilon_x)\upsilon_{xx} + g(x,\upsilon_x)$, *J. Math. Phys.*, **32**, 2988-2995.

[4]. Ibragimov, N. H. and M. Torris, (1992). A Simple Method for Group Analysis and its Application to a Model of Detonation, *J. Math. Phys.* **33**, 3931-3937.

[5]. Ibragimov, N. H. and M. Torrisi, (1994). Equivalence Groups for Balance Equations, *J. Math. Anal. and Appl.*, **184**, 441-452.

[6]. Ibragimov, N. H. (Editor), (1994). *CRC Handbook of Lie Group Analysis of Differential Equations, Vol.* **1**, Symmetries, Exact Solutions and Conservation Laws, CRC Press, Boca Raton, Ann Arbor, London, Tokyo

[7]. Ovsiannikov, L.V., *Group Analysis of Differential Equations,* Academic Press, New York, London (1982) [Russian edition published as *Gruppovoi analiz differentsial'nykh uravnenii*, Nauka, Moscow (1978)].

[8]. Şuhubi, E. S., (1991) Isovector Fields and Similarity Solutions for General Balance Equations *Int. J. Engng Sci.*, **29**, 133-150.